HUMAN FACTORS

IN **TRANSPORTATION**

SOCIAL AND TECHNOLOGICAL EVOLUTION
ACROSS MARITIME, ROAD, RAIL,
AND AVIATION DOMAINS

INDUSTRIAL AND SYSTEMS ENGINEERING SERIES

Series Editor
Waldemar Karwowski

RECENTLY PUBLISHED TITLES:

Human Factors in Transportation: Social and Technological Evolution Across Maritime, Road, Rail, and Aviation Domains
Giuseppe Di Bucchianico, Andrea Vallicelli, Neville A. Stanton, and Steven J. Landry

Ergonomics and Human Factors in Safety Management
Pedro Miguel Ferreira Martins Arezes and Paulo Victor Rodrigues de Carvalho

Manufacturing Productivity in China
Li Zheng, Simin Huang, and Zhihai Zhang

Supply Chain Management and Logistics: Innovative Strategies and Practical Solutions
Zhe Liang, Wanpracha Art Chaovalitwongse, and Leyuan Shi

Mobile Electronic Commerce: Foundations, Development, and Applications
June Wei

Managing Professional Service Delivery: 9 Rules for Success
Barry Mundt, Francis J. Smith, and Stephen D. Egan Jr.

Laser and Photonic Systems: Design and Integration
Shimon Y. Nof, Andrew M. Weiner, and Gary J. Cheng

Design and Construction of an RFID-enabled Infrastructure: The Next Avatar of the Internet
Nagabhushana Prabhu

Cultural Factors in Systems Design: Decision Making and Action
Robert W. Proctor, Shimon Y. Nof, and Yuehwern Yih

Handbook of Healthcare Delivery Systems
Yuehwern Yih

HUMAN FACTORS
IN **TRANSPORTATION**

SOCIAL AND TECHNOLOGICAL EVOLUTION
ACROSS MARITIME, ROAD, RAIL,
AND AVIATION DOMAINS

EDITED BY
GIUSEPPE DI BUCCHIANICO · ANDREA VALLICELLI
NEVILLE A. STANTON · STEVEN J. LANDRY

CRC Press
Taylor & Francis Group
Boca Raton London New York

CRC Press is an imprint of the
Taylor & Francis Group, an **informa** business

CRC Press
Taylor & Francis Group
6000 Broken Sound Parkway NW, Suite 300
Boca Raton, FL 33487-2742

First issued in paperback 2019

ISBN-13: 978-1-4987-2617-7 (hbk)
ISBN-13: 978-0-367-87322-6 (pbk)

Library of Congress Cataloging-in-Publication Data

Names: Di Bucchianico, Giuseppe., editor.
Title: Social and technological evolution across maritime, road, rail, and aviation domains / editors, Giuseppe Di Bucchianico, Andrea Vallicelli, Neville A. Stanton, and Steven J. Landry.
Description: Boca Raton : Taylor & Francis, CRC Press, 2017. | Series: Human factors in transportation | Series: Industrial and systems engineering series ; 11 | Includes bibliographical references and index.
Identifiers: LCCN 2016021879 | ISBN 9781498726177 (hard back)
Subjects: LCSH: Transportation engineering. | Human engineering. | Transportation--Social aspects.
Classification: LCC TA1145 .S64 2017 | DDC 629.04--dc23
LC record available at https://lccn.loc.gov/2016021879

Contents

Section II   Rail Domain

Section III   Road Domain

Section IV   Aviation Domain

Preface

The development of the transport industry, as a whole and in all its main domains (rail, road, maritime, and aviation), is essential for the well-being of our planet: it not only can facilitate the mobility of citizens and goods, but also has a significant impact on economic growth, on social development, and on the environment.

In recent decades, moreover, the movement of people and goods in the world has seen unprecedented expansion. The overall increase in traffic at the global level has been both a cause and consequence of many contemporary phenomena, often interconnected: increasing globalization, with the abolition of many import duties and borders; the liberalization of markets, which has in many cases led to substantial price reductions; the change of social structures and demographic evolution, which have fueled the increase in travels and trips; increasing urbanization, which favored the commuting phenomena; the process innovations in production systems and delivery of goods, which saw the elimination of warehouse stocks, the development of "just-in-time" supplies, and the rapid growth of mail order sales; the technological advances in energy and ICT (information and communication technology), which have revolutionized the same systems and the means of transportation.

So the transport sector has a growing role in the economic and social development of contemporary societies. This also related to the frantic evolution of technology, especially electronic and information, which led to inevitable and profound changes, not only in economic activities, but also in our daily lives. If, however, the sector as a whole on the one hand continues to have an unprecedented dynamism and economic prosperity, it determines ever-increasing social and ecological costs. Many problematic aspects are due in large part to a substantial gap between recent technological and social evolution and some endogenous and substantially inefficient characteristics of transport systems, already largely anachronistic, that tend to slow down its evolution to the idea of an overall system of integrated and sustainable mobility.

As a particular "critical" feature of the entire transport sector, it is important to first consider that it is largely characterized by small and microenterprises: the majority of companies have fewer than 50 employees, while large companies are considered typical mostly for air and rail transportation. This is certainly a hindering factor to the idea of systems integration and of the creation of combined and intermodal transport chains.

This assumption impacts directly or indirectly also on other factors, in many ways unfavorable, that crosswise characterize the whole sector.

This starts from the workforce, which continues to be predominantly male: in Europe, for example, men represent more than 80% of workers in the transport sector, although the gender difference is not so great in all subsectors. Actually the number of women working in the transport sector is growing, but on terms that do not conform to this changing reality. In fact, even for the jobs and the sectors with a growing number of female workers, ergonomic investments in these workplaces still continue to extensively refer to a work environment considered for men only, in terms of infrastructure, culture, and conditions of work. In this situation, women working in the transport sector must still continue to adapt to an organization focused mainly on male labor.

Again with reference to the personnel, it is necessary to reflect also on the age of employees, because the transport workforce is aging at a faster rate than the general active population in all other sectors. Again it can be useful to refer to Europe: although it is a continent notoriously "old," it is significant that a recent study highlighted how only about 6% of workers in the transportation sector are less than 25 years old (versus more than 11% that instead represents the youth labor force in the total EU workforce). The aging phenomenon, which also affects the transport sector, raises the issue of how to ensure better working conditions for older workers and how to contribute to the updating of their skills in relation to ongoing organizational and technological changes that characterize the sector while responding to the ever-increasing needs of innovation expressed by customers. In fact, the pervasive diffusion of technical applications, such as electronic remote control systems and mobile means of communication, has changed the content of the workload: the transport workers have to operate using this complex equipment, and it is not always a simple task, especially for the elderly, who therefore need ongoing and adequate training in step with these changes.

The need for training is also expressed by the workers with part-time contracts, widely used in this industry, especially for more monotonous activities and those requiring less professional skills.

The transport sector is also one of the productive sectors employing more migrant workers, mainly employed in jobs that typically are the most dirty, dangerous, and tiring ones: primarily as drivers on long distances in road transport; for baggage handling and cleaning in air transport; in maintenance and services on ships. Their work is often characterized by uncertainty, poor working conditions, part-time jobs, and low salaries. The growing size on a planetary scale of workers' migration puts in the foreground, next to the theme of safety and health in the workplace, the theme of cultural diversity and, therefore, of different habits, practices, and approaches to the development of tasks and activities. This raises, therefore, the need to adopt cross-cultural and universal procedures and practices easily transmitted through user-friendly devices, organizational systems taking into account human diversity (especially cultural and linguistic), and lifelong learning programs.

Further, secondary phenomena are flanked to the above features characterizing the whole transport sector related to the transformation of this sector in recent years and which tend to amplify their critical aspects: the rapid growth of air transport, especially low cost, with infrastructure subject to new intense activities, in relation to many more users, in environments which must be tailored to continuous change; the impact of the change in consumer habits, for example, relating to food, which affects the infrastructure related to the food industry; the increase in requests of dedicated transport for specific users or for private journeys of a short distance; the change of travel mode as far as distances, locations, destinations, and duration of the trips; the effect of an aging population on the demands of transport and on its related infrastructure; the rapid growth of public short distance commuting transport, even for specific user groups (school children, employees, etc.).

All of the above-mentioned characteristics and phenomena, finally, in the last few years have necessarily had to, and will more and more in the future, confront the environmental issues posed by the rapid growth in emissions due to transport systems. It is notoriously one of the largest global challenges to be faced in the fight against climate change. Over the past decade, in fact, emissions from transportation increased at a faster rate than any other industry that uses energy, and about a quarter of the gas emissions are mainly due to road vehicles. Achieving sustainable development will require a radical change of culture and of the overall infrastructure of transportation with the involvement of many other sectors in addition to the energy and environmental ones, such as the technological areas of information and communication, agriculture, and trade. This will require a multidisciplinary approach to issues that are increasingly interconnected.

In this complex and evolutionary scenario related to the transport sector, we wonder what role ergonomics may have for the sustainable development of the sector, this starting from a clarification on what the new paradigms of "well-being," which ergonomics refers to, will be.

In this regard, Ezio Manzini in 2004 summarized very clearly the transition from a "possession-based" well-being typical of preindustrial and industrial societies, to an "access-based" well-being in the transition to postindustrial societies, to, finally, a "context-based" well-being. And it is precisely this last conception of well-being, which will accompany the sustainable development of contemporary societies, that allows the recognition of the value of diversity of physical and social contexts, such as of diversity among individuals, in the idea that their valorization could carry to an overall improvement in the quality of life.

It is a transformation that will invest all the spheres of life and actions of man more and more, including those relating to transportation.

As a matter of fact, the new applied ergonomic issues related to transportation are required to deal with some emerging topics, such as growing automatism and manning reduction, advances in and pervasiveness of ICT, or new demographic and social phenomena, such as aging or multiculturalism.

New scenarios and vision for the years to come require new interpretation and development, including those issues and thematic areas that have already been well established in ergonomics literature and expertise. We refer, for example, to safety and wellness of workplaces, in the changing work scenario; to information and human–machine interaction onboard and in the infrastructures and systems connected with all means of transportation, in the ICT pervasiveness scenario; to human diversity and environmental design, in the new socio-demographic phenomena scenario.

Probably only renewing the interpretation of traditional topics or facing emerging ones related to ongoing social and technological phenomena in an innovative way, ergonomics can contribute to the development of an overall system of truly integrated and sustainable mobility.

This book aims to provide a first possible contribution in this direction, through a collection of reflections, experiences, and researches related to the relationship between human factors, recent social and technological developments, and the main areas of transportation: maritime, rail, road, and aviation.

Giuseppe Di Bucchianico

Editors

Giuseppe Di Bucchianico (Pescara, 1967), architect, earned a master's degree in ergonomics from Politecnico di Milano in 2000 and a PhD in technological culture and design innovation from University of Chieti-Pescara in 2001.

He currently is an associate professor in industrial design at the University of Chieti-Pescara. He conducts his research mainly in the field of relationships between individuals, artifacts, and environments, for the development of a holistic and user-centered approach to design, themes with which he has participated in several national and international conferences. The application areas are those of yacht design, design for all, and ergonomics for sustainable development.

Dr. Di Bucchianico is co-chairing several scientific and technical boards such as the technical committee "Ergonomics and Design for Sustainability" at IEA (International Ergonomic Association); the scientific advisory board at the International Conference on Human Factors in Transportation, and at the International Conference on Design for Inclusion, both in the ambit of the AHFE international conferences. Currently he is also president at Design for All Italia and vice president administration at EIDD-Design for All Europe.

Dr. Di Bucchianico has won numerous international architecture and design competitions and awards.

Professionally, he worked mainly as an industrial designer, with an approach geared primarily to social inclusion, working with numerous companies (including Gattocucine, Valcucine, Foster, Abis, and Tecnolam), exhibiting at major exhibitions and events in Milan, Verona, Brussels, Moscow, Paris, Frankfurt, and New York, and receiving reviews in prestigious magazines.

Andrea Vallicelli (Rome, 1951), architect, is professor of industrial design at the University of Chieti-Pescara and teacher in the master of yacht design at the Politecnico di Milano.

He has undertaken wide design activity in different areas. For the boating industry he has designed numerous pleasure boats built in Europe and America including Brava (1979 New York–1986 Limington); Azzurra (America's Cup 1983, 1987); Orsa Maggiore (the training ship for the Italian Navy, 1994); Virtuelle (in collaboration with Philippe Starck, 2000) and the family lines "Comet" (1985–2010) and "ISA" (2004–2010). In the furniture sector, he has collaborated with the Busnelli Industrial Group.

Dr. Vallicelli received several awards and recognitions and an overview of his works has been exhibited in the Pavilion of Contemporary Art in Milan (PAC) as part of the exhibition "It's Design," in 1983; in the Car & Yacht Designers' exhibition "The Birdhouse Project" in Osaka (Japan), in 2000; in the exhibition for SMEs in Canton (China), in 2006; and in the exhibition "Sport Design in Creative Interaction" in Jinan (Shandong, China), in 2008.

Dr. Vallicelli was the scientific coordinator for the unit of Chieti of the PRIN national researches (co-financed by the Italian Ministry of University and Scientific Research): "Sistema Design Italia" (which was awarded with the Compasso d'Oro ADI XIX 2001), in 1997/1999; "The design for the industrial districts," in 2001–2002; and "Me-Design," in 2002/2003.

He is co-chair of the scientific advisory board at the International Conference on Human Factors in Transportation, in the ambit of the AHFE International Conferences; and he was a member of the National Design Council of the Ministry of Cultural Heritage and he is a member of the Executive Committee of the Scientific Italian Society of Design (SID).

Neville A. Stanton earned his BSc in psychology from the University of Hull, Hull, UK, in 1986; an MPhil in applied psychology in 1990; a PhD in human factors engineering from Aston University, Birmingham, UK, in 1993; and a DSc in human factors engineering for the University of Southampton, Southampton, UK, in 2014.

Dr. Neville is both a chartered psychologist registered with the British Psychological Society and a chartered engineer registered with the Institution of Engineering and Technology in the United Kingdom. He has held a chair in human factors engineering since 1999, joining the University of Southampton in the Engineering Centre of Excellence in 2009. His research interest includes development and validation of ergonomics and human factors methods, analysis, and investigation of accidents, design of human–machine interaction, and modeling and investigation of human performance in highly automated systems. The Institute of Ergonomics and Human Factors awarded Dr. Neville the Otto Edholm Medal in 2001 for his contribution to ergonomics research, the President's Medal in 2008, and the Sir Frederic Bartlett Medal in 2012 for a lifetime contribution to ergonomics research. In 2007, The Royal Aeronautical Society awarded him the Hodgson Medal and Bronze Award with colleagues for their work on flight-deck safety.

Steven J. Landry is an associate professor and the associate head in the School of Industrial Engineering at Purdue University, and associate professor of aeronautics and astronautics (by courtesy). He has a BS in electrical engineering from Worcester Polytechnic Institute, an SM in aeronautics and astronautics from Massachusetts Institute of Technology, and a PhD in industrial and systems engineering from Georgia Tech. At Purdue, he conducts research in air transportation systems engineering and human factors. He teaches undergraduate and graduate courses in air transportation, human factors, statistics, and industrial engineering. Prior to joining the faculty at Purdue, Dr. Landry was an aeronautics engineer for NASA at the Ames Research Center, working on air traffic control automation. Dr. Landry was also previously a C-141B aircraft commander, instructor, and flight examiner with the US Air Force. He has over 2500 heavy jet flying hours, including extensive international and aerial refueling experience.

Dr. Landry's research has been funded by, among others, NASA, the Department of Transportation, and the FAA. Dr. Landry has published over 80 peer-reviewed journal articles, conference papers, and book chapters, including book chapters on human–computer interaction, automation, and human factors. He coauthored the second edition of the book *Introduction to Human Factors and Ergonomics for Engineers* (Lawrence Erlbaum Associates) with Dr. Mark Lehto, and published book chapters on Aviation human factors, in Salvendy, G. (Ed.), *Handbook of Human Factors and Ergonomics*, 4th edition; Human–computer interaction in aerospace, in J. Jacko (Ed.), *The Human Computer Interaction Handbook*, 3rd Edition; and Flight deck automation, in Nof, S. (Ed.), *Handbook of Automation*, Springer.

Dr. Landry is a senior member of the American Institute of Aeronautics and Astronautics and is a member of Sigma Xi; the IEEE Systems, Man, and Cybernetics Society; the Human Factors and Ergonomics Society; and the Institute of Industrial Engineers.

Contributors

Naseem Ahmadpour
Centre for Design Innovation
Swinburne University of Technology
Melbourne, Victoria, Australia

Bettina Bajaj
Centre for Transport Studies
Imperial College London
London, United Kingdom

Victoria A. Banks
Transportation Research Group
University of Southampton
Southampton, United Kingdom

Vanessa Beanland
Centre for Human Factors and
    Sociotechnical Systems
Faculty of Arts, Business and Law
University of the Sunshine Coast
Maroochydore, Queensland, Australia

Todd J. Callantine
Human-Systems Integration Division
San José State University/NASA Ames
    Research Center
Moffett Field, California

Stefania Camplone
Department of Architecture
University of Chieti-Pescara
Pescara, Italy

Miranda Cornelissen
Monash Injury Research Institute
Monash University
Melbourne, Australia

and

Griffith Aviation, School of Biomolecular
    and Physical Sciences
Griffith University
Brisbane, Australia

Jacquelyn Crebolder
Defence Research and Development
    Canada
Dartmouth, Nova Scotia, Canada

Christian Denker
OFFIS—Institute for Information
    Technology
R&D-Division Transportation
Oldenburg, Germany

Massimo Di Nicolantonio
Department of Architecture
University of Chieti-Pescara
Pescara, Italy

Tamsyn Edwards
Human Factors Research Group
Faculty of Engineering
University of Nottingham
Nottingham, United Kingdom

Sebastiano Ercoli
Dipartimento di Design
Politecnico di Milano
Milan, Italy

Emre Ergül
Dipartimento di Design
Politecnico di Milano
Milan, Italy

Florian Fortmann
OFFIS—Institute for Information
    Technology
R&D-Division Transportation
Oldenburg, Germany

Hartmut Fricke
Air Transport Technology and Logistics
Dresden University of Technology
Dresden, Germany

Marco Furtner
Department of Psychology
University of Innsbruck
Innsbruck, Tyrol, Austria

Katja Gaunitz
Air Transport Technology and Logistics
Dresden University of Technology
Dresden, Germany

Ashley Gomez
Human-Systems Integration Division
San José State University/NASA Ames
    Research Center
Moffett Field, California

Elizabeth M. Grey
Centre for Human Factors and
    Sociotechnical Systems
Faculty of Arts, Business and Law
University of the Sunshine Coast
Maroochydore, Queensland, Australia

and

Human Factors Team
Transport Safety Victoria
Melbourne, Victoria, Australia

Vimmy Gujral
Human-Systems Integration Division
San José State University/NASA Ames
    Research Center
Moffett Field, California

Axel Hahn
OFFIS—Institute for Information
    Technology
R&D-Division Transportation
Oldenburg, Germany

Eric Holder
Man-Machine-Systems
Fraunhofer Institute for Communication,
    Information Processing and
    Ergonomics
Wachtberg, North Rhine-Westphalia,
    Germany

I. C. MariAnne Karlsson
Division Design & Human Factors
Chalmers University of Technology
Göteborg, Sweden

Barry Kirwan
EUROCONTROL Experimental Centre
Bretigny/Orge, France

Dirk Schulze Kissing
Aviation and Space Psychology
German Aerospace Center/Deutsches
    Zentrum für Luft- und Raumfahrt
Hamburg, Hamburg, Germany

Michael Kupfer
Critical Networks
Harris Orthogon GmbH
Bremen, Germany

Michael G. Lenné
Monash University Accident Research
    Centre
Monash University
Melbourne, Victoria, Australia

Gitte Lindgaard
Centre for Design Innovation
Swinburne University of Technology
Melbourne, Victoria, Australia

Monica Lundh
Division of Maritime Human Factors
    and Navigation
Chalmers University of Technology
Department of Shipping and Marine
    Technology
Gothenburg, Sweden

Arnab Majumdar
Centre for Transport Studies
Imperial College London
London, United Kingdom

Yemao Man
Department of Shipping and Marine
    Technology
Chalmers University of Technology
Gothenburg, Sweden

Peer Manske
Center of Competence Safety
IABG mbH
Ottobrunn, Bavaria, Germany

Alastair Manson
Institute for Infrastructure and Environment
Heriot-Watt University
Edinburgh, Scotland, United Kingdom

Markus Martini
Department of Psychology
University of Innsbruck
Innsbruck, Tyrol, Austria

Pierangelo Masarati
Dipartimento di Scienze e Tecnologie
    Aerospaziali
Politecnico di Milano
Milan, Italy

Roderick McClure
Monash Injury Research Institute
Monash University
Melbourne, Australia

Rich C. McIlroy
Transportation Research Group
University of Southampton
Southampton, United Kingdom

Joey Mercer
Human-Systems Integration Division
NASA Ames Research Center
Moffett Field, California

Lothar Meyer
Air Transport Technology and Logistics
Dresden University of Technology
Dresden, Germany

Ann Mills
Professional Head of Human Factors
RSSB
London, United Kingdom

Alice Monk
Human Factors Specialist
RSSB
London, United Kingdom

Maria Carola Morozzo della Rocca
Department of Science for Architecture
Polytechnic School, University of Genoa
Genova, Italy

Florian Motz
Man-Machine-Systems
Fraunhofer Institute for Communication,
    Information Processing and
    Ergonomics
Wachtberg, North Rhine-Westphalia,
    Germany

Vincenzo Muscarello
Dipartimento di Scienze e Tecnologie
    Aerospaziali
Politecnico di Milano
Milan, Italy

Massimo Musio-Sale
Department of Sciences for Architecture
Polytechnic School, University of Genoa
Genova, Italy

Ursa Katharina Johanna Nagler
Analyses and Further Development
Federal Office of Bundeswehr Personnel
    Management
Cologne, North Rhine-Westphalia,
    Germany

Gina Netto
School of Energy, Geoscience
    Infrastructure and Society
Heriot-Watt University
Edinburgh, Scotland, United Kingdom

Lena Nilsson
The Swedish National Road and Transport
    Research Institute (VTI)
Linköping, Sweden

Marco Michael Nitzschner
Aeromedical and Psychological
    Assessment of Aircraft Accidents
Air Force Centre of Aerospace Medicine
Cologne, North Rhine-Westphalia,
    Germany

Helen Omole
School of Energy, Geoscience
    Infrastructure and Society
Heriot-Watt University
Edinburgh, Scotland, United Kingdom

Marie-Christin Ostendorp
OFFIS—Institute for Information Technology
R&D-Division Transportation
Oldenburg, Germany

Katherine L. Plant
Transportation Research Group
University of Southampton
Southampton, United Kingdom

Thomas Porathe
Department of Product Design
Norwegian University of Science and
    Technology (NTNU)
Trondheim, Norway

Giuseppe Quaranta
Dipartimento di Scienze e Tecnologie
    Aerospaziali
Politecnico di Milano
Milan, Italy

Andrea Ratti
Politecnico di Milano
Milan, Italy

Gemma J. M. Read
Centre for Human Factors and
    Sociotechnical Systems
Faculty of Arts, Business and Law
University of the Sunshine Coast
Maroochydore, Queensland, Australia

and

Monash University Accident Research
    Centre
Monash University
Clayton, Victoria, Australia

Jean-Marc Robert
Department of Mathematics and Industrial
    Engineering
Polytechnique Montréal
Montréal, Québec, Canada

Pierre Sachse
Department of Psychology
University of Innsbruck
Innsbruck, Tyrol, Austria

Joshua Salmon
Department of Psychiatry & Psychology/
    Neuroscience
Dalhousie University
Halifax, Nova Scotia, Canada

Paul M. Salmon
Centre for Human Factors and
    Sociotechnical Systems
Faculty of Arts, Business and Law
University of the Sunshine Coast
Maroochydore, Queensland, Australia

Kirsten Schreibers
INTERGO Human Factors &
    Ergonomics
Utrecht, The Netherlands

Sarah Sharples
Human Factors Research Group
University of Nottingham
Nottingham, United Kingdom

Ian Stevens
Suicide Prevention Programme Manager
Network Rail
Milton Keynes, United Kingdom

Niklas Strand
The Swedish National Road and Transport
    Research Institute
Göteborg, Sweden

Ailsa Strathie
School of Energy, Geoscience,
    Infrastructure and Society
Heriot-Watt University
Edinburgh, Scotland, United Kingdom

Richard van der Weide
INTERGO Human Factors & Ergonomics
Utrecht, The Netherlands

Guy Walker
Centre for Sustainable Road Freight
and
School of Energy, Geoscience,
    Infrastructure and Society
Heriot-Watt University
Edinburgh, Scotland, United Kingdom

Colete Weeda
INTERGO Human Factors & Ergonomics
Utrecht, The Netherlands

John Wilson
Human Factors Research
    Group
University of Nottingham
Nottingham, United Kingdom

Kristie L. Young
Monash University Accident Research
    Centre
Monash University
Melbourne, Victoria, Australia

Section I

Maritime Domain

Introduction

Social evolution today is closely connected with technological development more than in any other period in history. In fact, the continuous and frenetic process of technological evolution and, in particular, of information technology, entail inevitable and profound changes not only in economic assets, but also in the daily life of the relationship.

On the other hand, in this evolutionary context, the question is what the new paradigms of well-being are, which ergonomics deals with.

This section of the book, related to maritime transportation, tries a first attempt to relate some traditional ergonomic issues to a selection of emerging phenomena and questions posed by contemporary society. For this reason, the nine chapters can be informally divided into three main groups, comprising of certain issues, which, in general terms, could be defined as the design tools and methods (Chapters 1 through 3), the management and integration of information in navigation (Chapters 4 through 6), human diversity and accessibility (Chapters 7 through 9).

In particular, through the experiences described in these chapters, expert researchers, professionals, and doctoral students, aimed to explore the role of human factors in order to relate safety, security, and sustainable development in specific fields of the transportation domain; put forward new studies, tools, and design concepts related to spatial perception and comfort on board; develop new concepts for the management of the increasing amounts of information and communications onboard; propose reflections in reference to the growing reduction of crews and team sizes; highlight the potential relations and strategies between transportation, aging, and disabilities; wonder about new possible scenarios referring to growing human diversity and the need for social inclusion also in the transportation domain.

We hope that this joint effort represents the launch of an overall critical reflection on the way present socio-technical environment changes will affect the whole maritime transportation domain.

Giuseppe Di Bucchianico
Andrea Vallicelli

1

Ergonomics and Modern Technology in the Restoration of Historic Vessels: A Challenge for a New Life

Maria Carola Morozzo della Rocca

CONTENTS

1.1 Introduction

Historic ships are part of the cultural heritage of every state or nation. The protection, the valorization, and the restoration of navigating boats of historic interest are recent practice in the phase of aging or encoding.

Year after year the practice of nautical restoration grows and becomes stronger and stronger, but we need clear rules to achieve good recovery projects.

We can certainly say that the longevity of a vessel is directly proportional to the frequency and quality with which the operations of permanent maintenance are carried out. At the same time, the functionality of a vessel is closely related to the adjustment of the equipment and the necessary instrumentation to its navigation.

So, there is a multiplicity of aspects that are crucial to the renovation of historical craft, including primarily ergonomics as key to the design of how to operate boats as "places to live and voyage in" and technology (or instrumentation) as a fundamental element of navigation systems.

Technological development, miniaturization of instrumentation, and growth of technologies to increase the automatisms and to reduce the complexity of the maneuvers are an important aid in the restoration of historical boats.

The restoration of classic yachts with historical and cultural value must correctly consider all ergonomic, functional, and technological aspects in reference to both conservational issues and the need to upgrade existing designs to ensure that comfort and navigational safety standards are met.

1.2 Restoration of Classic Yachts and Contemporary Expectations

To evaluate the ergonomics of classic designs in terms of operational technology and livability of interior spaces is no banal aspect. Classic yachts show profiles, dimensions, technology, interior décor, and comfort of a very different standard level than today. These rétro features contribute to the cultural aura of classic designs. There are difficulties with the definition of whether insuperable limits exist to the upgrading of vintage craft according to ergonomics, technology, and functionality.

Current trends in yacht design are a continued design evolution and a research for habitability to meet increasing comfort and safety standards, independent of a vessel's technical sea keeping state. Yachts are complex products that are highly individualized and articulated. Design and manufacturing aspects that converge to make a yacht involve a variety of professional specializations. As the result of these different professional specializations a yacht follows their evolution in a transformation process that continues post manufacture, based on technical, technological, and functional upgrades that characterize periodical maintenance under applicable regulations.

We need a larger framework to analyze classic yacht renovation, because the innumerable possibilities of a contemporary yacht market must be evaluated in terms of their compliance under applicable legislation and in respect of preservation issues that characterize the recovery of historical artifacts.

Excluding a museological perspective as an extreme case for restoration, we have various degrees of operational freedom in the technological and ergonomic upgrade of historical yachts, for which many more years of service are expected post restoration, be it a conservational, philological, or critical-creative intervention. The only constraint in this respect is that the yacht's character is preserved, and as much as possible of the remains of its origins.

It is also key to this that a restoration project that respects the yacht's style language enables such upgrades as are required to ensure unlimited operational safety post restoration.

The big difference between the yachting industry and parallel sectors including the automotive industry is that no restrictions apply to the navigation of historical vessels, while there are restraints on classic cars under the law that only enable a sporadic, nonroutine utilization. Also, exceptions to the regulations disciplining the automotive sector do not apply to the yachting industry because classification societies require that periodic surveys are conducted to verify that vessels are maintained to class standards. Vessels are subsequently certified* (RINA) in conformity with laws and regulations. To maintain a yacht's efficiency unaltered is key to its preservation. To do this, there is a need to both rescue and update at the same time.

1.3 Critical Analysis of Case Studies

The restoration of historic boats performed over the past 50 years are many and very different. The work on boats and recovery related to historical ships, in fact, are the result of

---

* Periodic surveys of newly built RINA classed yachts that are CE compliant are conducted no later than 8 years of RINA's original classification for A and C categorized yachts, and no later than 10 years of RINA's original classification for C and D categorized yachts. Subsequent inspections are on a 5 year basis. RINA's surveys consist in a series of analysis that normally focus on hull, helm, and shafting line, pump room, piping and safety valves, engine cooling and exhaust system, fuel system, bilge system, electrical system, fire safety systems, and kitchen gas supply systems.

the passion and ability of a few individuals, researchers or scholars, most often specialized operators and masters of the yard.

JClass Shamrock V,* for example, was the only J that was maintained to continued service since launching of all JClass yachts. Derelict Endeavour and Velsheda were both salvaged in the 1980s. In a history of ownership changes, Shamrock V underwent repeated restorations that preserved its efficiency, while purists of its original design may have considered them controversial at times.

That total conservation *per se* that was eluded in this respect has enabled the continuity to this day of a historically significant exemplar that is technically representative of a particular formal value. This reasoning is limited to a particular example and may not be generalized in absolute terms.

So, the possibility that historical boats come back to shine their own lives is closely linked to their efficiency. Because a yacht generally represents capital expenditure that generates no return, this is accentuated if its operational perspective is extremely limited, due to difficult usability and livability. From a utilitarian, realistic standpoint, it is very difficult to find sponsors who have a disposition to exclusively safeguard cultural heritage unless it is perfectly efficient.

Classic yachts, those ocean ladies that fascinate habitués on the dock, can no longer afford large crews to maneuver them, and at the same time their wealthy owners may want to find all or almost all the comforts of contemporary living below deck.

Human factors and technology upgrades are key aspects of a restoration project to achieve full usability of the recovered good. It is not by chance that only among the numerous Js that were key players in the 1930s, Shamrock V only continued uninterrupted service while capitalizing on the possibilities of a technological upgrade as compliant under applicable regulations, which included the possibility to replace its original gaff versus Bermudan rig. Technology and passion for racing, seconded by the finance of the owners, also contributed to this result.

We look to the methodology of this upgrade and its impact on the yacht's existing general plan to possibly unveil a correct philosophical interpretation of a restoration project.

1.4 Demands of Modern Life and Contemporary Technology Evolution in the Recovery of Historic Boats

With reference to vintage designs, and to exemplify major criticalities of historical yachts largely, we may argue that operational upgrades address issues primarily in reference to a yacht's upper deck plan in terms of new maneuverability and compliance with existing safety standards, and to the redesign of interior spaces to new habitability practices in respect of earlier tradition.

Upgraded sails plan, rigging and deck equipment enable facilitated maneuvers and reduce crews as required for operational purposes, also in compliance with regulations that were not in force and binding at the time that vintage yachts were launched. There is a need to introduce new technology to enhance sail plan control that affects the upper deck

---

* Shamrock V is a JClass sailing yacht. A design of Charles Ernest Nicholson for Sir Thomas Lipton's fifth and last America's Cup challenge, lost to US design *Enterprise*.

morphology while maintaining the solid and void architecture of deckhouse and cockpit to their original design (Figure 1.1).

For vintage yachts older than 75 years, it is rare to find exemplars in their original designed state. Also, there may be uncertainty about a yacht's original designed plan based on documentation as available. Designers would personally engineer post launch upgrades to their original JClass race plans to improve their performance. Among others, this is the case with JClass Astra, a 1927 design of Camper and Nicholson, who reconfigured it to a new sail plan and water ballast ratio 20 years later. Because of this, there are difficulties in establishing a philological order and identifying whether Astra's authentic configuration is its original built condition or the condition after it was upgraded to an optimized performance. Because this is in issue, we argue that there is consistency in utilizing design skills and tools to upgrade and enhance a yacht's usability features to keep it alive, the same as the yacht's original designers did.

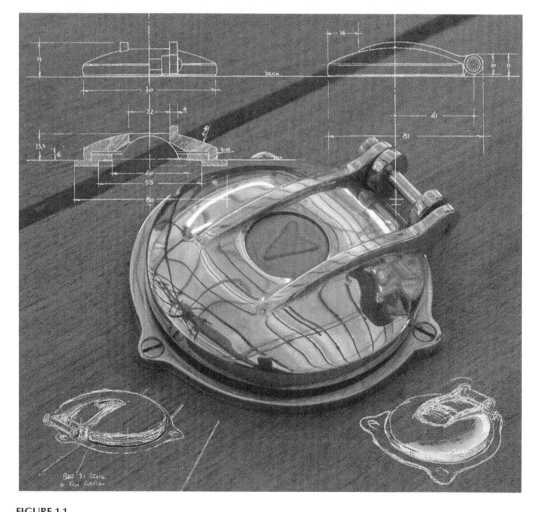

FIGURE 1.1
Lulworth (the world's largest surviving gaff rigged cutter) restored by Studio Faggioni Yacht Design of La Spezia, Italy. In the image, the design and the realization of the caps for covering the electric winches. (Photographs by Francesco Rastrelli, drawings by Studio Faggioni.)

For purebreds designed to be raced and subsequently owned to serve new purposes in cruise or charter work, there is a need for domestication to enable a new usability with smaller crews and adjusted parameters of guest spaces below deck. Because navigational safety and comfort have found an organic definition in an ergonomic approach, it is also indispensable to comply with those norms that discipline the relation of human work and the goods that are instrumental to it. This includes all elements that are in a relation with the operations that are routinely performed as part of human work. Applied to yacht design, this discipline focuses on all aspects where an interface exists between instruments and users.

One particular field of application of ergonomics to the redesign of sailing craft is a yacht's deck, where multiple activities take place in restricted spaces. For sailing yachts, propulsive functionality is a responsibility of crews who simultaneously interact with the yacht's equipment integrally. Cockpit and deck on board racers and cruisers alike epitomize a testing ground for human factors and safety of operations in terms of spaces and sizes, posture and coordinated motion, load levels on muscular and skeletal systems in various maneuvering positions.

During navigation a boat is a space where contemporary activities happen that are coordinated to be performed in a tight, hectic timeframe, such as course changes, tacking and jibing maneuvers, and complementary, more static actions consisting in maintaining pace and minimal maneuver regulation. Deck gear plus the kinematic mechanisms associated to them undergo continual upgrade that is incorporated in a variety of ergonomic assessment tools internationally, including OWAS (Ovaco working analyzing system), NIOSH (National Institute of Occupational Safety and Health), and RULA (rapid upper limb assessment).

Modern design and engineering ensure safety margins on deck that are absolutely superior to those in the past, both based on recently defined regulations and an in depth analysis of functions, operational dimensions, range, correct general logistics positions, reciprocal distances of various maneuvering positions in respect of winches, rails, gearings, halyards, sheets, cordage, etc. Winches, rails, and controls are located to optimize performances aligned to human factors considerations as earlier discussed. Because classic yachts came earlier than these concepts, the upgrade of their sail plan and maneuvrability can only be consequential to a larger operational rationalization. Safety in operations and onboard livability may not be secondary when design has a focus on a rationalization of the potentialities of an existing exemplar.

The restoration of historical craft affects both their operational efficiency and aspects in relation to better onboard livability, involving comfort levels of interior spaces and of all activities that are in association with routine organization below deck.

Preserving the original yacht layout and bulkhead partitions, interior décor upgrades are possible at a variety of levels to serve the purposes that generated them. In a comprehensive work logic, restoration enables limited changes that may not compromise the original yacht design in the name of better habitability, to the extent necessary to adapt interiors to human factors considerations as contemplated under current legislative trends on marine safety and livability.

Also, there is a need to consider an evolution of home habits and habitability requirements.

Specific style and form criteria condition improvements to classic designs, such that additions must be completely integrated in the general development of the original décor. Interventions typically focus on critical kitchen, restroom facilities, cabin comfort, and operational facilities and equipment.

For example, interior spaces that are also noncompliant in respect of current standards are restroom facilities. Spartan, undersized spaces, peripheral to the overall yacht's design, restroom facilities hardly accord with current standards and need substantial updating to meet the needs of a changed culture and lifestyle.

Facilities are an important component for which update is indispensable. This includes the redesign of kitchen technology (Figure 1.2).

Contemporary kitchen technology is a must both at home and on board a classic yacht. To utilize kitchen appliances and equipment that guarantee efficiency in terms of the yacht's energy, balance justifies the replacement of obsolete elements that are both non-compliant with current safety legislation and with the standards of primary classification societies internationally.

There is a need to replace or else integrate, in a careful mimesis, each single element, from dishwasher to refrigerator, to fit in the original vintage décor of these charming

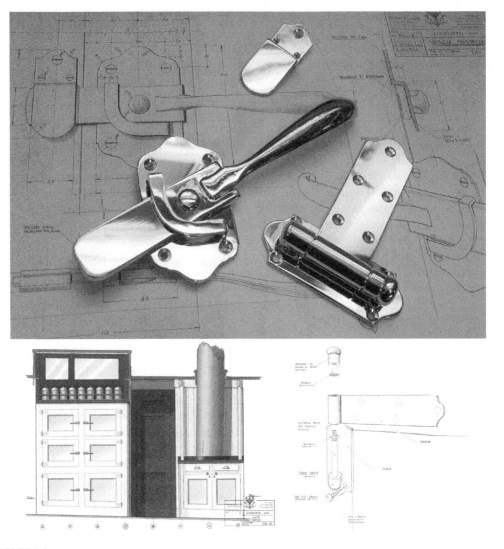

FIGURE 1.2
Lulworth (the world's largest surviving gaff rigged cutter) restored by Studio Faggioni Yacht Design of La Spezia, Italy. In the image, a detail of the icebox realized using the modern technologies but at the same time respecting the original galley's style. (Photographs by Francesco Rastrelli, drawings by Studio Faggioni.)

"ladies of the ocean" yachts. For particular home appliances, such as an icebox, individual-ized redesign enables upgrades to refrigeration technology in conformity with applicable regulations while preserving the original cabinet design.

Also at the same time that interiors are modified, aligned to current hygrothermal comfort standards, there is the possibility to modify existing interiors and install—rigorously integrated in the original interior design and hidden from sight—heating and air conditioning systems.

Contemporary habitability styles strongly characterize the offer of technology for basic utilities in fields that include the recreational marine industry. Home automation is increasingly a determinant of interior comfort levels as a key aspect of buying decisions in the yachting industry, in regard to both new and historical craft. Indispensable onboard equipment includes such utilities that enable the same livability standards as a residential setting. This involves control technology for environmental comfort, multimedia, information and entertainment, auxiliary instrumentation including bow thrusters, and tension inverters in support of portable devices as are ordinarily indispensable.

Technological implementation interfaced with evolving customs affects a yacht's habitability and its maneuverability and navigational instrumentation. It is necessary for advanced multiple device panel instrumentation that controls onboard functionalities—engines, trim controls or stabilizers, lightning, environmental conditioning, hydraulic systems, geolocation systems, communication—to be integrated and aptly concealed on board classic yachts.

The organization of advanced navigational systems configured in accordance with current cognitive ergonomics is no "impossible thing" for classic yachts. To upgrade a yacht design to current requirements is both a good design practice and obligatory under applicable legislation, particularly in such cases as, for example, when private yachts for charter are involved. There are safety issues in this respect that are a responsibility of owners toward yacht charterers, under normal circumstances.

1.5 Conclusions

We believe that the talent, knowledge, and cultural sensitivity of designers, through an ability to skillfully integrate "the old and the new," will successfully enable better upgrades to classic yachts, that are both in accordance with the original yacht design and functional to avoid their anachronistic unsuitability and inevitable dereliction.

The continuous technological development and the increasing miniaturization of instrumentation are a big help for designers to improve all nautical restoration projects. The latest technologies, in fact, can be easily adopted and integrated on historical boats without altering the spirit of the ancient yachts.

Further Reading

Ciocchetti, D., Rosa, A. 2006. I Relitti Galleggianti. Riccardo I. L'ultima Comacina di Comacchio. Storia di un recupero, in: *Il MiBAC ricerca e applicazioni a confronto, Atti del convegno*, Venezia, dicembre 1–3, X Salone dei Beni e delle Attività Culturali.

Di Maggio, P. 2005. La dichiarazione delle imbarcazioni storiche: Il recupero di Pianosa, in: *Yachting Library*, Library Editore srl, Milano.

Faggioni, S. 2002. *Storia, Tecnologia e Committenza: restauro di una barca d'epoca*, in: *DDD—Rivista trimestrale di Disegno Digitale e Design, anno 1 n°3, Edizioni Poli. Design*, Milano, lug/set 2002, p. 47.

Giacinti, R. 2007. L'imbarcazione storica come bene culturale, in: *Seminario III° Raduno Vele Storiche Viareggio*, Viareggio, Italy.

Giacinti, R. 2008a. Il bene culturale barca visto dalle associazioni, in: *Nautica, n°552*, Nautica Editrice, Roma.

Giacinti, R. 2008b. La barca storica come bene culturale, in: *Nautica*, n°553, Nautica Editrice, Roma.

Giacinti, R. 2008c. Norme per le imbarcazioni d'epoca e di interesse storico e collezionistico, in: *Convegno Associazione Vele Storiche di Viareggio*, Viareggio, Italy.

Giacinti, R. 2009. DDL di regolamentazione delle imbarcazioni d'epoca, in: *Yacht Digest, n°155*, Nautica Editrice, Roma.

Giacinti, R. 2012a. Il Redditometro della nautica, in: *BEC–Barche d'Epoca e Classiche*, anno 2 n°4, gen/feb/mar.

Giacinti, R. 2012b. Inquadramento fiscale dei servizi di ormeggio svolti da circoli nautici, in: *BEC-Barche d'Epoca e Classiche*, anno 2 n°6, lug/ago/set.

Giacinti, R. 2013a. Il Redditometro per le barche d'epoca, in: *BEC-Barche d'Epoca e Classiche*, anno 3 n°11, ott/nov/dic.

Giacinti, R. 2013b. Il Registro delle imbarcazioni storiche con la Legge di Stabilità 2012, in: *Convegno Associazione Vele Storiche di Viareggio*.

Giacinti, R., Cambi, S. 2011. Inquadramento fiscale dei servizi di rimessaggio di imbarcazioni svolti dai circoli nautici, in: *Terzo Settore, Il Sole 24 Ore, n°10, Grupp 24 Ore*, Milano.

Giacinti, R., Fani, A. 2008. Proposta di regolamentazione delle imbarcazioni d'epoca e di interesse storico, in: *Bozza disegno di legge, IV Raduno Associazione Vele Storiche Viareggio*, Viareggio, Italy.

Giacinti, R., Fani, A. 2009a. Proposta di legge sull'istituzione del Registro delle Imbarcazioni di Interesse Storico, in: *Convegno Vele Storiche Viareggio*, Viareggio, Italy.

Giacinti, R., Marino, G. 2006. Tutela e valorizzazione di imbarcazioni storiche, in: *Terzo Settore, Il Sole 24 Ore, n°12, Gruppo 24 Ore*, Milano.

Moretti, M., Liberati, S. 1995. *Guida alla stima di un'auto d'epoca. Metodologia di analisi e valutazione*, Palombi Editori, Roma.

Morozzo della Rocca, M. C. 2006. Un'esperienza assoluta, Lulworth raccontata da Stefano Faggioni, in: Editrice A. (Ed.), *GUD, Genova Università Design*, n°6, Firenze, pp. 82–93.

Morozzo della Rocca, M. C. 2009a. Le tipologie contemporanee, definizione dei recenti modelli nautici, in: Musio Sale M. (Ed.), *Yacht design, dal concept alla rappresentazione, Tecniche Nuove editrice*, Milano, pp. 39–55.

Morozzo della Rocca, M. C. 2009b. Nuovi scenari per il design nautico, il design per il restauro occasioni di riflessione, in: Musio Sale M. (Ed.), *Yacht design, dal concept alla rappresentazione*, Tecniche Nuove editrice, Milano, pp. 39–55.

Morozzo della Rocca, M. C. 2010. Le radici del design nautico contemporaneo. Rapporti disciplinari e ambiti progettuali. The roots of contemporary naval design. Interdisciplinary elements and fields of design, in: Falcidieno M.L. (a cura di), *Le scienze per l'architettura. Frammenti di sapere/Architectural sciences. Fragments of knowledge*, Allinea, Firenze, pp. 340–359.

Morozzo della Rocca, M. C. 2014. *Yachts Restoration*, Umberto Allemandi and C., Torino.

Nardella, D. 2004. Uscita e ingresso nel territorio nazionale, in: *AA. VV., Il codice dei beni culturali e del paesaggio*, il Mulino, Bologna.

Petragnani Ciancarelli, A. 2005. Il formale riconoscimento di interesse culturale per un'imbarcazione storica, in: *Nautica n°516*, Nautica Editrice, Roma, p. 246.

Prud Homme, G. in press. Le label Bateau d'interet patrimonial, in: *Atti del Convegno "Censimento e tutela del naviglio storico e tradizionale in Italia. Le imbarcazioni da lavoro,"* Cesenatico, giugno 17, 2009, Museo della Marineria di Cesenatico e di ISTIAEN.

Riccardo, A. 2006. L'Ultima delle Comacina di Comacchio. Storia di un recupero, in: *Archeologia e Territorio, Atti del Convegno*, Paestum, novembre 16–19, IX Edizione, Borsa Mediterranea del Turismo Archeologico.

Robinson, J. in press. Censimento della flotta storica: l'esperienza britannica, in: *Atti del Convegno "Censimento e tutela del naviglio storico e tradizionale in Italia. Le imbarcazioni da lavoro,"* Cesenatico, giugno 17, 2009, Museo della Marineria di Cesenatico e di ISTIAEN.

Rodrigo De Larrucea, J., 2009. La protection del patrimonio flottante: hacia un estatuto juridico del buque historico, Communication for the "Jornada tècnica sabre la gestión de embarcaciones históricas," on the occasion of the Salone Nautico Internazionale di Barcellona 2009, Museu Maritim de Barcelona, Spain.

Rosa, A. 2006. Barche tradizionali dell'Adriatico. Tutela del Patrimonio Storico Artistico Etnoantropologico, in: *Atti del Convegno*, Paestum, novembre 16–19, 2006, IX Edizione, Borsa Mediterranea del Turismo Archeologico.

Rosato, G. 2011. *La tutela e il restauro delle imbarcazioni storiche*, Ministero per i Beni e le Attività Culturali, Soprintendenza Beni Storici Artistici, Etnoantropologici della Liguria.

Zaccagni, E. 2006. Vintage Yachts: il lavoro del project management, in: *Yachting Quarterly, n°10*, Fabio Ratti Editoria, Milano.

Zignego, M. I. 2012. *Yacht Refitting. Nuove frontiere dell'allestimento nautico*, Aracne Editrice, Roma.

2

Assessing the Fitness of Information Supply and Demand during User Interface Design

Christian Denker, Florian Fortmann, Marie-Christin Ostendorp, and Axel Hahn

CONTENTS

2.1 Background and Driving Forces

Accident reports reveal that human error is the number one cause of accidents in transportation. For example, a review of accidents in the aeronautical domain states that 71% of the flight accidents investigated by the U.S. National Transportation Safety Board (NTSB) between 1989 and 1992 were caused by human error (Jones and Endsley, 1996). Further, a review of accidents in the maritime domain, which was performed on the basis of data from the U.K. Marine Accident Investigation Branch (MAIB), the Transportation Safety Board of Canada (TSBC), and the Australian Transport Safety Bureau (ATSB), showed that more than 82% of accidents in shipping were associated with human error, and in 46% human error was even the main cause (Baker and McCafferty, 2005) of which 71% were caused by degraded situation awareness (SA) (Hetherington et al., 2006).

SA can be seen as a state of mind, which contributes essentially to the human decision-making process. According to Endsley's very common SA model, SA is composed of three levels: the perception (level 1), comprehension (level 2), and projection (level 3) of information (Endsley, 1995). Reports from human factors research show that 60%–77% of SA-induced errors were errors on level 1, besides 20%–30% on level 2, and 3%–9% on level 3 (Grech et al., 2002; Jones and Endsley, 1996). Thus, errors related to the perception of information are clearly the most frequent source of error within this three-level taxonomy. Furthermore, errors on level 1 can cascade to errors on level 2 and level 3 (Endsley, 1995). This implies that the elimination of level 1 errors could lead to a significant reduction of accidents and an increase of overall safety in the maritime domain.

Level 1 errors can have various causes, which cannot be strictly attributed to humans but is a severe problem with regard to information distribution and human–machine interaction on board which is closely related to the ICT (information and communication technologies) pervasiveness scenario. This can also be reasoned from the SA error taxonomy which states five causes for level 1 error (Jones and Endsley, 1996): "data is not available," "data is hard to discriminate or detect," "monitoring or observation of data failed," "misperception of data occurred," and "memory loss."

In this chapter, we focus on the two level 1 error causes "data is not available" and "data is hard to discriminate or detect." We present a formal method for the assessment of the fitness between information supplied by a user interface (UI) and the information demanded by a user. We implemented the method in an integrated systems modeling environment. We demonstrate the method with a course change task on a ship bridge. We performed an interview with system engineers from the maritime domain and a human factors ergonomist as an initial evaluation. Overall, the feedback was very positive and warrants further investigation of the method.

2.2 Related Work

In computer science formal verification methods from mathematics are used to prove the correctness of an algorithm in accordance to a defined system specification or property description. A common method is model checking. Model checking allows us to verify whether a given system model fulfills a specification. As Meolic et al. state the "method requires that a system is given with a graph, which describes the system behavior in terms of states and actions" (Meolic et al., 2000). The specification is expressed as logical propositions. The verification is done by checking the system models compliance to the propositions (Clarke et al., 1999; Meolic et al., 2000). In terms of model checking, human and machine could be defined as individual system models. In our approach the human's demand for information can be defined as the required specification. The information demand specification is used to verify the existence of the machine's information supply. Therefore, the states of the machines model contain supplied information as properties.

To establish SA and incorporate new information humans search for required information in their environment. They interact with machines and other humans to exchange information during task execution. Koreimann defined three types of information exchange between human and machine in information systems: dialog, report, and information retrieval (Koreimann, 2000). A dialog describes a bidirectional information exchange, for example, in a request and response pattern. The report is a unidirectional flow where information is transferred from a computer to the user or from the user to the computer (e.g., displaying information about the system status). An information retrieval is also a unidirectional flow, but it differs in the direction of the information flow, since information is taken from one of the information system's parts (e.g., looking for information on computer's display). The concepts of dialog, report, and information retrieval are abstractly used in our model to describe the direction of an information flow and set the initiator of an information exchange.

Besides the concrete interaction, a human has to set herself/himself into a position in which it is possible to take part in the information exchange. Here, the spatial distribution of information supply is important; because it influences the time it takes to build up SA. In the domain of human factors there exist a variety of methods considering the spatial distribution of information. One example is a method called link analysis (Chapanis, 1965; Wilson and Corlett, 2010). Link analysis allows identifying "links" between interface components (or functions) and human operators. The "links" are constructed out of a human's gaze movements between the components or a sequence of use of components or functions (Stanton and Young, 1999). For instance, the sequence of pressing button *A* and afterwards pressing button *B* would construct a link between buttons *A* and *B*. The analyst records the frequency and execution times of the links during a task under investigation. Based on the records the links are drawn onto a schematic representation of the interface to construct the so-called link diagram and a link table is created, which contains the same information in tabular form. The results of a link analysis are used to optimize the interface by reducing the spatial distance between linked components (Stanton and Young, 1999).

2.3 Information Gap Model

A concept which encompasses our view on level 1 errors is the so-called information gap introduced by Endsley and Jones (Endsley, 2000; Endsley and Jones, 2011). Their concept describes the information gap as an inconsistency between data produced and information needed. There are various definitions and meanings on what data and what information are and how they differ. The data–information–knowledge–wisdom discussion gives an insight into that field (Fricke, 2008; Rowley, 2007). Since we follow a human's task-oriented perspective, we solely focus on information. The reason is that data "has no meaning or value because it is without context and interpretation" (Rowley, 2007). In contrast, information has a format, is structured and organized, has a meaning, and a value feature (Rowley, 2007). Since information has a meaning and value feature, we will use the term "supplied information" instead of "produced data" in the remainder of this paper. Furthermore, we use the term "information demand" instead of information need.

During task execution in transportation systems, humans and machines are demanding and supplying information. This basic concept of supply and demand facilitated in our approach is taken from business studies. In business studies' controlling for instance a set theoretical concept of information supply, demand, and requirements exists (Weber and Schäffer, 2006). Within that concept information requirements describe all information which are necessary to the management, for example, for making a decision. An information demand is issued to fulfill the management's information requirements. The information demand describes information which is requested from the information supply. In the "ideal situation" the sets of required information, demanded information, and supplied information overlap (Weber and Schäffer, 2006). We used the basics of this concept and transferred it to investigate gaps between information supply and demand between humans and machines. The result is the information gap model (Figure 2.1). In the model an information gap is defined by the two complements of the intersection of information supply and demand. This means that an information gap can have two

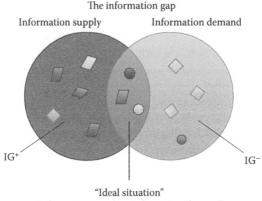

FIGURE 2.1

The information gap model describes the relation between information supply and demand. We model information supply and demand as two sets of atomic IE. IG⁺ represents the subset of oversupply and IG⁻ the subset of undersupply of IE. In the ideal situation, information supply and demand are well-balanced.

characteristics: (1) supplied information is not demanded or (2) demanded information is not supplied. The former is also part of the previously stated definition of the information gap by Endsley. We call this part information gap⁺ (IG⁺), since there is more information available than demanded. The second part is called information gap⁻ (IG⁻), because there is less information available than demanded. This information gap definition is the baseline of our concept. In the ideal case, information supply and demand are well-balanced.

2.4 Method

In this section we present our method to detect information gaps and assess the fitness of information supply and demand during the UI design phase. The method consists of the four steps *definition*, *modeling*, *detection*, and *assessment* of information gaps and can be integrated into system design processes, such as the human-centered design process (ISO9241-210, 2009). In the following sections, we describe each step of the method in detail.

2.4.1 Step 1: Definition of Human Tasks and Machines under Investigation

In the first step of the method, the scope of the system investigation is defined. This means that the human tasks and the UI to consider are specified in detail. Therefore, hierarchical task analyses (HTA) of the humans' tasks can be conducted. An HTA typically results in a hierarchical tree-structure, where the task is decomposed into multiple subtasks (Hollnagel, 2003). Based on the HTA, atomic information elements (IEs) are extracted. IEs represent the smallest unit of information within a task which provides a meaning to the human. A digital speedometer for instance typically contains an IE (information element) that can be called "current_speed." In this approach the sufficient grade of detail is reached, when all IEs, which are necessary to complete the task at hand, are identified.

Separated from conducting the HTA, the UI is analyzed to gather the contained IEs. Contained IEs are in this case only relating to information which is shown to the users in one modus of the UI. In some cases the UI's size may exhaust the effort for gathering all IEs. Then, depending on the tasks under investigation, the grade of effort can be scaled by neglecting parts of the UI in the analysis, which would not influence the perception and SA of the humans.

During human task and machine modus definition, the analyst has to mind safety aspects. Missing or disregarding safety-relevant tasks and corresponding IEs can have a huge impact on system safety. During task definition questions like "what can go wrong?" and "which evasive tasks need to be executed?" must be considered to derive safety-relevant IEs. Another point referring to IEs is that they may be integrated with other IEs so that a new IE is derived. IEs extracted from HTAs may be integrated. By using integrated IE during modeling also parts of level 2 or level 3 SA will become the subject of the assessment's result. For instance the speedometer could be used in two different tasks to demand both the IE "current_speed" and also the IE "target_speed_deviation" as integration of, for example, "target_speed" and "current_speed." In such a case the analyst can determine whether his machine can support other SA level. The integration of IEs can be considered in modeling and in the detection and assessment of the information gap. In the remainder of the concept description, level 1 SA is considered.

2.4.2 Step 2: Modeling of Human Tasks and Machine States

In the second step, the information is gathered and an information supply and demand model (ISDM) is instantiated. The ISDM is a set of classes, which allows detecting information gaps both in a static comparison of IE sets and in a dynamic simulation of task execution. The class *InformationElement* represents an IE and is the smallest and atomic entity of the ISDM. Every IE is unique and distinct from other IEs. An *InformationElementPool* (IEP) is a global container for all unique IEs, which holds both the human tasks' and the machines' IEs. There further exist three roles (*InformationRole*) called *InformationSupplier*, *InformationDemander*, and *InformationHybrid*. An *InformationSupplier* is a role, which is capable of emitting/supplying IEs via a so-called *InformationSupplySide*. For example the whole machine under investigation can have the role of an *InformationSupplier*. The *InformationSupplySide* would then be, for example, a console or display. *InformationDemander* is a role which is capable of receiving/demanding IEs from *InformationSupplier*'s *InformationSupplySide*s. The source of an information demand is an *InformationDemandSide*, which is part of the *InformationDemander*. To enable for dynamic analysis, the *InformationDemanders*' *InformationDemandSides* and *InformationSuppliers*' *InformationSupplySides* are ordered according to their temporal execution of demand and supply. Referenced IEs of both *InformationMarketSides* can be ordered in the same manner, to reflect, for example, detailed sequences of eye movements (saccades). The *InformationHybrid* is a combination of both *InformationSupplier* and *InformationDemander* and can thus contain *InformationSupplySides* and *InformationDemandSides*. Both the *InformationDemandSide* and the *InformationSupplySide* have a location in the system under analysis. Furthermore, the *InformationMarketSides* can be enriched with properties corresponding to their contained IEs. For example a value of importance or a value representing the saliency of IEs could be defined for each IE. This enables not only for spatial analysis of the human–machine interaction, but also allows considering further human or machine factors in the later analysis. An *InformationFlow* is used to connect the *InformationDemandSides* with *InformationSupplySides*. This concept allows modeling the fulfillment of information demand via distributed *InformationSupplySide*s. The *InformationFlow* class can be used to model dialogs, reports, and information retrievals. With the model it is also possible to express information which is not perceivable on the UI, but existing in the internal model of the human or machine. The class representing this is called *InformationPotential*. In case of a machine this class can contain, for example, information produced by sensors, which are currently not showing on the display. For humans the *InformationPotential* can contain all IE which were already perceived during a task's execution. The construct can also be used to model cognitive limitations, for example, to set a max amount of information in a human's memory. The *InformationPotential* can increase or decrease in the amount of contained IEs during task execution.

2.4.3 Step 3: Detection of Information Gaps

In the third step information, gaps are detected by analyzing the ISDM. Therefore, information supply and demand is compared to check for existence of IG$^+$ and IG$^-$ in the system under investigation. The comparison of the properties of the *InformationMarketSide*s is also a part of the detection. As preparation to the detection and assessment, the *InformationDemanders'* *InformationDemandSide*s are inspected. For each *InformationDemandSide* connected *InformationSupplySide*s are resolved, which are connected via *InformationFlow*s. The *InformationSupplySides'* IEs are then aggregated to form a joint set of supplied IEs. IEs of the focused *InformationDemandSide* are considered as a set of IEs as well. Of course properties of the *InformationDemandSide* and *InformationSupplySide*, which have been previously annotated, need to be referenced to their corresponding IEs, if they have to be considered in the analysis as well. A possibility for annotating the sets is to construct tuples, which consist of the IE and its properties.

Having the two sets of information supply and information demand set up, the detection of the information gap may begin. Let A be the information supply set and B be the information demand set, then $f(A - B)$ results in IG$^+$ and $f(B - A)$ results in IG$^-$. The function f is called the matching function. It maps the referenced IEs of sets A and B to each other. The function allows scoring differences in properties of IEs between both demand and supply side. The results of the detection step are the sets of IG$^+$ and IG$^-$ including properties' scores.

2.4.4 Step 4: Assessment of Information Gaps

In the fourth step, the information gaps are assessed. The assessment is based on the two sets A and B and their properties which were created in the third step. The assessment is done by application of Tversky's ratio model similarity (Tversky, 1977)

$$S(a,b) = \frac{f(A \cap B)}{(f(A \cap B) + \alpha f(A - B) + \beta f(B - A))}, \quad \alpha, \beta \geq 0 \tag{2.1}$$

It is a mathematical model which allows comparing two sets and results in a ratio. The ratio indicates the similarity of the given sets A and B and expresses it as normalized real number between 0 and 1. Here again, the complements of the intersection of A and B represent the information gap consisting of IG$^+$ and IG$^-$. α and β are weightings to the complements and hence allow for changes of the influences of IG$^+$ and IG$^-$ to the metric's result. When $\alpha = \beta = 1$ the model reduces to $f(A \cap B)/f(A \cup B)$ (Gregson, 1975). Again, the function f is called the matching function. The matching function allows to integrate further mapping functions between both *InformationDemandSides'* and *InformationSupplySides'* IE-referencing properties. The implementation of the matching function depends on the particular property and its metric. A simple example is the mapping of an IE's value of importance for the *InformationDemandSide*.

When calculations for each *InformationDemandSide* were carried out, the arithmetic mean is used to rate the entire tasks in relation to the machines. When the metric result equals one, no information gap exists. A result of zero would indicate that there exists no information exchange. Instead of the arithmetic mean, it is of course also possible to apply other, even more complex functions, which weight in a more meaningful manner, for example, in accordance to task priority.

2.5 Implementation in a Modeling Environment

In this section, the implementation of the concept is described. The aim of the implementation is to show how the concept can be integrated into existing integrated modeling environments to automate the static detection and assessment method. We chose to implement the concept as plugin to MagicDraw. The UML (unified modeling language) tool supports business process, architecture, software, and system modeling (NoMagic, 2014). The main advantage is its extensibility via the provided OpenAPI. The OpenAPI allows access to various internal modeling constructs and enables us to extend MagicDraw with custom plugins. The internal models are mapped to the ISDM and a plugin is developed which enables us to detect and assess information gaps. The static analysis disregards spatio-temporal aspects of the information gap. An IE property which states the priority within a task under investigation is integrated into the assessment.

2.5.1 Mapping the ISDM

The ISDM is mapped to existing constructs of MagicDraw. These constructs are SysML requirement diagrams, UI models, and traceability links. In systems engineering SysML requirement, diagrams are used to define requirements to a system and requirements' relations (Weilkiens, 2011). Here, the SysML requirement diagram is used to express the *InformationDemander* with one *InformationDemandSide* of the ISDM. The diagram's parts called "information requirements" represent IEs. The priority of an IE is settable via the information requirement's property attribute. UI models, which can be created with MagicDraw's UI Modeler, are used to constitute the *InformationSupplier* including one *InformationSupplySide*. Traceability links enable to connect SysML requirement diagram's information requirements with the UI model's elements and therefore surrogate the *InformationFlow*. *InformationElementPool*, *InformationPotential*, and *InformationHybrid* are not considered in this implementation.

2.5.2 Plugin Implementation

The requirement diagram, UI model, and traceability links are facilitated by the developed plugin extension. The plugin is implemented in Java and consists of the two classes

IGMetricPlugin and *IGMetric*. These extend the abstract classes *Plugin* and *Metric* of the MagicDraw OpenAPI. The *IGMetricPlugin* instantiates *IGMetric* in its *init*-method and adds it to the *MetricsManager* of MagicDraw. This makes the metric available in MagicDraw. The *IGMetric* contains the logic for the automated detection and assessment. In its *calculateLocalMetricValue*-method, the ratio model similarity is calculated. During calculation IEs of the information requirements diagram are mapped to the priority. The result of the calculation is shown as a report in MagicDraw. The report states IG⁺, IG⁻, IEs of the "ideal situation," and the result of the ratio model similarity calculation. The method *acceptModelElement* is used to specify permitted input elements to the metric. Here all required diagram elements are permitted. An example of the results is presented within the use case in the next section. As completion of the implementation, the plugin was added to MagicDraw.

2.6 Use Case

We applied our method to assess the fitness between an autopilot UI (defines the information supply) and a course change task on a ship (defines the information demand). The course change task under investigation is artificial, since it only focuses on operational changes to keep the ship on its track and is not capable of extraction of all required information. For instance, in such a task a target course is identified through other nautical means, which are not considered in this use case.

The course change task was examined in a HTA which was conducted and evaluated in an expert interview with a master-licensed seafarer. The resulting HTA is shown in Figure 2.2 and was used to identify the IEs demanded during the task. To change the course, the autopilot needs to be configured, before maneuver execution is possible. A maneuver is executed by setting a new heading to the autopilot and monitoring the ship's corresponding behavior. The identified IEs were added to a requirement diagram in MagicDraw (see Figure 2.2). The diagram contains nine IEs called *current_rate_of_turn*, *new_rate_of_turn*, *current_rudder_limit*, *new_rudder_limit*, *current_heading*, *new_heading*, *current_reference_compass*, *new_reference_compass*, and *current_rudder_position*. Every IE got a priority of 1 assigned. Exceptions to this are the IEs *current_heading* and *new_heading*, which got a priority of 2 assigned, only for demonstrative purposes. Next, a commercial autopilot's UI was recreated in an abstract manner with MagicDraw's UI modeler. The abstract UI model of the autopilot is shown in Figure 2.2 as well. There IEs of the UI model are represented as *GroupBoxes* and a variety of control buttons were aggregated to *GroupBoxes* called "controls" to minimize the textual output of the plugin. This of course influences the metric's result. The *GroupBoxes* were then linked to the information requirements with MagicDraw's traceability functionality. Finally, the *IGMetric* was executed to calculate the ratio model similarity between the information supply and the information demand. The textual output containing IG⁺, IG⁻, and the ratio model similarity's result is shown in Table 2.1.

One information requirement, or IE respectively, was not implemented in the UI model (IG⁻) and seven supplied IEs were not demanded within the task (IG⁺). The ratio model similarity results in 0.578947. This results from the mapping function to the information requirements' priority property. In fractional notation, the result equals (11/(11 + 7 + 1)).

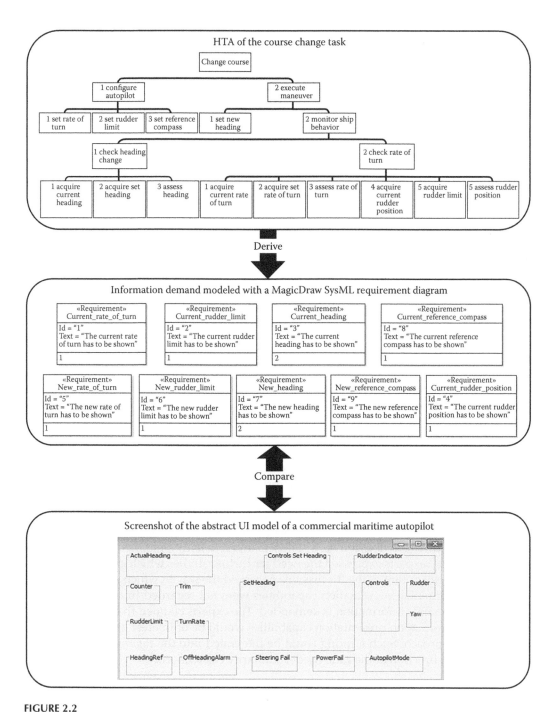

FIGURE 2.2
Overview on derivation of information requirements from HTA and comparison to the abstract UI model of a commercial maritime autopilot.

TABLE 2.1

Calculation Results of the Information Gap Metric Plugin

WARNING (IG$^-$): Requirement current_rate_of_turn is disregarded in GUI!
WARNING (IG$^+$): GroupBox AutopilotMode is not derived from information requirement
WARNING (IG$^+$): GroupBox Yaw is not derived from information requirement
WARNING (IG$^+$): GroupBox PowerFail is not derived from information requirement
WARNING (IG$^+$): GroupBox Steering Fail is not derived from information requirement
WARNING (IG$^+$): GroupBox Controls is not derived from information requirement
WARNING (IG$^+$): GroupBox Counter is not derived from information requirement
WARNING (IG$^+$): GroupBox Trim is not derived from information requirement
Ratio model similarity: 0.578947

2.7 Evaluation

We evaluated our approach in collaboration with system engineers from the maritime domain and with a human factors ergonomist. Therefore we presented the concept of the method to them and demonstrated its application. Then we asked the experts to estimate the applicability, benefits, and shortcomings of the method. Overall, the feedback was positive. The experts found that the method is a good complement for system design processes. Especially the seamless integration into an existing systems engineering tool was stated as beneficial. The experts agreed that the spatiotemporal resolution of the information gap needs to be addressed in future work. Furthermore, the ratio model similarity metric delivering a normalized estimation of the severity of information gaps was considered as helpful. They estimated the metric would have its strength in the assessment for comparison of various UIs and different UI modes during design time. An engineer mentioned that the approach requires a huge initial modeling effort and that detailed task analyses may consume much time. During development this can have a negative bias for the time to market. However, another expert with background in task analysis reasoned that gathering IEs would cause little additional effort to a typical task analysis. In the end a comparison of cost and benefits will drive the decision on whether to invest in additional modeling effort. The discussion with the experts expressed the need for further studies which examine the costs and benefits. Another point in the discussion concerned the way in which IEs are supplied. For instance information presentation capabilities were not regarded in our use case, but have an impact on how information is demanded. The experts claimed that the integration of a rating for information presentation capabilities would be of interest as an extension to the presented approach. Such a rating could be integrated into the *InformationMarketSide*s and the mapping function of the ratio model similarity calculation.

2.8 Conclusion and Outlook

In this chapter, we have demonstrated a method to assess the fitness of information supply and demand on the UI during the design phase to improve the information distribution and thereby the human–machine interaction on board with regard to the ICT pervasiveness scenario. As basis of our method, we derived the information gap model comprising information supply and information demand. Our method consists of the four steps: (1) definition of human tasks and machines under investigation, (2) modeling of human tasks and machine states, (3) detection of information gaps, and (4) assessment of information gaps. We applied the method to assess the fitness of information supply and demand of an autopilot component in a course change task on a ship bridge. We presented our method and its application to system engineers and a human factors ergonomist. Then, we interviewed them to get an initial evaluation. In the interview, we asked the experts to estimate the applicability, benefits, and costs of the method.

The overall result is positive and warrants further research. A study which investigates the applicability of the method with respect to benefits and costs would be of interest for industrial stakeholders. Another research demand concerns the extension of our method by including further properties and the investigation of distributed information supply and demand. In our implementation we considered an IE's priority out of the demand side's perspective. As extension to this work, further properties of IEs which influence the information flow between information supply and information demand side could be considered. For instance an integration of the properties of Wickens' SEEV-Model (salience, effort, expectancy, and value) from applied attention theory may be considered (Wickens and McCarley, 2008). In future work, the effort property of this framework can be used to compensate for the spatiotemporal aspects which are not considered within our implementation. Furthermore, we conclude that analysts have to consider IG^+ and IG^- during assessment with the presented implementation, since the arithmetic mean of multiple ratio model similarity metrics may mask important information gaps. However, our method enables us to integrate more powerful aggregation functions, which may compensate that shortcoming. We further identified that the presentation of the assessments' results could be improved. Analysts applying our implementation to optimize complex systems consisting of multiple humans and machines might struggle in finding major system problems in the overloaded textual results. The results could be improved through visualization, for example, as a graph visualization (Herman, 2000).

Acknowledgments

This research was carried out with support from the EU FP7 project CASCADe, GA No.: 314352. Any contents herein are from the authors and do not necessarily reflect the views of the European Commission. The authors would like to thank our partners in CASCADe and the participating domain experts who supported this work.

References

Baker, C. C., and McCafferty, D. B. 2005. Accident database review of human element concerns: What do the results mean for classification? In: *Human Factors in Ship Design, Safety and Operation* (p. 4), February 23–24, London: Royal Institution of Naval Architects.

Chapanis, A. 1965. *Research Techniques in Human Engineering* (316pp). Baltimore: Johns Hopkins University Press.

Clarke, E. M., Grumberg, O., and Peled, D. A. 1999. Model checking. *Journal of the American Statistical Association*, 962, 314. Doi: 10.1109/43.845084.

Endsley, M. 2000. Theoretical underpinnings of situation awareness: A critical review. In M. R. Endsley and D. J. Garland (Eds.), *Situation Awareness Analysis and Measurement* (pp. 3–28). Mahwah, NJ: Lawrence Erlbaum Associates Inc.

Endsley, M. R. 1995. Towards a theory of situation awareness. *Human Factors*, 37(1), 32–64.

Endsley, M. R., and Jones, D. G. 2011. *Designing for Situation Awareness: An Approach to User-Centered Design* (2nd Edition). Boca Raton, FL: CRC Press.

Fricke, M. 2008. The knowledge pyramid: A critique of the DIKW hierarchy. *Journal of Information Science*, 35(2), 131–142. Doi: 10.1177/0165551508094050.

Grech, M., Horberry, T., and Smith, A. 2002. Human error in maritime operations: Analyses of accident reports using the leximancer tool. In *Proceedings of the Human Factors and Ergonomics Society Annual Meeting* (Vol. 46, pp. 1718–1721). Baltimore, Maryland: SAGE Publications.

Gregson, R. A. M. 1975. *Psychometrics of Similarity* (262pp). New York: Academic Press.

Herman, I. 2000. Graph visualization and navigation in information visualization: A survey. *IEEE Transactions on Visualization and Computer Graphics*, 6(1), 24–43.

Hetherington, C., Flin, R., and Mearns, K. 2006. Safety in shipping: The human element. *Journal of Safety Research*, 37(4), 401–411. Doi: 10.1016/j.jsr.2006.04.007.

Hollnagel, E. (Ed.) 2003. *Handbook of Cognitive Task Design* (Vol. 20031153). Boca Raton, FL: CRC Press. Doi: 10.1201/9781410607775.

ISO9241-210. 2009. *ISO 9241-210: Ergonomics of Human-System Interaction—Part 210: Human-Centred Design for Interactive Systems* (ISO 9241-2).

Jones, D., and Endsley, M. 1996. Sources of situation awareness errors in aviation. *Aviation, Space, and Environmental Medicine*, 67(6), 507–512.

Koreimann, D. S. 2000. *Grundlagen der Softwareentwicklung* (3rd Edition). Munich: Oldenbourg.

Meolic, R., Kapus, T., and Brezocnik, Z. 2000. Model checking: A formal method for safety assurance of logistic systems. In: *2nd Congress Transport—Traffic -Logistics Portoroz*, Slovenia, pp. 355–358.

NoMagic. 2014. NoMagic Website. Retrieved February 22, 2014, from http://www.nomagic.com/products/magicdraw.html

Rowley, J. 2007. The wisdom hierarchy: Representations of the DIKW hierarchy. *Journal of Information Science*, 33(2), 163–180. Doi: 10.1177/0165551506070706.

Stanton, N. A., and Young, M. S. 1999. *A Guide to Methodology in Ergonomics: Designing for Human Use*. London: Taylor and Francis.

Tversky, A. 1977. Features of similarity. *Psychological Review*, 84(4), 327–352.

Weber, J., and Schäffer, U. 2006. *Einführung in das Controlling*. Stuttgart, Germany: Schäffer-Poeschel.

Weilkiens, T. 2011. *Systems Engineering with SysML/UML: Modeling, Analysis, Design*. Heidelberg, Germany: Elsevier Science.

Wickens, C. D., and McCarley, J. S. 2008. *Applied Attention Theory*. Boca Raton, FL: Taylor and Francis.

Wilson, J. R., and Corlett, N. 2010. *Evaluation of Human Work*. Boca Raton, FL: Taylor and Francis.

3

Visual Pleasantness in Interior Yacht Design: A Case Study of the Pleasure-Based Approach Application

Massimo Di Nicolantonio, Giuseppe Di Bucchianico,
Stefania Camplone, and Andrea Vallicelli

CONTENTS

3.1 Introduction

The quality of the environment is closely related to the control and availability of natural light and possible views. In the field of yachting, however, the design solution of the hulls and the internal environments often makes the relation between the availability of natural light and views inside the boat especially critical. Anyway, the demands of representation and sociality required by final users needs the introduction of new systems of windows, terraces, and new layouts, to relate the interior of the boats with the deck, attributing new meanings within the social idea of "going out to sea." This is how the small portholes, with the original function of air environments, can aspire to become large openings for dialog with sky and sea. However, the unstable horizon of the boat and the dynamic changing nature of views and natural light generate very difficult conditions with regard to the control of the factors which can help determine the good size of domestic interiors and their relationship with the environment.

3.2 New Ideas of "Transparency" in Yacht Design

In recent years, the yachting industry has had an unprecedented expansion and evolution. Technological development has been complemented by experimental research of new layout solutions, new product concepts, and new morphologies, where the methodologies, techniques, and tools of the project have responded with unprecedented dynamism, research, and solutions that guarantee the improvement of each performance, putting the end user at the center of the whole creative process. Today it is known that yachts are not only for transportation, for competition, or for exploration, but increasingly assume the role of status symbols, places of representation with high social value. It is therefore inevitable that the spaces on board need substantial, technical, and functional changes. Referring to the range of the sailing yachts of medium and large size, the attention of yacht designers today takes care of both the pleasure of sailing conditions and the living conditions, changing spatiality, and shapes. Actually, to obtain the best organization of the interior living spaces, the designers tend to preserve the most typical layouts, to introduce new materials and advanced technologies, to improve the performance values of the boat, and make new ways of living the boat possible.

The new technological solutions improve the operating conditions of the crew, so it is possible to reduce the number of crew members and to reduce the crew area. It is now possible to create larger external living spaces, separated and organized better to increase the levels of comfort and hospitality. One of the most significant innovations is related to the research about the interiors "transparency." The interiors are no longer small, dark, cramped, protective. The interiors tend to "open-up" to the surrounding environment, giving new social meaning to the idea of "going out to sea." The small portholes, originally designed to ventilate the interior, now become large openings, so the crew can really dialog with the sea and the sky from the interiors of the boat. Hulls and decks structures are cut from large windows. The walls can be more transparent, and the crew can live in dialog with the sea and nature. New physical and visual relationships between the interior and exterior of the boat are established, through the inclusion of unpublished terraces and openings.

In this context of technological innovation and space control experimentation, this new visual attractiveness represents an interesting and strategic project proposal, both referring to the "views" and the possibility for the crew to relate visually from the surrounding natural environment, both reported to control the quantity and quality of natural light that invests on a daily basis the inside of the boats, which often must be shielded in order to obtain acceptable levels of visual comfort.

3.3 Verify and Assess the Visual Pleasantness on Board

The visual comfort in any indoor environment is the result of several factors.

The natural and artificial lighting, the colors, the views, the shape of the space, which must be carefully related to each other in order to produce well-being and not malaise. Most of these aspects are derived directly from the sizing, the shape, and positioning of any openings to the outside.

The research was conducted with an initial investigation of the general nature on the subject of visual comfort through natural and artificial lighting and through the use of color and its effects on humans, also investigating those activities the user is going to do into these specific environments.

The study was preliminary and extremely useful for defining the research objectives, and to obtain qualitative data on the attractiveness of the apertures toward the outside and the colors in the living area of the yacht.

The first issue was the choice of the size of a 15 m sailing yacht, due to the fact that the interior layout of this type of boat can be considered a good compromise between a marine space and architectural civil space. The smallest sailing yachts have internal spaces similar to cockpits. On the contrary, the environment of this type of boat can be compared to those of the civil architecture, which lead to a different spatial perception from that of the interior of a boat.

The second issue was the choice to conduct the experiment on the living area, because it is the common and the most experienced space of the yacht. This particular environment, defined "square," represents the social area of the yacht, generated by the union of the saloon and the kitchen. Thus it is possible to prepare and eat food, while relaxing or making conversation, often also in special spatial conditions (a boat tilted, in continuous movement, etc.).

3.4 The Pleasure-Based Approach

The research is part of the most famous experiments in applied ergonomics that make reference to the pleasure-based approach (PBA). The designers use this methodology to know the needs and desires of the target audience through tests made with specific groups of individuals at all stages of design.

In particular, the articulation of the experimental phases refers to the SeQUAM* (Bandini Buti, 1997), sensorial quality assessment method, with some modifications due to the specific contexts of use.

The objectives are

- To create approval rating indexes associated with the objective aspects of the product, to point out the satisfaction ranges for each parameter from which the guidelines of pleasantness will result
- To define guidelines of pleasantness for designers
- To formulate design specifications about object pleasantness and every individual parameter
- To obtain transmittable and certain data, based on certainty basis and applicable to the design, making predictive assessments related to objective parameters

The method involves that the operative scheme will be repeated thrice during the sequential survey:

- Survey on the present
- Innovation survey
- Prototype check

---

* The SeQUAM method (Method for sensory quality assessment evaluation) was developed by L. Bandini Buti and L. Bonapace, since 1992, as part of a project for FIAT Auto aimed at increasing the levels of pleasantness perceived by users during eye contact, touch, and body parts and components of motor vehicles.

In the first phase (survey on the present), objects selected from those offered by the market will be investigated, which are particularly interesting to analyze for aspects of attractiveness. The purpose of the first phase is to derive quantitative assessments on the appreciation of the members of the stimuli and to guide the design of experimental maquettes necessary for a systematic study of the individual component of agreeableness.

The advantages to investigate the series products are

- To allow rapid retrieval of samples to be tested
- To be able to test all the variables because the object is finished and is functioning

The disadvantages are

- That the object can recall the interviewee stereotypes which are difficult to eliminate (e.g., known or mythical objects, or objects that are better known than others, etc.)
- It is only possible to test common and not innovative parameters

This first phase investigates by using the observation matrixes M1 and M2.

The establishment of pleasantness criteria inserted into the matrixes, combined with the reading of the results, allows the research group to obtain useful data for the design of maquettes to be included in the test of pleasantness in phase 2.

Phase 2 (innovation survey), starts from the indications of pleasantness obtained during phase 1. It refers to those solutions which are currently in production and are therefore significant for today, but do not say anything about innovative trends, especially if it is taken into account that what is offered by the market has been conceived a long time before. It is therefore necessary to introduce a stage in the investigation that analyzes objects specially designed for the research, which should allow us to

- Investigate parameters that in the first phase showed trends that should be deepened
- Analyze trends emerged from the survey but not verified due to a lack of suitable objects
- Investigate innovative trends

This second phase involves the design, the construction, and the test of a homogeneous series of maquettes which lead the survey of the subjective response related to the changes of individual components in "low noise" situations, that is not polluted by past prejudices. The series of maquettes can be used to check the performance's limits related to the variability of the components investigated by evaluating models with features also esasperated (e.g., grips of excessively large size or excessively small).

The advantages to investigate maquettes are

- To analyze very well the individual parameters by comparing maquettes, equal in all ways except for the variation of the parameter
- The time required for the development of this phase is not much longer than that of the previous phase, because it requires the development of projects, and the execution of models

The disadvantages are

- Timing and costs of production enforce that the research can be made generally by using maquettes just not working

The innovation survey is conducted by a test subject in a representative group of potential users. The results of the tests crossed to the results of the matrixes allow us to obtain guidelines based on reliable data for the realization of a concept.

In the first two phases, the research is carried out in conditions that should enable us to get the maximum information with content development time, as required by the logic of production marketing.

Under experimental conditions, the subjects pay attention only to the objects themselves, which must be used in the presence of observers. In a real condition of use, subjects pay attention only to the results of their actions.

The final and third phase of SeQUAM is called "verification on the prototype."

It requires us to carry out research on working prototypes that have all the characteristics of the finished product and which are included in the correct environment, from the functional point of view, and from that formal point of view, and last, that can be used in real conditions by potential users.

In this case the third phase did not take place.

3.4.1 Application of M1 and M2 Matrixes

In this case study, to carry out the "analysis on the current" in a structured way, the research developed two matrixes or cards classification.

The matrix M1 refers to the "analysis of the views" and is a table that organizes the "elements" and "criteria" to analyze the views to the outside.

In particular, among the "elements" are considered some aspects of identification of the vessels analyzed, and especially the vertical and horizontal elements having the ability to control the stimuli coming from the outside. They are the four "directions" in which apertures are inserted (at the bow, to the stern, aft to the bulwarks, on the ceiling).

The "criteria" have been shown: position of the openings/transparent surfaces, with which you can benefit from the view posture (standing, sitting, half-lying, lying), and finally what kind of view of return openings (circumscribed, elongated, fragmented, overview).

The use of the matrix M1 has allowed us to analyze a sample sufficiently significant of boats.

The matrix M2 refers to the "analysis of the use of colors" and it is a table that organizes the "elements" and "criteria" to analyze the use of color in the living spaces of the yachts.

In particular, among the "elements" are considered some aspects of the yachts analyzed identifiers (name, images of the square) and especially the opaque surfaces of the living area such as the deck, the sides, the bulkheads, and the top.

Among the "criteria," however, have been shown, each part has its own color with high gloss levels, those characterized by colors that convey similar feelings, and those colors that convey different sensations.

Also in this case the use of the matrix M2 has allowed us to analyze and systematize the data referring to a significant sample of yachts.

This phase of the research has had the aim of obtaining qualitative data on the attractiveness of the views to the outside, and the colors, related to the living area of a sailing boat.

3.4.2 Investigation with Maquettes and Render Scenes

An evaluation test was designed to investigate the "new tendencies." For the experiment a questionnaire was used accompanied by some "maquettes," in respect of which, the judgment of pleasantness was asked for.

The variables evaluated (and on which were carried out maquette/render scenes) were considered with respect to (1) placement of transparent surfaces; (2) posture of the individual for the views; (3) kind of fruition views; and (4) dyes for coloring matt surfaces.

The sample was composed of 26 members, divided into three age groups between 21 and 65 years, 13 of them with experience in sailing and 13 with no experience of sailing.

In particular, users had to answer a questionnaire based on some cards containing render scenes describing an "environment type," of which we must make a judgment. The reproduced scenes on each card are different only for the feature that is intended to be investigated.

The ratings of pleasantness judgments were expressed using a simple rating scale (1: not pleasant; 10: very pleasant), with the possibility to indicate intermediate values. Each judgment of pleasantness expressed in numerical form (quantitative assessment) was also required to be associated with an explicit "motivation" (qualitative judgment).

The scenes for which we asked to give the judgments of pleasantness were 61, divided into 8 tabs. Each tab on the back reported two questions to answer. Assuming an average rating for each scene of 15 s, a minute and a half to meet every couple of questions, and adding about 5 or 6 min to fill out the Getting Started tab and be informed on the tasks to be performed, it was previously calculated that the overall average length of the test was about 40–45 min, a time short enough to hold the attention and get answers that were instinctive enough, and at the same time sufficiently long enough to avoid a hasty completion of the questionnaire.

Even the conducting of the trial was designed in detail, describing in detail the activities and roles of the researcher to obtain sufficiently objective results. The test was carried out in the laboratory.

3.5 Data Analysis and Guidelines

The results have been organized in respect of two aspects.

The first aspect concerns the system of guidelines for the awareness design of portholes in the living nautical spaces, and the guidelines for the usage of colors.

The second consists of a further study that reported the possible control systems of the quality and quantity of light through natural apertures which could be transparent or translucent.

The data collected were allowed to organize guidelines in respect to placement of the apertures (and the respective openings that allow it), postures to enjoy the apertures, and the type of use of the views.

The data collected were allowed to organize, in general, that the requirements relating to internal living spaces for sailing boats of similar size to those analyzed are attributable to the following concepts: highest availability of free surfaces on the sides;

convertibility, mobility, and flexibility in the use of equipment and furnishings to allow variability in exhibition set up environments; maximum spaciousness, and maximum usability views.

The same comparison was conducted with respect to the use of colors. In this case, the guidelines have been organized with respect to dyes to color the side surfaces of matt colors for opaque surfaces at the bottom (deck); colors for opaque surfaces at the top (top).

The requirements related to the use of the colors were: transformability (the possibility to change the colors also instantly depending on the prevailing activities) and high gloss.

3.6 Daylight Control Systems

Another interesting study of the research has focused on the identification of possible smart systems to control the quality and quantity of natural light that passes transparently or translucently into the yachts. Our attention was focused on the identification of a useful smart system for screening:

- To control the amount of natural light that penetrates inside at certain times of the day and according to specific conditions and changing orientation of the yacht relative to the position of the sun (mostly because of the so-called "greenhouse effect")
- To choose the quality of natural light, even filtered through special screening systems that can help to determine new shades for specific environmental conditions and activities

The research has been focused on chromogenic devices that also allow you to greatly reduce the energy consumption for cooling and lighting the indoor environment. In marine applications, the control of the solar factor referred to in the transparent or translucent parts is very important for maintaining microclimate comfort, as well as to provide specific visual effects, high levels of security and privacy that may differ when related to internal and external factors regarding the yacht.

The control of "natural light" or "sunlight" depends on the type of yacht, the weather, the seasons, the different times of the day, and on the characteristics of use. The use of chromogenic materials is more effective than the use of traditional solar shading systems such as blinds, curtains, drapes, and many others. The systems chromogenic materials are characterized by the ability to modify the optical properties with a reversible effect, following the application of an electrical stimulus, heat, or light. According to their behavior we identify four types of chromogenic materials: liquid crystals, electrochemical, photochromic, and thermochromic. In this case the chromogenic and photochromic devices proved potentially more effective compared to the specific conditions of use. In particular, these devices that are sensitive to environmental parameters and placed in special "smart windows" react with the gradual and reversible change of color when exposed to light stimuli of varying intensity (UV exposure). With the same technology is also possible to control the tint of the light, turning the transparent surface into a real color filter.

3.7 Concept Design: A Study Case of PBA Approach in 15 m Sailing Yacht Interiors

The system of guidelines that referred to the design of the apertures to the outside has allowed a deepening in terms of the design of possible responses. The starting point of view, related with the innovative relationship between interior and exterior spaces of the sailing yacht, through the combination of the data produced by research and the new concept referring to the relationship between the product and the marine environment, has had great influence on the final concept. At last, the solution was found, through two ways of action:

- New interior concept
- New exterior concept

With reference to the interiors the choice has been to reconfigure the living area, through the use of large apertures on both sides of the hull and on the deck of the yacht. This choice, obtained with the adoption of smart materials and control systems of the intensity of natural light and with the aid of the control systems of the staining, allows us to optimize the relationship with the natural environment.

New technological and formal solutions have been found for the exterior design that could highlight the innovation represented by large apertures. The first step was to redesign a structural framework that would guarantee the stability of the entire system under the effect of stress to which a yacht is constantly subject. Second, it is necessary to rethink the yacht starting with the elements that characterize the morphological forms of the hull.

All the design choices implemented have allowed the research group to develop the project of a boat with extremely innovative solutions in morphological, structural, and distributive terms. All the solutions allowed, especially in the middle of the square, a visual relationship with the external environment, characterized as extremely flexible, instantly reconfigurable, and really unusual for the sailing yacht industry.

3.7.1 Design of Apertures toward the Outside

The first insight of the concept took care of the design and reconfiguration of an original deck layout, thanks to an original design of topsides related to the introduction of these great apertures.

At the same time new solutions for a new internal layout were found, related to the choice of introducing great apertures on the top and on the sides, and introducing the new technology of the chromogenic panels, with the purpose of obtaining these large apertures and color variations of sensory character.

The correct height of the washboard has allowed the insertion of these large portholes in the hull, which acquire a great architectural value, due to their size, and due to the fact that the entire system is redesigned on the outer side, so unusual and original, enhancing the form and function through the use of views. The final result has been a new and original design of the hull profile, strongly characterized by the system of architectural openings embedded in a dynamic way (Figure 3.1).

3.7.2 The New Concept Design of the Hull Structures

The inclusion of a large number of apertures into the deck and into the hull of the sailing yacht led the designers to analyze the functioning of the structures of the entire boat.

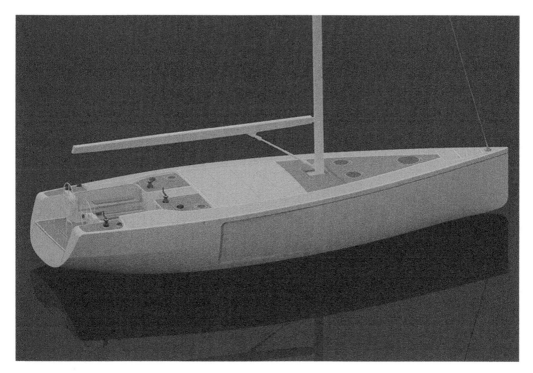

FIGURE 3.1
General view of the sailing yacht.

As known, the sailing boat is by nature subject to a number of high stresses that are absorbed by a complex system of structural stiffening of the body. In this case, the stress increase is due to the inclusion of large apertures. It has therefore become necessary to optimize the structural system of the floor plates and transversal beams. The structural grid has been reinforced with the right number of beams that serve the dual function of supporting the chromogenic panels. The reinforcement system increases its mechanical performance with the use of the bulkheads in a carbon single-skin laminated with vinyl ester resin, which can block the movement of the structural reference cage. The rigidity of the cage is also increased through the right dimensioning of the high current, placed in the vicinity of the top side line of the hull, and the low current, which also absolve the dual function of support and base for the coupling of the chromogenic panels inside. The construction material is carbon fiber, resin laminated and the structural omega and at the hull with vinylester resin, and subsequently worked with the system of the vacuum bag to clear any possible delamination problems. The entire system of hull lamination is based on a sandwich with a honeycomb core and internal and external layers of carbon fiber material.

3.7.3 Design Solutions "to Live" the Living Area

The two last insights involved the insertion of two sitting systems in the living area and the transformation system of the living area into a dining area.

In the first chase, the concept proposal was the insertion of two symmetrical and linear components, equipped with low seats inside.

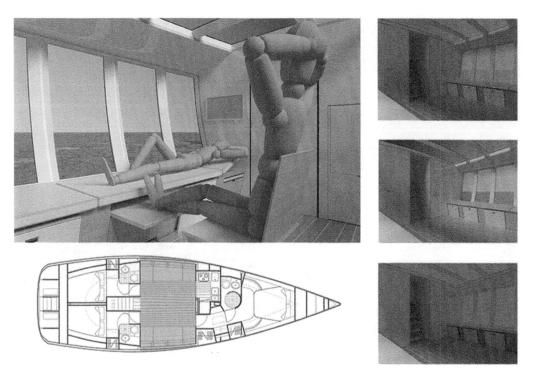

FIGURE 3.2
Interior layout and views of the living area.

This sitting system puts the user in a position to take advantage of the space in different ways, depending on the different useful functions and criteria of the management system of the apertures outside, in the hull, and on the deck. This new solution introduces a new way of feeling the outside context, and creating a new type of relationship between the user and the internal living space, like a scenic surrounding (Figure 3.2).

In the second case, as the square is a multifunctional environment concept, it is possible to transform the living area into a hybrid space involving the function of dining through a mechanism that moves a folding table in the forward bulkhead. The table, with characteristics of modularity, can accommodate a minimum of two guests to a maximum of six guests, according to the organization of the general layout.

3.8 Conclusions

This research aimed to investigate how the PBA could give reliable data on subjective aspects of the project, so far left only to the individual sensitivity and culture of the designers.

The PBA looks at design for the individual. What could be considered quality for one final users could be considered the opposite for many other final users.

Numerous methods to transfer subjective aspects into the projects have been studied, such as the SeQUAM, which has been used in this research to investigate the issue of visual pleasantness as a way of attractiveness.

Such a study was conducted in the field of sailing yachts, considered a key sector of Italian industry.

A first result of the research was to identify the most subjective aspects that could generate healthy sight in the living area of the boat resulting in the criteria that were used to analyze a number of case studies on the market today by using the principles of SeQUAM, given us by the results of matrixes M1 and M2. It is true that in this way it was possible to design and implement a questionnaire, which allowed us to investigate aspects of visual attractiveness. The test was administered to a representative set of potential users. The results were extrapolated to the guidelines and requirements for the design of a 15 m sailing yacht concept.

The design requirements of the yacht had to immediately deal with those problems related to the structural nature of the hull, along with those concerning the external layout, and then those related to the distribution and organization of the external and internal living areas, where the different activities based on different needs were considered. The first result is a concept which has several innovative ideas and can potentially be engineered according to a PBA to the project.

The methodological approach of this research can be considered a starting point for further developments and insights, not only in a yachting living space, but in any interior housing.

Credits

Here, we have detailed the results achieved in a master's degree thesis "Yacht Design_ PBA pleasure-based approach" (Advisor: Professor G. Di Bucchianico, PhD; co-advisor: Professor M. Di Nicolantonio; technical consultants: S. Camplone, A. Vallicelli; candidate: F.P. Salvemini), developed in the "Interior Design of Sustainable Living" Degree Laboratory, at the Department of Architecture of the University of Chieti-Pescara (Italy). All the images reported in this chapter are taken from the abovementioned MSc thesis.

This chapter can be considered the consequence of common discussion and collective review among authors. In particular, the writing of the various sections can be attributed to Giuseppe Di Bucchianico (Section 3.1), Andrea Vallicelli (Section 3.2), Stefania Camplone (Sections 3.3 through 3.5), and Massimo Di Nicolantonio (Sections 3.6 through 3.8).

Reference

Bandini Buti, L., Bonapace, L., and Tarzia, A. 1997. Sensorial quality assessment: A method to incorporate perceived user sensations in product design. Applications in the field of automobiles. In: *IEA '97 Proceedings*. Helsinki: Finnish Institute of Occupational Health, pp. 186–189.

4

Implementing Information and Communication Technology Onboard: An Example for the Integration of Information Received via Communication Equipment with Onboard Navigation Systems

Eric Holder and Florian Motz

CONTENTS

4.1 Overview

In the maritime domain communications between ship and shore, ship and other ships, shore entities with other shore entities, and other stakeholders are now a regular part of operations. Furthermore, the needs and expectations for information sharing are expanding rapidly. This includes the exchange of navigational and voyage planning information (i.e., weather information, optimized routing, and voyage details), emergency information, business and administration information, and personal communications such as emails for passengers and crew among others. The costs and availability of communications and connections (mobile networks, satellite, wi-fi, etc.) vary dramatically depending on the ship's location, also bringing the need for strategic decisions on when and how to use, or allow the use of, these services. The increasing needs and expectations also elicit the emergence of new ideas, such as the Maritime Cloud and Maritime Service Portfolios (Danish

Maritime Authority, 2015). Further, there is the need for affordable safety-based communication services, such as the components of the Global Maritime Distress and Safety System (GMDSS), and hybrids such as the now established VHF (very high frequency)-based Automatic Identification System (AIS) that can transmit limited ownship and safety information.

Alongside these trends is an increase in both automation and technological complexity onboard. Therefore the ship, and especially the bridge, is becoming a prime test case for information and communication technology (ICT) applications with the need for human factors and a systems thinking philosophy to ensure integration in both a user-friendly and safe manner. This chapter will describe the results from a series of projects sponsored by the German Ministry of Transport and Digital Infrastructure (BMVI) demonstrating how to integrate systems-based human factors thinking in the design of an onboard ICT application.

4.2 Information and Human–Machine Interaction Onboard and an ICT Application

Modern ship bridges are highly complex man–machine systems. As such, the safety and efficiency of their handling is dependent on the interaction between humans and machines during the accomplishment of tasks. Humans can fulfill their assigned monitoring, control, and decision tasks (e.g., collision avoidance, conning, and navigation) most effectively, if the information flow between them and the systems on the bridge is adapted to the human's needs, skills, and abilities.

Although not standard practice in the development of maritime equipment (Cyclades, 2015), this concept of human-centered design is receiving increased attention and acceptance within the maritime domain. The International Maritime Organization (IMO) identified the need to equip shipboard users and those ashore responsible for the safety of shipping with modern, proven tools optimized for good decision making in order to make maritime navigation more reliable and user-friendly. Within this framework the IMO decided at the 81st session of the Maritime Safety Committee (MSC) on the proposal of several member states to develop an e-navigation (e-Nav) strategy to integrate and utilize new technologies in a holistic and systematic manner and to make them compliant with the various navigational and communication technologies and services that are already available. The strategy was approved at the 85th session of the MSC.

The e-navigation strategy aims to enhance berth-to-berth navigation and related services by harmonizing the collection, integration, exchange, presentation, and analysis of marine information onboard and ashore by electronic means (IMO, 2008). The e-navigation strategy is supposed to be user-driven, not technology-driven, in order to meet present and future user needs (IMO, 2008).

One of the major issues that was identified through the process of identifying e-navigation user needs (IMO, 2009; Motz et al., 2009) was the need for user-selectable presentation of information received via communication equipment on the navigational displays of the ship's bridge. The availability of safety-relevant information in real time with possible presentation on the navigational displays is considered as an advantage for mariners regarding informed decision making and safety of navigation. This solution is the focus of the project described here.

The current separation of communication systems and navigational systems does not support this need. Onboard, the hydrographical, meteorological, and safety-related information is presented on the communication equipment without filtering or solely as paper printouts with minimal options for efficient integration with the information presented on navigation displays. Technical as well as legal conditions (e.g., separation of responsibilities for radio-communications and safety of navigation in the *Safety of Life at Sea Convention—SOLAS*, Chapters IV and V) hinder the integration of information provided by communication equipment in the navigational systems, which reduces their utilization (IMO, 1974).

The implementation of effective solutions for integrating communicated information into navigational displays required gaining practical experience and methodologically examining the workflow and user needs in detail. This has been a priority research focus for BMVI-funded projects conducted by FKIE (Motz et al., 2009, 2011, 2013). This chapter extrapolates the work presented by Holder et al. (2014) to provide an overview of the methodological human factors process and activities undertaken, and required, to implement effective ICT onboard.

4.3 Goal and Guiding Concepts

Implementation of these solutions onboard was envisioned through an Integrated Navigation System, or INS. The concept for INS as specified in MSC.252(83) (IMO, 2007) is being considered as a dominant factor for the development of e-navigation onboard. The INS concept provides a task-oriented approach which offers the possibility to integrate further functionalities specified with the e-navigation process and present those in a manner that gets the right information to the right person when and where it is needed. A task-oriented integration and presentation of information, when all necessary information for the respective task and situation is available in a fast, reliable, consistent, and easily interpretable format, will support the officers onboard and personnel ashore in their decision making and enhance the safety of navigation. The solutions for integration should improve the safety of navigation and enhance data transfer both between vessels and also between vessels and shore-based authorities without increasing the workload of the user or producing information overload.

The context and constraints on receiving and displaying these types of information items that are received via communication equipment on navigational displays, both current and anticipated for the e-navigation future, need to be understood and considered when defining solutions.

The following considerations were identified as ones with significant impact. Large amounts of information are already available, and this amount will likely increase in the foreseeable future of e-navigation through advances in data production, storage and transmission technologies, and general trends in information sharing. Not all of this information should be presented on navigational displays (e.g., irrelevant, inaccurate, etc.). Not all of this information is relevant to every task display (route monitoring, collision avoidance, etc.). Not all information is relevant to every situation (e.g., location, vessel type, etc.). Not all information will arrive formatted for integration with navigation displays. The point to take away from this list is that it is very important to consider the content, timing, workflow, presentation format, and relationship to the mariner's tasks and overall bridge resource management when defining solutions.

It is also important to learn from the mistakes made in the past and to avoid common human-centered mistakes made in the maritime bridge environment. These include problems such as lack of standardization, inconsistent presentation of information or control–display relationships, unnecessary levels of complexity, and information and alarm overload.

4.4 Methodological Overview

The first step toward defining the design space was to determine what information is out there, or could realistically become available in the future, to be integrated, the context that the information is delivered and used in, and the related problems that mariners were currently experiencing. This would provide the foundation and starting point for understanding the situation and user requirements. This information was then used to develop concepts for managing and integrating information, the user requirements and functional requirements, and a prioritized list of information items to focus on for detailed requirements. A risk identification and risk-control options analysis for the sending and receiving of safety-related information was also performed. A test-bed consortium was assembled to implement selected solutions. Then candidate ships, operating on test-bed supported routes and using manufacturer partner-supplied equipment, were selected to support the conduct of an evaluation and additional data collection activities undertaken to understand these vessel-specific task and information requirements and preferences. Further details describing some of the methodologies used and output at the various steps are described in the following sections as a test case demonstration for what is effectively required to implement ICT solutions onboard.

4.5 Defining the Information, Context, and Problems

Various methods were employed to understand the information that is currently available, and could be available in the future, along with how that information would ideally be used and the problems and constraints in doing so. These methods included reviewing the literature available (including regulatory requirements), reviewing existing and expected communication equipment and content, conducting observations, and a series of mariner interviews and surveys to understand this information and the needs and current problems mariners are experiencing in context. These steps are described in detail in Holder et al. (2014) and Dalinger and Motz (2011).

4.6 Understanding the Onboard Architecture

In order to effectively implement potential solutions onboard it is also necessary to consider, and possibly enhance, an onboard architecture for the ship bridge that allows a user-friendly integration, presentation, and use of the information received via communication equipment. Based on the preceding research, the following objectives were defined and can be used to evaluate existing systems:

- Task-oriented integration and presentation of information received via communication equipment in shipboard navigation systems
- User-selectable filtering and routing of information to prevent information overload
- Data evaluation (quality assurance) and storage
- Provision of source and channel management (selection of best connection according to criteria, e.g., content, integrity, costs)
- Data safety regarding cyber security issues
- Increased availability and reliability of information due to efficient use of different communication channels

The realization of these objectives needs to be supported by the management of the information on both the ship and shore-side in a manner that is designed to allow access, selection, sorting, filtering, and presentation of the information. Fraunhofer FKIE, in cooperation with the members of the national e-navigation working group, developed a proposal for a detailed shipboard bridge architecture based on the modular concept with INS as a core element. Figure 4.1 provides an overview of the architecture for the onboard integration.

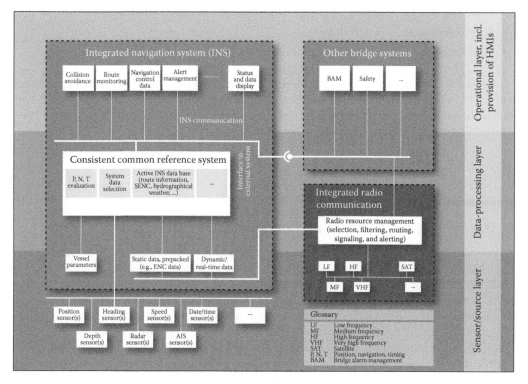

FIGURE 4.1
Shipboard bridge architecture.

4.7 Risk Assessment for the Transfer and Integration of Safety-Relevant Information (Objects)

The risks and risk control options were detailed separately for the sending and receiving process in relationship to the major attributes of the process. For the sending process, the major attributes were risks and errors related to

- Initiating request or reporting requirement
- Filtering, sorting, and quality assurance of information to transmit
- Interaction via the message/information handling HMI (human–machine interface)
- Interaction with onboard databases
- Assimilated content for transmission
- Shipboard source/channel/equipment management

For the receiving process, the major attributes were risks and errors related to

- External source
- Shipboard source/channel/equipment management
- Filtering, sorting, and quality assurance of external information received
- Selection for presentation (message/information handling HMI)
- Interaction with onboard databases
- Presentation on INS displays and user experience

The risks and errors were described in both figure (interactive mindmaps) and textual form (with expanded descriptions) to map out the risks and potential errors. A table was created to match risk control options to risks. The risks and risk control options were then included in further design discussions and decisions.

4.8 Organizing and Prioritizing Communicated Information

The previous steps determined the types of communication information that are received on the ship's bridge from various sources (e.g., vessel traffic services, other ships, coastal authorities, company, etc.) via various media (e.g., AIS, radio, satellite, etc.) and the INS as the best bridge system option to integrate this information. To allow for an initial classification of the safety-relevant information, we utilized categories of generic information types based on the overall type of information (i.e., emergency information, meteorological information, hydrographical information, navigational information, traffic information, communications with office and authorities, and security information) and then subcategories based on the general properties of the message that impacted presentation (priorities and graphical presentation qualities). Specific examples that required special presentation consideration due to priority or presentation properties were also identified. This procedure was chosen due to the abundance of specific information items that can be referred to in a communication that would have similar display properties (e.g., all the updates concerning specific chart-referenced features of the same priority level can use the same workflow and utilize preexisting symbology for graphical presentation).

A matrix was developed to list the information items, details concerning the information included, the INS task supported (and how), suggested presentation format and workflow, timing information, sources of the information, additional considerations, and references.

The matrix was used as the base to begin the process of prioritizing items and safety-related information objects for detailed evaluation. The goal was to focus on a limited number of items (4–7) with the highest compelling need for the users and to produce detailed workflows and display requirements for these items. These items would be evaluated by representative end users using a variety of methodologies, as appropriate for the situation.

The high-priority list of seven items is described briefly below but additional detail along with the logic of that item's selection can be found in Holder et al. (2014). Key considerations in selection included user requested features, relevance to INS, practicality of development and evaluation during the project timeline, and inclusion of items that demonstrated a range of the features and functions of a communication management system.

Item 1—Message/information handling HMI (MIHI): This item is the HMI for message handling and display selections.

Item 2—Alterations to ownship route: This item addresses communications containing alterations to the ownship route (e.g., recommended for safety, economy, or required).

Item 3—Geo-referenced locations to avoid or with special procedures: This item concerns communications containing updates with geo-referenced locations to avoid (hazards or regulations) or with special procedures (e.g., speed or fuel restrictions).

Item 4—Safe-depth information: This item concerns communications containing safe-depth information (tidal information and under keel clearance—UKC).

Item 5—Dynamic air gap information: This item concerns communications (possibly obtaining information directly from on-site sensors) containing safe clearance information (between an overhead object and the water level).

Item 6—Collision avoidance information: This item concerns the exchange of route, maneuver, or intention information for collision avoidance.

Item 7—Meteorological and hydrographic data that impacts conning: This item concerns the near-term information that impacts route and steering (conning). This includes the representation of set and drift (current) and leeway (wind) and their impact on steering and ship handling. This may also include high wave prediction, as well as parametric rolling condition monitoring.

Items not included in the priority list: The following items were identified as interesting but only to be included in evaluations as time and resources allowed. These items included storm information that was not captured in items 3 or 7; piracy information; general meteorological; and hydrographic data not covered in priority items; ice information; search and rescue information; pilot information/pilot services; the use of textual information for collision avoidance; and security-terrorism information.

4.9 Detailed Evaluation of Priority Items

The priority items were then evaluated in detail to further define the user requirements, functional requirements, workflow, and the optimal presentation parameters for effective solutions. These definition steps included iterative human factors review, as well as interviews with a variety of stakeholders and simulator evaluations of design concepts and mock-ups. An overview description of one of the additional site visits that was not reported in Holder et al. (2014) is provided in the following section. Complete, detailed results are not included due to space constraints.

4.9.1 Canada 2014: St. Lawrence Pilots and Coast Guard

Although prior research activities had looked at test beds and prototypes, it was desired to also look at how similar concepts were being used in current operations. Therefore, interview sessions were arranged with representatives of the St. Lawrence Pilot group (March 31, 2014) and the Canadian Coast Guard (April 1, 2014) based out of Quebec City, Quebec to gain experience about the exchange and integration of information in the St. Lawrence seaway system. The groups were chosen not only for their in-depth experience and industry-leading application of e-navigation solutions but also specifically as a group that had significant experience with ice and ice portrayals. The interviewers explained the project and the concepts that were being evaluated. The participants were given an overview of the project and its goals and informed that the purpose of the interviews was to collect their feedback and input on the priority items from communication equipment to be presented on an INS and for the best manner to integrate that information. The participants were asked to describe how they currently exchanged and used these types of information and then were shown example portrayals, as warranted, to further the discussion. The participants were encouraged to suggest additional information items or use-cases.

The participants provided detailed feedback on the services offered concerning safe-depth information, ice information, meteorological and hydrographic information, and geo-referenced information items. Notable points included that they are sharing information via their online portal and broadcasting other information via an AIS relay system. It was not seen as feasible to model and depict the constantly moving river ice in a useful way on navigation devices. Instead a system of cameras was implemented and could be accessed online to allow a view of the current ice conditions before making the voyage or when internet access was available. Detailed results can be found in Motz et al. (2013).

4.10 Test-Bed Implementation

The vision throughout the project was to conduct research and then implement and evaluate the selected solutions in real-world use in a national test bed. To cover the whole process and workflow including data production, processing, and preparation of the data for dissemination, the exchange of information between ship and shore, the integration onboard, and finally the presentation on the navigational systems, it was essential to form

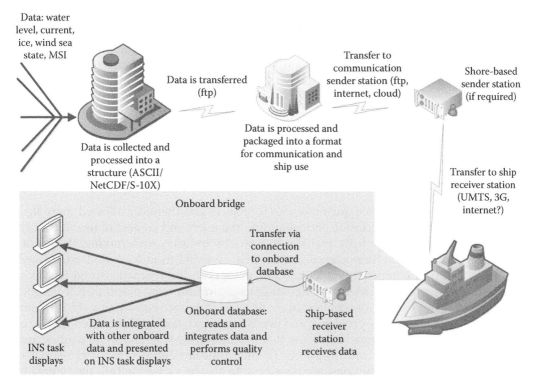

FIGURE 4.2
Communication and infrastructure overview.

a consortium of diverse partners for the test bed (see Figure 4.2). These include The German Federal Maritime and Hydrographic Agency (BSH) and German Weather Service (DWD) as data providers; Jeppesen as a data processor, preparer, and disseminator; Sam Electronics (partner) and Raytheon Anschütz (observer and consultant) as onboard equipment manufacturers; and vessel and shore participation supported by partners such as TT-Line and the German Waterways and Shipping Administration (WSV). Detailed requirements were provided to the equipment manufacturers and other partners and extensive discussion took place concerning the information, formats, features, and portrayal options.

The vessel Neuwerk operated by the WSV was selected as a candidate test vessel and therefore was also visited for observation and interviews to define the task and information needs in more detail. The data collected included the crew structure, planning procedures, information sources in use and desired, and a detailed analysis of relevant tasks focusing on the information currently used and also desired, and how the proposed test-bed information could support these specific tasks.

It was clear from the discussions with the Neuwerk, and other representative users, that safe-depth information was a top priority and also one of the most difficult items to deal with in terms of processing and portrayal options. Therefore it was chosen as the first step for implementation in the test bed and this implementation is currently in progress. It is not possible to go into detail but challenging considerations include processing time when integrating prediction models and detailed bathymetry, matching the updated data to ENC (electronic navigational chart) cells to allow time-based presentation at smaller scales than ENC cells, systematically dealing with the error and safety margins in the data to still reap the benefit of improved data while ensuring safety, integrating new content with existing INS features and functionality, integrating alarm functions, selecting between a more push or pull-based information requesting system, and several others. The test data sets created were portrayed on Sam Electronic's (now Wärtsilä) Platinum System demonstrating that the data could be integrated and displayed.

4.11 Summary

This research has taken place during research projects and therefore allowed more flexibility and time and resources for understanding the users and context of use. Although these resources may not be fully available in a strictly industry undertaking, the results hopefully emphasize the importance of why it is essential to at least understand and take these end-user components into account, along with the regulations and technical details that are already considered in regular practice. In summary, maritime ICT product development should adopt a systematic human factors approach considering the different intended user groups, tasks, problems, and needs in context; taking a larger perspective of the communication management network that includes other ship and shore entities; and the potential risks and risk control options. The maritime environment is a very regulated one and therefore all of these considerations have to also be viewed within the context of the applicable rules and guidance, either following these guidelines or providing compelling evidence for a need to change them. Further it is recommended to assemble a broad stakeholder group to bring in the various perspectives required to implement a successful maritime ICT system.

References

CyClaDes. 2015. Best practices in the maritime industry. BALance Technology Consulting. http://www.cyclades-project.eu/CyClaDes/results/best-practices-in-the-maritime-industry;jsessionid=01a3c63cd6a3072f12898b16a9ae?title=Best+practices+in+the+maritime+industry (Accessed March 3, 2015).

Dalinger, E. and Motz, F. 2011. Design of a communication management system on board in the framework of e-navigation concept. In: *Proceedings of the ERGOShip Conference on Maritime Human Factors*, Gothenburg, Sweden. http://conferences.chalmers.se/index.php/ergoship/ergoship11/schedConf/presentations (Accessed May 9, 2016).

Danish Maritime Authority. 2015. An overview of the "Maritime Cloud" proposed information exchange infrastructure for e-navigation. https://dma-enav.atlassian.net/wiki/display/MCCT/Maritime+Cloud (Accessed August 31, 2015).

Holder, E., Motz, F., Horoufchin, H., and Baldauf, M. 2014. Concept for the integration of information received via communication equipment with on-board navigational systems. In: Ahram, T., Karowski, W., and Marek, T., (eds.) *Proceedings of the 5th International Conference on Applied Human Factors and Ergonomics AHFE 2014*, Kraków, Poland July 19–23. Louisville, KY (USA): AHFE Conference, pp. 203–214.

IMO 1974. *International Convention for the Safety of Life At Sea*, November 1, 1974, 1184 UNTS 3. London: International Maritime Organization.

IMO 2007. *Revised Performance Standards for Integrated Navigation Systems MSC.252(83)*. London: International Maritime Organization.

IMO 2008. *Report of the Maritime Safety Committee on its Eighty-Fifth Session*. MSC 85/26. London: International Maritime Organization.

IMO 2009. Development of an e-navigation strategy implementation plan—Results of a world-wide e-navigation user needs survey. *Submitted by Germany to the 55th Session of the IMO Sub-Committee on Safety of Navigation*. NAV 55/11/3. London: International Maritime Organization.

Motz, F., Dalinger, E., and Holder, E. 2011. *Communication Elements in Integrated Navigation Systems (CE-INS)*. Final Report for the Ministry of Transport, Building and Urban Development. Wachtberg: FKIE.

Motz, F., Höckel, S., and Dalinger, E. 2009. *On-Board User Requirements for the e-Navigation Concept*. Final Report for the Ministry of Transport, Building and Urban Development. Wachtberg: FGAN-FKIE.

Motz, F., Holder, E., and Horoufchin, H. 2013. *Safety Relevant Information Objects for INS (SINFO)*. Final Report for the Ministry of Transport, Building and Urban Affairs. Wachtberg: Fraunhofer FKIE.

5

Using Eye-Tracking and Mouse Cursor Location to Examine Visual Alerting in a Multi-Display Environment

Jacquelyn Crebolder and Joshua Salmon

CONTENTS

5.1 Introduction

With increasing levels of complexity being designed and implemented into many of today's socio-technical systems, the assistance of automation is integral to enable users to sort through and handle the enormous amount of information, data, and choices available to them. One style of automated assistance comes in the form of alerts or alarms delivered to direct the user to aspects of interest or significance. Appropriate alerting can be critical in complex, high workload environments, where individuals use multiple information displays to conduct several tasks concurrently. Under pressure of high workload, and in the absence of effective automated alerting, operators may become focused on a task and miss important details or information. One example of this kind of environment is the operations room of a navy frigate. The frigate operations room is the processing hub for all sensory information pulled in from the world outside. Incoming information is collated here and formed into a global picture that provides the command team with situation awareness to support operational decision making. Operators in the operations room use multiple displays to perform their jobs and they are heavily tasked, being required to quickly read and interpret incoming information while monitoring for new information and changes to existing data. Speed and accuracy are fundamental to timely decision making and the operations room is at times a noisy, distracting, and intense arena.

Because of the concentrated effort and decision-making pressure within the operations room, operators could be prone to attentional tunneling (Wickens, 2000, 2005) where focus on a particular task or area of interest on the display is so concentrated that critical information is missed. This is one reason that effective modes of automatically alerting operators to specific conditions, states, and points of interest are critical. Most of the alerting in the operations room is currently provided through the auditory modality which is heavily taxed even when the alerting component is excluded. Thus, the visual realm may be an alternate channel for supplementing automated alerting in this complex environment.

5.1.1 Visual Automated Alerting

In a multiyear research program, we have been investigating visual forms of alerting to supplement the fully loaded auditory modality in the navy frigate operations room. Two forms of visual alerting have been examined: (i) a short red bar 2 cm wide appearing at the side, top, or bottom of the operator's display and (ii) a red border 2 cm wide around the perimeter of the display. Using a three-display workstation (see Figure 5.1), location of the alerts has also been studied, by presenting the alert on a single display (either the left, middle, or right display), or on all three displays simultaneously (Crebolder et al., 2010; Nakashima and Crebolder, 2010). We have also investigated whether detecting alerts was affected by their appearance as static or flashing (Crebolder and Beardsall, 2008, 2009).

These studies were conducted using a task in which participants were required to detect the visual alerts while performing a secondary task that required their full attention. The results consistently showed that the bar alert was responded to more quickly than the border form of alert, and, in contrast to other research (Goldstein and Lamb, 1967; Li et al., 2014) that flashing alerts did not improve response time over static presentation. We have

Display type

Status   Tactical   Reporting

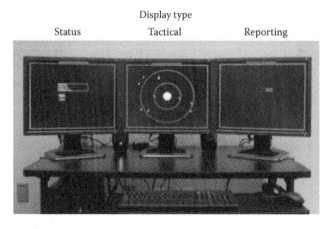

FIGURE 5.1
Red bar alerts located at the top of the screen on three displays. Border alerts were similar in width and color but surrounded the entire perimeter of the display.

also found that responses to bar alerts were fastest when alerts appeared simultaneously on all three displays, but for border alerts the left and middle displays produced response times that were equal to the "all" display condition.

The consistent result that bar alerts were detected faster than border is a somewhat non-intuitive finding since one might expect that the larger surface area of the border surrounding the entire perimeter of the display would be easier to detect that a smaller short bar specifically positioned on the display. One hypothesis for this result is that the bar alert is detected more rapidly because of its more compact, concise form and its consequent ability to fall into a defined attentional radius or spotlight, as compared to the larger spread of the border. If this is the case, bar alerts should be detected faster when participants are attending to the same display the alert appears on because the alert is being captured in the same radius of attention.

The next step in the research program then was to capture the data that showed the display the alert appeared on and the display the participant was looking at when the alert appeared. These data could verify whether eyes on the same display as the alert affected detection time. We used the location of the mouse cursor (the display the cursor was on) as a basic estimation of where the participant was looking (Crebolder, 2011). Findings were varied and they did not provide solid evidence for or against the spotlight of attention theory. Results showed that, for the border alert, response times were in fact slower when the alert appeared on the left display and the participant was attending to that display. This finding suggests an effect of attentional tunneling, or tunnel vision, whereby the user is so immersed in a task on, in this case the left display, and is focusing on a particular area of the display to the point that they miss other critical information on that display (Wickens, 2000), which in this case was the alert. This is not an uncommon phenomenon in environments where an individual is required to perform multiple simultaneous tasks under time pressure and where the consequences of inaccuracy are severe, such as for operations room personnel, and air traffic controllers, maritime helicopter flight deck operators, and nuclear power plant operators (Rubinstein and Mason, 1979). Furthermore, the abrupt onset of a visual stimulus has often been used successfully to capture attention, but in some cases, where attentional resources are allocated to other information and tasks the effect of abrupt onset can be significantly reduced (Yantis and Jonides, 1990).

The results further showed that eyes on the right or middle displays resulted in no difference in response time when the alert was on either of those displays. In fact response time was faster when eyes were on the left display.

On the other hand, for the bar alert, response to an alert was considerably slower when it appeared on the right display and eyes were on the left as compared to when eyes were on the right display.

Generally the findings showed that alerts presented simultaneously on all displays were attended to fastest but that in some cases attentional tunneling may have been evident whereby alert detection was hurt by having the alert on the same display that the participant was looking at.

Using the cursor position as an estimate of where participants were looking is a relatively elementary method but one that others have used and regarded as valid (Bieg et al., 2010; Guo and Agichtein, 2010; Huang et al., 2011). Much of the work using cursor position as an assessment of where people are looking and attending is found in the web-based applications research, where the interest is in where users are focusing on a webpage or how the cursor is used to help a user read a web page. For example, looking at eye movement and cursor movement, Rodden and Fu (2007) found a strong relationship, as did Chen et al. (2001), in web browsing tasks.

There are a number of advantages if the cursor can be used as an estimator of where users are looking as compared to using an eye tracker. Using the cursor is of no cost as compared to investing in an eye-tracking system, there is no setup required, a cursor is nonintrusive, and perhaps of most importance in the case of web-based research, users do not have to be physically present in order to track their web-based behavior. For our needs, being able to use the cursor as a means of estimating where a participant is looking would be advantageous for all those reasons, but particularly because the cursor is nonintrusive. Eye-tracking equipment attached to a participant's head may impede their ability to easily turn and attend to all information displays without affecting the sensitive calibration of the system and without discomfort. Crebolder et al. (2010) have examined the relationship between cursor and eye tracking and found the cursor to be a reasonable assumption of where the eyes are looking. The present study is an opportunity to provide further validation to that assumption for the kind of detection task used in our studies. This detection task is one that emulates the kind of task a sensor operator might be performing in a frigate operations room.

Thus, one objective of the study reported here was to validate use of the cursor against eye-tracking data in a multiple display detection task. The other primary objective was to gather in-depth information about where the participant is looking when an alert appears in order to delve more deeply into the spotlight of attention theory and why a bar alert has consistently proven to be more quickly detected than a border. As in previous experiments, data were collected on human performance as a function of alert type (border, bar) and alert location (left, middle, right).* The experiment was a within-subjects design. If the results show that the bar alert is better detected when the alert location and the display the eyes are on are the same, as compared to the border under the same criteria, one inference could point to the bar alert falling within the spotlight of attention created as the participant performs the secondary task on the same display, making it easier to detect.

5.2 Method

5.2.1 Task

As in previous studies in this series, the participant's primary task was to detect alerts that appeared randomly on a three-display workstation while performing a secondary task that required detecting and categorizing targets as they appeared on one of the displays. The task was designed to emulate in simple form the workstation and the kind of tasks a sensor operator might do in the frigate operations room.

5.2.2 Apparatus

The task was presented on a workstation consisting of three 20.1" liquid crystal display (LCD) computer monitors, running Windows XP Professional (Service Pack 2), with a single keyboard and mouse input device. The displays were configured with the middle display directly in front of the participant, and one display on either side (see Figure 5.1).

---

* Alert location level "all" was removed in this study because eyes and cursor could never be on all displays at the same time.

Participants were fitted with eye-tracker apparatus, Viewpoint® EyeTracker PC-60 SceneCamera System by Arrington Research, which was used to monitor where the eyes were looking while completing the task.

5.2.3 Tasks and Alerts

Primary task—The primary task was to detect alerts that appeared randomly on the displays. Responses to detection were made by pressing the spacebar on the keyboard as quickly as possible.

Alerts were presented as static (i.e., not flashing)* and were either in the form of a border (a red, 2 cm continuous band around the display perimeter); or a bar (a red, 2 × 10 cm strip placed at the top of the display.[†]

Alerts could appear on any single display. The display the alert appeared on (left, middle, right) was randomized, with the condition that alerts appeared at all possible display locations an equal number of times.

Secondary task—The secondary task was a categorization task that required using the three displays to categorize targets as neutral or hostile. Failure to correctly categorize a target within a limited time period resulted in the assumption it was hostile and the destruction of the participant's ship, followed by a restart of the task. Details of the task and display setup were as follows, beginning with the middle display which describes the fundamental task.

Middle display—tactical display—Depicted the participant's ship (ownship), represented as a gray filled circle (60 pixel radius), that remained stationary in the center of the display, as well as other vessels (contacts) represented as yellow triangles that originated in the periphery of the tactical display and advanced toward the ownship in incremental steps at 2 s intervals. Thus, there were a number of contacts on the screen at once, all moving incrementally toward the participant's ownship in the center. The task was to categorize contacts on the display as hostile or neutral by moving the mouse cursor over a chosen contact. This action generated attribute information about that specific contact that appeared on the left display (status display).

Left display—status display—Showed attribute information about each contact after it had been moused-over on the middle (tactical) display. The information on the status display was required to classify contacts as hostile or neutral. Three categories of information were provided:

Speed:	Fast = hostile	Slow = neutral
Size:	Small = hostile	Large = neutral
Weapons on board:	Yes = hostile	No = neutral

Based on the above, a score of >2 attributes in one of the hostile or neutral categories resulted in the contact being classified as such. After making a decision as to whether the contact was hostile or friendly, the participant was required

---

* In some previous experiments a flashing component of 3.333 Hz was used (Crebolder and Beardsall, 2008, 2009).

[†] Previous work has examined placing the bar at the side, and on the bottom of the display (Crebolder and Beardsall, 2008, 2009).

to use the cursor to select and click on one of two text boxes representing hostile and neutral.

Right display—reporting display—Participants entered their response of neutral or hostile in a text box on the reporting display located to the right of the tactical display. The cursor was used to click on the text box before a response could be entered. A correct response resulted in the contact of interest disappearing from the tactical display. An incorrect response required repeating the mouse-over contact process and reviewing the status display attribute information once again, subsequently going through the action of reporting the choice on the reporting display and using the cursor to highlight the reporting box before entering a response.

The entire task was very interactive, employing all three displays as equally as possible and ensuring that use of the cursor was required on all the displays.

Participants were instructed that there was a time limit to categorize incoming targets and that contacts coming within a predetermined radius of the ownship would result in the ship being destroyed (accompanied by an audio file sound effect "kaboom" with a JPEG picture of an exploding ship displayed on the tactical display). If the ownship was destroyed the task was paused for 3 s after which time it was automatically restarted, with contacts once again originating in the periphery of the tactical display and moving incrementally toward the ownship.

Participants were instructed that detecting alerts was their primary task and that they were to make their response as quickly as possible after by pressing the spacebar.

5.2.4 Participants

After explanation of the task, the participant put on the eye-tracking headwear and the eye tracker was calibrated. Four blocks of practice trials and 18 experimental blocks (six alerts per display per block) followed. The alert condition of type was held constant throughout each block of trials and the order of blocks was counterbalanced across participants.

Participants were instructed to reduce head movement as much as possible so that the eye-tracker calibration would remain stable. The restriction was somewhat artificial considering the task involved several displays but it was a necessary request based on limitations of the head-mounted eye-tracking equipment. A key variable of interest was the comparison of performance between the bar and border alerts, with the addition of the eye tracker allowing for more in-depth analysis. Thus it was deemed reasonable to accept the limitations of the system. The effect of performance on the task itself would remain to be seen.

5.2.5 Performance Measures

Performance measures on several factors were collected but those of most relevance to this study were response time to alerts and position of alert, cursor, and eyes with respect to the display.

Response time was automatically collected via keystroke. For the eye-tracking and cursor data the initial plan was to divide each display into a 3×3 grid and record the grid in which the eyes were focused or the cursor was placed, but this approach proved to be problematic. First, the border and bar would not naturally fit in the same space. The bar would occupy a single grid, while the border would fill eight grid locations on the display with the middle (center) grid unoccupied. We chose the eye distance as the distance of the eye from the lines making up the border (EyeXdist, EyeYdist) as the best way to measure distance from the eye location to the border location. Second, if frame of reference moved (e.g., head movement), then the association with the correct grid location might be compromised. Finally, how best to represent these grid locations in an economic way for analysis was an issue. Rather than using discrete eye locations, a program was written to capture eye space as a continuous spatial value in the X and Y domains (independently). The advantage of this approach was that the data could be represented as a function of distance from the alert over time (similar to the way event-related potential waveforms are represented during electroencephalography analysis).

An additional constraint was that eye location could only be sampled every so often since pure continuous tracking would result in an infinitely long data file. Consequently a manageable sampling period pre- and post-alert was determined. A resolution of 30 ms was used, being shorter than average fixation length (50 ms, cf. Nuthmann et al., 2010) and about equivalent to two screen refreshes on a 60 Hz display. Eye-location data were recorded from 300 ms before alert onset, while the alert was visible to the participant, and continued for 300 ms after the alert disappeared.

5.2.6 Data Preparation

5.2.6.1 Comparison of Eye to Cursor

The comparison between eye points to cursor points would be easier to interpret if the two were in the same space (i.e., the range of values were equivalent). By default, eyes were tracked from values 0.000 to 1.000, which, when multiplied by 1000 equaled values of 0–1000 (in both the X and Y dimensions). The cursor, on the other hand, was measured by pixel coordinates, which on three 1280 × 1024 displays were 3840 pixels wide by 1024 pixels tall. In order to compare eye position to cursor position, these numbers were translated so that the same values corresponded to the same locations (i.e., a cursor position of 100 equals an eye position of 100). This space translation was computed by calculating the range of the spaces and calculating how much the data should be shifted to overlap each other. Within Excel, a method of calculating both the slope/scalar value and intercept was required in both X and Y dimensions to perform a linear transform ($y = mx + b$) on the data. These values were then applied to generate a position of where the eyes were in pixel space.

5.2.7 Data Extraction

For the eye-tracking data, a "parser" file was used to manipulate and retrieve relevant data points. The parser was embedded in a Microsoft Excel® file with underlying VBA code/macros that could be accessed through pull-down menus. Because the measure for the eye-tracking data was a continuous one, the amount of data per trial was much larger than earlier experiments. For example, previous experiments generated approximately 3000 lines of data while the current experiment, with a sampler of 30 ms, generated about 17,000 lines of data for a 50 min period. To reduce the amount of data the cursor versus eye data was limited to the X dimension since distance measures were not of value in the Y dimension. Thus the data analyzed for the eye cursor was cursor X dimension, eye X dimension, for border and bar.

5.3 Data Analysis

5.3.1 Response Time

Response times to alerts detected correctly were analyzed. Cell means for alert type and alert location, for each participant, were entered into a repeated measures Analysis of Variance (ANOVA). Significant effects of alert type [$F(1, 23) = 25.205$, $p < 0.001$, $MS_e = 16{,}597.283$] and alert location [$F(2, 46) = 10.889$, $p < 0.001$, $MS_e = 5293.428$] were present, with no interaction. Overall, responses were significantly faster to the border alert ($M = 806.38$ ms) than to the bar alert ($M = 874.33$ ms). Furthermore, responses were fastest when the alert appeared on the middle display and were slowest to the right display.

The alert type result is contrary to previous work in the series where the bar alert has been consistently detected faster than the border. As such, further analysis of alert type is not extended here, but this contradictory finding is discussed in the discussion section.

5.3.2 Eye Tracking/Cursor

A primary objective of the study was to validate the use of cursor location as a means of inferring eye location. In the eye-tracking/cursor analysis, the data were collapsed across the variable alert location. The mean for the eye-tracking and cursor data was calculated over all participants and plotted as a function of alert type (border/bar) and distance from the alert (in pixels) over time, in the X dimension. Not surprisingly, the average distance from the alert for participants' eyes varied because of differences in eye calibrations and head positions. To correct for this variance, a baseline correction was applied by subtracting each participant's average distance pre-alert from all their data points. The correction had the effect of recentering each individual's average distance at 0, with any significant deviations from 0 representing a significant deviation from the average. Figure 5.2 shows the data for eyes and cursor in the border and bar conditions plotted after baseline correction.

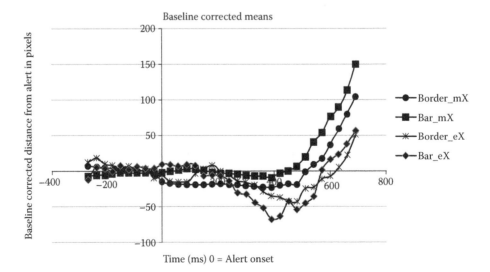

FIGURE 5.2
Distribution of baseline corrected mean eye location (eX) and cursor location (mX) for the bar alert in the X dimension, over time (ms).

The Y axis represents pixels, 0 being the alert. The X axis represents time, 0 being alert onset. Data points below 0 on the Y axis indicate that the eyes are moving toward the alert, and above 0 that they are moving away from the alert. For the bar and the border the eyes moved toward the alert and then moved away. Participants looked toward the alert approximately 200 ms after onset, and subsequently the eyes looked away at about 450 ms. Border data was a little more variable showing an initial move away at about 100 ms before changing toward the alert.

Generally, the cursor and eyes were well aligned, the Pearson product-moment correlation coefficient for the bar alert was $r = 0.55$ [$p < 0.0011$] and for the border $r = 0.67$ [$p < 0.0001$], with a combined correlation of $r = 0.62$ [$p < 0.001$]. The values represent positive correlations of eye and cursor location, showing that where the cursor moved so did the eyes. Thus, the data support the assumption that the location of the cursor is a reasonable estimate of where the eyes are looking in the task that was used in this experiment and in previous studies in this series.

5.4 Discussion

The objectives of this study were to replicate previous findings in a series of studies where one form of alert has been found to be more quickly detected than another, and to validate the use of cursor location against the location of the eyes in a computer-based task. Alerts were presented randomly while participants performed a categorization task that required their full concentration.

The bar alert, which had been detected faster than the border alert in all previous studies in the series, was in fact slower to detect in this experiment. The initial rationale for that result might be that wearing the eye-tracking equipment impeded performance. As such, the attentional spotlight, that may have served to capture the bar in previous studies, was not as effective in this particular experiment. One theory is that the instruction to limit head movement, brought on by the head-mounted eye tracker, forced participants to expand their attentional beam. Consequently the advantage of the smaller concise bar falling into an attentional spotlight was reduced, resulting in the border being detected faster than the bar. The theory sounds plausible, although the same eye-tracking equipment had been used as part of a previous experiment in which the tracker, being new to the lab, was tested. No eye-tracking data was collected for formal analysis in that experiment but the bar–border comparison was consistent with all previous ones, with the bar being responded to faster than the border (Crebolder, 2011). In that study, participants were not instructed to limit head movement, because the eye tracker was an add-on for initial testing purposes. As such, expansion of the attentional beam remains a plausible theory and worthy of further investigation.

We had hoped to use the eye-tracking data to delve more deeply into the previously shown superiority of the bar alert to shed light on why such an alert would prove to be faster to respond to than a full border. However, because of the unexpected result it is not possible to analyze the data as hoped. Overall, responses were fastest when the alert appeared on the middle display and slowest on the right display, which are findings in line with previous work.

Nevertheless, it is of interest to use the eye-tracking data to examine behavior with respect to alert onset. Participants looked toward the alert on average 200 ms after onset and subsequently looked away at approximately the 400–450 ms mark. The cursor and eyes were quite well-aligned except during the short toward-away alert behavior. This pattern is to be expected since the eyes are capable of quick saccadic movement and the cursor was not required to contact with the alert as the eyes were. The speed at which the eyes moved toward the alert is consistent with typical reflexive saccade movement (Purves et al., 2001). Saccades to an unexpected stimulus are usually about 200 ms and last up to about the same amount of time. So the time frame is as expected. Wood (1995) has noted that attention to alerts of any kind does not come from a neutral impartial state but requires actually shifting attention from an existing event to a new and relevant one. Participants looked away from the alert relatively quickly which was most likely a consequence of the task, where lingering on the alert after detection would be detrimental to performance in the secondary task. Note that overall mean response time was 840 ms showing that participants continued to process the alert into the motor action of response (pressing the spacebar) after their focus had moved away from the visual appearance of the alert on the screen. This pattern is also in accordance with the literature in which eye fixation is followed by processing toward response (Just and Carpenter, 1976; Carpenter, 1998).

5.5 Conclusions

Overall, this study demonstrated that cursor location is an accurate estimate of where an individual is looking. Many times the outcome of basic research is generalizable to other tasks and domains. However, results may be more specific and may change when associated with a different context or domain, particularly in cases like in this series of studies where it appears that small changes to the stimulus or experimental paradigm can bring about very different results. Therefore, care must always be taken in generalizing results to other tasks where research has originally been designed for a particular context, as this one was.

There was an indication in this study of the possible effects of wearing a head-mounted eye-tracking system and, as a consequence, we feel that more investigation is required to understand these implications. Further investigation may shed light on why the border alert proved to be easier to detect in this particular study which is in contrast to other studies in this series where the bar alert has consistently been detected faster than the border.

With the ever-increasing complexity of technology in the maritime transportation domain, as well as in other environments, there springs a rudimentary requirement for basic research to feed into the development of socio-technical systems design. Basic research is essential to a solid design foundation and the importance of providing input at the initial, grassroots level in the design process cannot be overstated. Here human factors and ergonomics expertise, effort, and input should be part of the overall design process from the start, rather than being appended or provided as a modification piece at the end of the production line.

References

Bieg, H., Chuang, L., Fleming, R., Reiterer, H., and Bülthoff, H. 2010. Eye and pointer coordination in search and selection tasks. In: *Proceedings of the 2010 Symposium on Eye-Tracking Research and Applications*, Austin, TX, pp. 89–92.

Carpenter, R. 1998. *Movement of the Eyes.* London: Pion.

Chen, M., Anderson, J., and Sohn, M. 2001. What can a mouse cursor tell us more? Correlation of eye/mouse movements in web browsing. In: *Proceedings of the 2001 ACM CHI Conference in Human Factors in Computing Systems,* Seattle, WA, pp. 206–212.

Crebolder, J. 2011. Investigating human performance in complex command and control environments. *Journal of Human Performance in Extreme Environments,* 10(1). http://dx.doi.org/10.7771/2327-2937.1000.

Crebolder, J., and Beardsall, J. 2008. Investigating visual alerting in maritime command and control. *Defence Research and Development Canada Atlantic Technical Memorandum TM 2008-281.*

Crebolder, J., and Beardsall, J. 2009. Visual alerting in complex command and control environments. In: *Proceedings of the 53rd Human Factors and Ergonomics Annual Meeting,* San Antonio, TX, pp. 1129–1133.

Crebolder, J., Salmon, J., and Klein, R. 2010. The cost of location switching during visual alerting: Effects of experience and age. In: *Proceedings of the 54th Human Factors and Ergonomics Annual Meeting,* San Francisco, CA, pp. 1655–1659.

Goldstein, D. A., and Lamb, J. C. 1967. Visual coding using flashing lights. *Human Factors,* 9, 405–408.

Guo, Q., and Agichtein, E. 2010. Towards predicting web searcher gaze position from mouse movements. In: *Proceedings of the ACM CHI Conference in Human Factors in Computing Systems,* Atlanta, GA, pp. 3601–3606.

Huang, J., White, R., and Dumais, S. 2011. No clicks, no problem: Using cursor movements to understand and improve search. In: *Proceedings of the 2011 ACM CHI Conference in Human Factors in Computing Systems,* Vancouver, BC, Canada, pp. 1225–1234.

Just, M., and Carpenter, P. 1976. Eye fixation and cognitive processes. *Cognitive Psychology,* 8, 441–480.

Li, G., Wang, W., Li, S., Cheng, B., and Green, P. 2014. Effectiveness of flashing brake and hazard systems in avoiding rear-end crashes. *Advances in Mechanical Engineering,* Vol. 2014. http://dx.doi.org/10.1155/2014/792670.

Nakashima, A., and Crebolder, J. 2010. Evaluation of audio and visual alerting during a divided attention task in noise. *Canadian Acoustics,* 38(4), 3–8.

Nuthmann, A., Smith, T., Engbert, R., and Henderson, J. 2010. CRISP: A computational model of fixation durations in scene viewing. *Psychological Review,* 117(2), 382–405.

Purves, D., Augustine, G., and Fitzpatrick, D. (Eds). 2001. Types of eye movements and their functions. *Neuroscience,* 2nd edition. Sunderland, Massachusetts: Sinauer Associates, pp. 361–390.

Rodden, K., and Fu, X. 2007. Exploring how mouse movements relate to eye movements on web search results page. In: *Proceedings of the 30th Annual International ACM SIGIR Conference—Web Information Seeking and Interaction Workshop,* Amsterdam, Holland, pp. 29–32.

Rubinstein, T., and Mason, J. 1979. An analysis of Three Mile Island. *IEEE Spectrum,* November, 37–57.

Wickens, C. 2000. Designing for stress. *Journal of Human Performance in Extreme Environments,* 5(1), Article 11. http://dx.doi.org/10.7771/2327-2937.1012.

Wickens, C. 2005. Attentional tunneling and task management. In: *Proceedings of the 13th International Symposium on Aviation Psychology.* Dayton, Ohio: International Symposium on Aviation Psychology.

Wood, D. 1995. The alarm problem and directed attention in dynamic fault management. *Journal of Ergonomics,* 38(11), 2371–2393.

Yantis, S., and Jonides, J. 1990. Abrupt visual onsets and selective attention: Voluntary versus automatic allocation. *Journal of Experimental Psychology: Human Perception and Performance,* 16(1), 121–134.

6

Seeking Harmony in Shore-Based Unmanned Ship Handling: From the Perspective of Human Factors, What Is the Difference We Need to Focus on from Being Onboard to Onshore?

Yemao Man, Monica Lundh, and Thomas Porathe

CONTENTS

6.1 Introduction

During the ship handling process, ship handlers have always strived for a continuous balanced effect by tuning the ship to the environment under different situations. Previous studies (Prison et al., 2009) have discovered that one tacit but indispensable gut feeling known as ship sense is intensively involved in ship handling for the safety of the vessel and people. When the bridge officer lacks visual reference, navigational instruments like the radar and the electronic nautical chart will be the main input source. However, when the weather gets rough, he will make use of ship sense to handle the ship in relation to the direction of the oncoming wave (Porathe et al., 2014). Sensing the ship's movements, the bridge crew will maneuver the ship to achieve the goal of safety.

Ship sense has never been the magic word from the perspective of perception and cognition. During the ship's maneuvering, information will first be gathered through the ship handlers' senses via different perception receptors. For example, the information could be the kinetic feeling of heaving, pitching, and vibration of the vessel, seeing the wave patterns, hearing the wind, wave slamming, engine sound (Prison et al., 2009), etc. Then the ship handler has to interpret the information and make sense of the situation with the mental model. By using his experience and skills, he will eventually make a decision. Since the dynamic information comes from the environment and the vessel whose physical state is constantly changing, the ship handler has to cope with fast-emerging tasks, such as slowing down the speed or adjusting the rate of turn when feeling the bank suction effect. Effective decision making and appropriate actions from personnel can only be achieved by successfully balancing task demands and the human's individual capabilities (Fuller, 2000). In the task of ship handling, there is a balancing act between the ship handler's capabilities (based on his personal prerequisites) and the task demand (made up by the environmental prerequisites) conducted through his vessel (the tool) (Prison, 2013). That is the "harmony" between the ship and environment that ship sense serves to continuously assure safety (Prison et al., 2013).

While the concept of ship sense and harmony is originally created for onboard ship maneuvering, this chapter extends it to the domain of shore-based unmanned ship monitoring and control from the perspective of changes in human factors. The 3-year 7th Framework EU Project MUNIN (Maritime Unmanned Ship through Intelligence in Networks) has been investigating the feasibility of autonomous unmanned ship and prototype implementation of its shore-based control center (SCC) since 2012. The motivation for MUNIN is presented as the striving for a better working environment, reducing costs of transportation, the global need of reducing emissions, and increased safety in shipping (Porathe et al., 2014).

In MUNIN, the unmanned ship is one 200 m long dry bulk carrier with the intelligent Autonomous Ship Controller (ASC) system. The slow-steaming ship conducts collision avoidance without human interference during intercontinental voyages. Meanwhile, the ship is also constantly monitored by a manned SCC. The operations in SCC includes remote monitoring and remote control (Rødseth et al., 2013), so SCC can decide when to intervene based on the status information sent from the ship, and also override ASC to make sure the ship is working under International Regulations for Preventing Collisions at Sea (COLREGS).

With the apparent changes made in the system, people are no longer maneuvering ship onboard but from ashore. Nowadays, the maritime industry is facing more human factor issues (Han and Ding, 2013). Unmanned ship might resolve fatigue problems but it also brings more questions concerning remote supervisory monitoring, as people need to be able to get into the control loop at any time. For example, how do operators in the SCC perceive the ship's movements and maneuver the ship without ship sense when the working environment has greatly changed in the SCC? There will be no physical connection between the human and the vessel, and no directly perceived information from the ship's environment. Specifically, the visual perception of the environment, a vital sense in ship handling for bridge officers, will be lost. The important questions will arise: Are there going to be new human factor issues? Will the same human factors be applied as they do for the manned ship? If no, what factors behind ship sense onboard need to be refactored to the shore side? How can we prioritize them to regain harmony?

In fact, the sense deprivation and new way of human machine interaction indicates the importance and necessity to reanalyze how human factors are applied in a distributed system. This chapter provides a preliminary exploration of human factor issues in shore-based unmanned ship handling and explores some influential aspects of human factors we need to focus on in order to facilitate shore-based ship handlers to regain harmony. Ten master mariner program students with experience at sea were invited to take part in a focus group interview. The purpose is to discuss the different tasks and actions onboard and ashore as a basis to explore underlying human factors requirements in the context of the MUNIN project. The results highlight several differential aspects in human factors that should be considered, refactored, and prioritized. It also suggests general approaches to user-centered design for SCC in practice.

6.2 Methodology

The study adopts the focus group interview (Kitzinger, 1995) as the main data collection approach. Ten undergraduate students in Chalmers University of Technology voluntarily took part in the focus group interview. The participants' background was similar: they were studying in the same master mariner's program and they all had sea experience prior to the focus group interview, however, not as officers. Their previous active time at sea varied between 9 and 33 months, average 16.5 months (the standard deviation is 7.2). Only one participant was Mexican-Swedish while the rest nine participants were all Swedish. Their ages range from 22 to 41 years old, average 27 years old (the standard deviation is 7.2). One of the participants was female (10%) while the rest were males (90%). Out of the 10 participants, only one person (10%) did not have ship maneuvering and navigation experience, the rest (90%) all had experience in ship handling from the bridge, either alone or under the supervision of the captain. Fifty percent of the participants had the experience of remote ship monitoring or controlling, including in a simulation environment. Besides, 50% of the participants had been previously involved in ship or workplace design work (ships, systems, tools). Forty percent of the participants mentioned that they also had working experience in maritime-related activities at the same time as they studied, mainly being able seaman and working for passenger vessels.

The focus group interview took place at Chalmers University of Technology, Gothenburg, Sweden. All participants signed a written consent about the anonymous and ethical usage of their data in the academic research. The interview process was recorded by a voice recorder for further analysis after the interview. The interview lasted for approximately 2 h. The focus group interview assistant took field notes on the participants' discussion. All participants were briefed about the MUNIN project with the idea of a dry bulk carrier sailing without a helmsman by remote ship monitoring and control. The discussion was based on the constraints and conditions in the project described earlier in this chapter.

The first questions asked the participants to discuss the possible actions to execute ship handling that would actually correlate with their past ship maneuvering experience: *What actions will it take to monitor and maneuver the ship onboard today?*

The replies from the participants were continually listed on the whiteboard. Then the second question asked the participants to envision an operators' possible action in an SCC: *What actions will it take to monitor and maneuver an autonomous unmanned ship from an SCC?*

With the actions and tasks being discussed in both onboard and onshore situations, the third question asked the participants to identify the changing aspects of human factors under these two circumstances: *From the perspective of human factors, what is the difference when we shift ship handling from being onboard to being onshore?*

Finally, the participants were asked to prioritize the key aspects of the human factors that would require special attention, especially in terms of designing work for the SCC.

After the focus group interview, the ordering scheme for the data with prioritized feature lists was initially created and summarized. Then the lightweight qualitative data analysis approach (Goodman et al. 2012) was taken through by analyzing the audio recordings together with the field notes as well as the lists.

6.3 Results

Based on each participant's own experience, the replies to Question 1 were as follows:

- Checking screen, radars, conning display, AIS for maneuvering
- Looking outside the window (to get a feel for weather, wind, speed)
- Feeling the sense of balance
- Feeling waves, rolling, pitching
- Getting an intuitive feeling of what the needs are and being less stressful
- Feeling the ship (e.g., the ship's performance when cargo is loaded, how the ship's sensitivity is when turning)

The majority of participants mentioned that they would use the navigational instruments in the bridge to see the status of the ship and the surrounding environment, for example, "checking screens," "radar," "conning display," or "AIS" to make sure the ship was safe. However the most discussed key word was "feelings" that they perceived by looking outside of the window and experiencing "standing wave," "rolling," or "sense of balance" with the vessel. The participants thought this was one important intuitive sense that kept their stress levels down and even helped them to take corresponding actions more efficiently with regard to the external environment, because "body reacted quicker than the instruments." In terms of maneuvering, they thought one important aspect of the feelings was to sense their ship, for example, "feeling the sensitivity of the ship" or "feeling the ship's performance when cargo is fully loaded or not."

When the discussion turned to remote supervisory monitoring (i.e., Question 2), the participants envisaged what the operators in the shore control center would probably do, but they suggested it was an unprecedented challenge for which they did not have a perfect solution (see Table 6.1).

Basically the actions that operators can do ashore were to observe the screens and perceive dynamic real-time information. Multiple human–machine interfaces ashore was discussed compared to onboard ship handling. Most participants deemed the simulator as the ideal human–machines interface used in the shore control center, as "They don't want a mouse button but a joystick handle." In terms of sense, they anticipated there would be gyros and other senses that could simulate the feelings onboard. As more assumptions were proposed in the focus interview, the participants turned to list the leading consequences being onshore as the unprecedented challenges, for example, maintenance work, economy cost, reliability of the system, etc. Meanwhile, the participants realized that "not the same human factors were needed" in both situations, the discussion moved on naturally to the main research interest of the interview (i.e., Question 3). The overview of the replies was presented in Table 6.2.

From the perspective of the majority of the participants, the most controversial question they would consider in priority was the possibility to build a "the full-proof system,"

TABLE 6.1

Actions and Confronted Challenges Discussion Concerning Shore-Based Ship Monitoring and Controlling

What the Operators Would Do	Consequences as Challenges
Observe multiple screens	It must be possible to display all-needed information and allow perceiving it as onboard but it would cause information overloading problem;
Use simulator as human–machine interfaces rather than mouse/keyboards	The operators must be considered as seafarers with expertise
Monitor incidents onboard Well prepare for emergency	How to handle maintenance work immediately and management (ordering spare parts)
Observe gyro and other sensors	Are they real-time sensors, if so, what the cost would be
Let system calculate risks and alternatives	Ensure more backup sensors and systems on the ship to prevent/handle severe technical failure (e.g., connection lost)
Trust in the system and sensors	How to guarantee the reliability of the system so people could really trust it

because they believed that it could be a big risk to solely rely on the shore-based monitoring system and therefore judge things from it. Except for the skepticism, they explicitly mentioned situation awareness as the most significant key to focus on when shifting ship handling from ship to shore followed by information overloading and organizational issues.

TABLE 6.2

Overview of Changing Human Factors from Ship to Shore for Ship Handling

Human Factors	Presentation of These Factors	Voices
Sense	Visual, auditory, sense of smell, kinetic feeling, sense of balance	*"Ship starts vibrating and pitching when changing the course a bit, but these senses are lost ashore."* *"Everything got closer ashore."*
Perception—cognition	Mental model, decision making, situation awareness, information overloading, stress, trust in the system	*"You may pay attention to parameters that don't matter or are wrong and you worry for nothing."* *"Receiving much more information but you can't discern what matters to you as you did onboard."* *"When you're onboard, fear is stimulating but you're less stressed ashore. Complacency. Maybe too relaxed."*
Work space	Working environment, ergonomics, hardware, software	*"Only rely on instruments ashore."*
Maintenance	Backup systems, maintaining approaches	*"A big part of the ship work is maintenance."* *"What happens if there is a malfunction or emergency?"*
Risk	Risk assessment, shifting risk	*"Risks for other boats around."* *"Not that risky being onshore."*
Organization	Expertise, structure, roles, education/training	*"Computer engineers for the operator ashore would be good since they monitor ships through computers. Seafarers would not need that"*
Legal perspective	Regulations, laws	*"Who is responsible if the ship is in international waters?"*

6.4 Discussion

The focus group is used as the data collection method and is also used partially for the data analysis in this research. The reason to choose the focus group is because it is suitable for identifying problems, seeking to solve problems from the stakeholders' view with an exploratory research manner (Ivey, 2011). More importantly, it can provide insights into the sources of complex behaviors and motivations (Morgan and Krueger, 1993). The purpose of this research is to explore the key aspects of human factors with regard to maintain ship sense when shifting people from onboard maneuvering to shore-based monitoring. The target audience are fourth year master mariner program students with a certain amount of navigational experience at sea. The average age is 27 years and they are comfortable using a daily digitalized device, for example, laptop, iPad, iPhone, etc. Therefore the focus group can provide multifaceted opinions by looking deeper into their working experience and maneuvering behaviors, and seek the affected human factors behind the explorative computerized solution in the SCC. Although the focus group cannot substitute a usability test and observation of the product in use to evaluate how efficiently people will use a certain product, it can underpin the research of human factors in complex systems and provide values on which design direction would be widely accepted.

Designing a focus group interview elaborately to ensure its structure is vital (Morgan, 1996) so the moderator controlled the topics to be discussed step by step and involved each individual participant as equally as it could. Following the discussion that went from actions onboard to ashore, a comparative analysis was conducted during the interview. It might not provide the "full picture" to all aspects that need to be covered within 2 h, but it indeed afforded valuable insights on some discernible "tips of the submerged iceberg," such as how the perception difference might shape the operators' behaviors in the SCC, and what the main factors were that hindered them from achieving a high level of situation awareness.

However, the focus group has limitations like other forms of data collection. First it is not a statistically significant sample (Goodman et al., 2012). Although the group participants have sea experience, they are far from experienced seafarers. None of them are bridge officers. Second, those whose viewpoint is a minority perspective may not be inclined to speak up or risk negative reactions (Patton, 2002). Besides, there are no fully fledged proven solutions to be provided to the participants, thus it is hard to envisage what the SCC would look like at the end of the day. After all, in the exploration-oriented group discussion, the lack of the reference frame may lead the participants to some "limbo" state and therefore affect the discussion results. There is also one possibility that the participants see the autonomous unmanned ship as one potential conflict with their career development, which might partially explain why the discussion was at once deviated to the unfeasibility of the whole concept.

The result from actions onboard provided by the participants indicates that "feeling" is very important in ship maneuvering. This tacit and gut feeling is interpreted as the ship sense which is seen as critical for seafarers in ship maneuvering (Prison et al., 2009). Such feelings are strongly related to the kinetics that can give the bridge crew vital information in a more straightforward manner than the instrument screens: how the ship is behaving now and under what circumstances. It means the sensitivity that the ship presents when reacting to the external environment (e.g., how the ship is reacting to the bank suction effect) and internal status (e.g., full cargo or not), as well as the constraints from the environment (e.g., weather, wind, wave) that the bridge crew must take into account in ship handling. Visual perception has significant weight in all the sensations. Ship status can be read and judged from the screens in the bridge in combination with the environment information that can be gained by looking outside. Figure 6.1 presents the analysis results on how seafarers get visual information from navigational instruments while perceiving feedback coming from the environment and the movement of the vessel.

From the perspective of human–machine interaction, the seafarers are constantly "interacting" with both the vessel and the environment. Although the participants were more stressed under such circumstances, they expressed that they have the ability to discern the priority of information and act intuitively. However, the ability would become inability when people are located far away from the conducted vessel in a shore-based center. There will be no physical connection between the human and the vessel, and no directly perceived information from the ship's environment. Specifically the visual perception and the kinetic feeling will be lost, which would truly jeopardize ship sense.

The participants acknowledged that operating ashore would put the inevitable "feeling" in jeopardy, "the connection with the outside world is lost," and therefore they proposed many ways to compensate such feelings or substitute such feelings in their replies. Some participants went for the simulation setup or for visualization solutions so that the SCC could mimic the ship sense. One typical suggestion is to use the simulator as human–machine interface, consequently the bridge crew can see the surrounded 3D (three-dimensional) visualized environment, provided that the sensors can transfer sensory information to shore in real time. One of the critical concerns is that visualization might not be able to provide enough situation awareness. For example, it does not

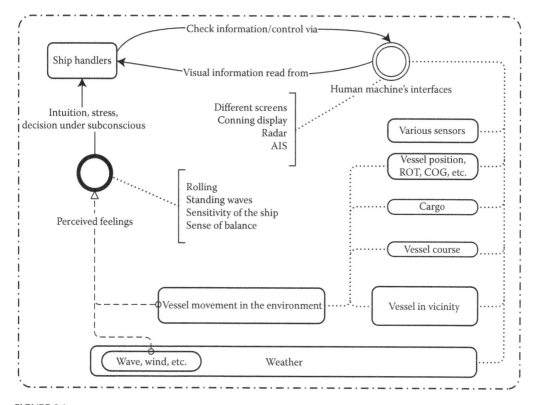

FIGURE 6.1
Ship handlers gain visual information from bridge instruments while perceiving feedback coming from the environment and the movement of the vessel.

resolve the problem caused by the loss of motion, as people do not move with the ship any more as they do onboard, thus they are not able to feel the tool they are operating. Some other participants turned to the sensors to seek alternatives for ship sense, as gyros can tell vibration, roll, and heave. However, there are also usability issues as too much visual information might cause information overloading for the operators ashore.

With the previous discussion (treating onboard and onshore situations separately) as the underpinning blocks, the participants seemed to understand the topic more comprehensively and had contributed something more valuable in Table 6.2—there are indeed several identified aspects of these changes in human factors that we must not ignore when shifting navigation from ship to shore. Except for the skepticism, the participants listed situation awareness as the most significant key to focus on. Situation awareness stands for three levels of information processing: perceiving information, understanding information, and anticipating information (Endsley, 1988). When fulfilling the task of maneuvering, the information is gathered through seafarers' senses via different perception receptors like the retina, which indicates the first level of situation awareness. Previous studies (Endsley, 1995) find that attention and working memory are the critical factors when people are interpreting things from the environment into their mind. The concern from the participants perfectly match these critical factors in situation awareness, for example, "you may pay attention to parameters that don't matter or are wrong and you worry for nothing," and the notorious "information overloading" problem.

The mental model is seen as an important sense-making tool to overcome such limits and decide the priority of information (Endsley, 2011). The participants felt that the monitoring process was generally full of "complacency and relaxation," until the occurrence of unexpected automation failures. "Receiving much more information but you can't discern what matters to you as you did onboard." It suggests the emerging challenges caused by degraded situation awareness and partial sensory deprivation ashore. It also indicates that operators who were assumed to be bridge officers need to develop new mental models as a higher level situation awareness enables the adaption to the working pattern ashore. Maintaining situation awareness would be even more challenging than achieving a certain level of situation awareness since it needs to keep users in the loop of the dynamic situation (Endsley, 2011).

Along with the described issues with respect to perception and cognition, the organizational problem is also considered as one prioritized aspect in the development of a shore control center. It raises questions like, what the role of the operators should be, what the difference would be compared with seafarers today, what regulations or rules were needed, how the training program should be tailored for them, etc. The puzzle needs to be solved from multifaceted views.

Noticeably, maintenance is identified as one of most serious issues with no one onboard. It explained partially why participants asked for backup solutions in a full-proof system in the first place. The trust from the operators is there only because of the reliability, resilience, and robustness of the system. However, there is hardly confirmed evidence to prove that an unmanned autonomous ship could function with "fail-safe" guarantee during the whole voyage. Could it even be managed by the shore-based operators at any time, the majority of the participants held a skeptical attitude toward the concept of an autonomous unmanned ship. Along with the LinkedIn group discussion (Unmanned Ships on the Horizon, 2014), there has always been a problem with the acceptance of the concept of an autonomous unmanned ship. The goal of the MUNIN project is not only to study the feasibility of an unmanned ship and SCC, but also to aim at improving sensor systems, cooperation work flow between ship and shore, maintenance procedures, and reliability and cost-effectiveness. Those studies and research may also be used in the future concept that only partially removes seafarers from ship to shore and makes the maritime industry more attractive and safer. Some key human factors onboard influencing safety have been identified, such as fatigue, automation, situation awareness, communication, decision making, and teamwork (Hetherington et al., 2006). Automation is often introduced to reduce human error and work load, but it also shapes crew assessments and actions (Lützhöft and Dekker, 2002). Automation surely cannot simply replace human work with machine work and MUNIN is just a first step towards a distributed human-centered automation framework for deep sea voyages. It provided opportunities to explore the different presentation formalities and facades of known human factors along with other emerging challenges under one new business model in the maritime domain.

6.5 Conclusions

The results from the focus group indicate the gap between the tasks that require adequate situation awareness to maintain ship sense and the inability of personnel ashore to do this due to the lack of conventional sensory cues and appropriate organizational regulations and management for SCC. The original "harmony" faces new challenges by reconstructing its constituents (Prison, 2013), that is, people, vessel, and environment. Ship handlers still strive for a continuous balanced effect by tuning the ship to the environment, but in the remote control pattern. On the one hand it might bring more risks in operations due to the lack of situation awareness and thus put harmony in jeopardy, but on the other hand, it suggests the approach to design an integrated system by studying the changes in various aspects of underpinning human factors. Through the deep analysis in the contextual nature of the onboard and shore-based environment, the intrinsic variability of those applied human factors can be exposed for further human factors refactoring. What is going on ashore, how it is different, and how it can be adapted to humans, the explorative research in the future unmanned ship presents to the industry unprecedented challenges as well as endless opportunities.

Acknowledgment

We gratefully acknowledge the financial support from MUNIN, which is funded by the EU's 7th Framework Program under the Grant Agreement Number 314286.

References

Endsley, M. R. 1988. Design and evaluation for situation awareness enhancement. *Proceedings of the Human Factors and Ergonomics Society Annual Meeting*, 32(2), 97–101. Doi: 10.1177/154193128803200221.

Endsley, M. R. 1995. Toward a theory of situation awareness in dynamic-systems. *Human Factors*, 37(1), 32–64. Doi: 10.1518/001872095779049543.

Endsley, M. R. 2011. *Designing for Situation Awareness: An Approach to User-Centered Design*, Second Edition. Boca Raton, Florida: CRC Press, Inc.

Fuller, R. 2000. The task-capability interface model of the driving process. *Recherche—Transports—Sécurité*, 66(0), 47–57. http://dx.doi.org/10.1016/S0761-8980(00)90006-2

Goodman, E., Kuniavsky, M., and Moed, A. 2012. *Observing the User Experience: A Practitioner's Guide to User Research*, Second Edition. USA: Morgan Kaufmann Publishers Inc.

Han, D. F., and Ding, S. 2013. Review of human factors in maritime system. *Applied Mechanics and Materials*, 397–400, 679–682.

Hetherington, C., Flin, R., and Mearns, K. 2006. Safety in shipping: The human element. *Journal of Safety Research*, 37(4), 401–411. Doi: 10.1016/j.jsr.2006.04.007.

Ivey, J. 2011. Focus groups. *Pediatric Nursing*, 37(5), 251.

Kitzinger, J. 1995. Qualitative research. Introducing focus groups. *BMJ*, 311(7000), 299–302.

Lützhöft, M. H., and Dekker, S. W. A. 2002. On your watch: Automation on the bridge. *The Journal of Navigation*, 55(01), 83–96. Doi: 10.1017/S0373463301001588.

Morgan, D. L. 1996. Focus groups. *Annual Review of Sociology*, 22, 129–152.

Morgan, D. L., and Krueger, R. A. 1993. *When to Use Focus Groups and Why Successful Focus Group: Advancing the State of the Art* (pp. 3–19). Newbury Park, California: Sage.

Patton, M. Q. 2002. *Qualitative Research & Evaluation Methods*. Thousand Oaks, California: Sage.

Porathe, T., Prison, J., and Man, Y. 2014. Situation awareness in remote control centres for unmanned ships. *Paper presented at the RINA Human Factors in Ship Design and Operation Conference* (pp. 93–101), London, UK.

Prison, J. 2013. *Ship Sense—Exploring the Constituents of Shiphandling*. Göteborg: Chalmers tekniska högskola.

Prison, J., Dahlman, J., and Lundh, M. 2013. Ship sense—Striving for harmony in ship manoeuvring. *WMU Journal of Maritime Affairs*, 12, 115–127.

Prison, J., Lützhöft, M., and Porathe, T. 2009. Ship sense—What is it and how does one get it? *Paper presented at the RINA Human Factors in Ship Design and Operation Conference* (pp. 127–130), London, UK.

Rødseth, Ø. J., Kvamstad, B., Porathe, T., and Burmeister, H.-C. 2013. Communication architecture for an unmanned merchant ship. Paper presented at the *OCEANS—Bergen*, MTS/IEEE. http://dx.doi.org/10.1109/OCEANS-Bergen.2013.6608075

Unmanned Ships on the Horizon. 2014. Retrieved on February 22, 2014, from http://www.linkedin.com/groupAnswers?viewQuestionAndAnswers=&discussionID=5818020799168155648&gid=44626&commentID=5826083705612943360&trk=view_disc&fromEmail=&ut=3tagxIxMVQM641

7

A Multi-Method Analysis of the Accessibility of the Izmir Ferry System

Sebastiano Ercoli, Andrea Ratti, and Emre Ergül

CONTENTS

7.1 Introduction

This chapter presents the planning and results of a multi-method research aimed at understanding critical and positive aspects of current designs of the Izmir ferry system in terms of accessibility for all users, encompassing people with temporary or permanent disabilities, pregnant women, elderly people, and children.

EMTA (association of European Metropolitan Transport Authorities) estimates that 12% of the European population has a disability (Fiedler 2007). This is mostly due to the aging of population, because of declining birth rates and increasing longevity. Disability is strongly connected to age: 80% of people with disabilities are 65 years old or more. Within Europe, Asia, and North America, the current percentage of the population over age 65 ranges from 6% to 22%. By 2030, these percentages are estimated to range from 17% to 29%. The fastest growing subgroup represents those over 80 years of age. Thus, an inclusive approach is becoming an urgent need in design. Universal Design of transport projects brings benefits in excess of investment costs (Odeck et al. 2010), benefiting not only the impaired, but all users of a facility. Marine transport system suffers from high manufacture costs, long design process, and red tape problems, therefore it is usually slow in adopting innovations, but inclusion issues can no longer be neglected.

The urban ferry system in service in Izmir was considered as a case study to assess users' opinions and expectations. The results of the research will be used to compile a guide on accessibility issues and user needs, addressed to vessel builders and operators, to be used as an input set before constructing new vessels or before refitting existing ones.

7.2 Research Problem

The research aimed at offering an answer to the following questions: (1) Who are the users in trouble with accessibility? (2) Which design elements cause accessibility problems, and to which users?

7.3 Context and Theoretical Premises

7.3.1 Izmir and İzdeniz A.Ş

Izmir is the third largest metropolis of Turkey, with a population of 4 million people. It spreads at the head of a long gulf with the same name. The city center and the most of the offices are located in the metropolitan districts of Konak, Pasaport, and Alsancak. Karşıyaka, on the opposite side of the bay, is the favorite place for dwelling. Thus, a great mass of people commute from the north side of the bay to the south. Due to various factors such as the layout of the city around the bay, the size of the city itself, road congestion, and limited land public transport, public marine transit service is very used in İzmir, with 125 million passengers in 2012 and 130 million in 2013.

The public marine transit service is operated by İzdeniz A.Ş. (Izdeniz), a company which is majority-owned by the Metropolitan Municipality of Izmir. Izdeniz operates 18 passenger boats, 2 passenger ships, 3 catamarans, and 3 car ferries (data updated to October 2014). The company is currently under the biggest renovation of the fleet in its history: it is acquiring 15 new passenger ferries, of which three are already in service and the others being supplied by 2017.

Starting from 2013, the Izmir Municipality has been developing a 5-year strategic plan—EngelsIzmir, "Izmir without obstacles"—aimed at improving the physical accessibility of the city, including its urban transit network. In this context, the municipality established the Red Flag Committee (Kirmizi Bayrak) with the purpose of inspecting and encouraging the improvement of accessibility. The design of the new passenger ferries was supervised and endorsed by this committee.

The fleet is operated through eight stopovers. All the routes lie in internal waters. The transit service is integrated with the urban land network of public transport, sharing the same ticket system and fare. Travel times range approximately from 10 min to half an hour, depending on the routes.

7.3.2 The Seamless Journey

Instead than focusing on ferries only, the research considered the system of water public transport in its whole. What is around the ferry, such as piers and berthing facilities, has the same importance in providing a smooth, seamless journey for the passenger. The hardest obstacles for accessibility are likely to be at the interfaces of the components of the transit system, that is, where the gangway touches the deck of the vessel.

An interesting model of ferry travel is described by the Department of Transport of Ireland (Department of Transport of Ireland and NDA 2012). According to this model, a single journey is divided in seven phases: (1) information; (2) travel to vessel; (3) wait for sailing; (4) board vessel; (5) sailing; (6) get off vessel; and (7) travel to destination. Specific studies have already been carried out on the different aspects of the journey, as pre-trip information needed by people with disabilities (Waara 2009, 2013), boarding, sailing, and getting off vessels (Chapman 2004, 2006, Ercoli et al. 2014), and terminal infrastructures (International Maritime Organization 1989). Other studies encompass all the phases (Minister of Public Works and Government Services Canada 1999). The authors adopted this model to plan field observations, survey, and interviews, and to organize findings.

7.3.3 Addressing the Needs of Real Users

Life expectancy is increasing all over the world, and the numbers of people with disabilities are rising. In fact, disabilities are strongly related to age, especially in most developed countries. All sectors need to consider the shift of the user base and the marine industry is no exception. This is why in 2010 the European regulation on the rights of passengers traveling by sea and internal waters (EU Regulation No. 1177/2010, articles 6 and 7) stated that "in deciding on the design of new ports and terminals, and as part of major refurbishments, the bodies responsible for those facilities should take into account the needs of disabled persons and persons with reduced mobility, in particular with regard to accessibility, paying particular consideration to 'design for all' requirements. Carriers should take such needs into account when deciding on the design of new and newly refurbished passenger ships."

In 2010, a guideline based on the International Maritime Organization's IMO "Recommendation on design and operation of passenger ships to elderly and disabled person's needs" dated June 24, 1996 and referred MSC/735 has been published by the ministry to regulate the related fleet in Turkey.

In the near past, the main way to address the issues of accessibility was through add-ons which were not connected with the main design, but which were thought of as a posteriori corrections. Disabilities were seldom considered in the first phase of the project, thus corrections were necessary in the last phase. An example is the use of stair-lifts to allow people on wheelchairs to overcome a flight of stairs, instead of designing a building without architectural barriers. In this approach, people with disabilities were considered as "special" users who needed "special" solutions. These dedicated, segregated solutions are often expensive, and usually unappealing. Designers and architects soon understood that many of the changes needed for people with disabilities benefited everyone, and that it is more convenient to consider them at the beginning of the design process. Recognition that such features could be commonly provided and thus be less expensive, unlabeled, attractive, and even marketable laid the foundation for the Universal Design movement (Story et al. 1998).

Moreover, in that period the social model of disability was developed: disabilities were not to be considered as intrinsic characteristics of the individual. As opposed to that, they are conditions determined by society through the construction of social and environmental barriers which isolate and exclude people with impairments from full participation in society.

The first principle of Universal Design (Connell et al. 1997) seeks to consider the breadth of human diversity across the lifespan to create design solutions that work for all users. In order to evaluate the needs of the variety of users, a classification system is needed.

EU Regulation No. 1177/2010 establishes rules for the nondiscrimination of "disabled person" and "person with reduced mobility," meaning any person whose mobility when using transport is reduced as a result of any physical disability (sensory or locomotor, permanent or temporary), intellectual disability or impairment, or any other cause of disability, or as a result of age, and whose situation needs appropriate attention and adaptation to his particular needs of the service made available to all passengers.

This classification is not usable for designers, because the categories are too wide. Instead of a minority model, which classifies people according to their impairments, a positive universal model which considers functional abilities seems more appropriate (World Health Organization 2001):

- Mental functions
- Sensory functions and pain
- Voice and speech functions
- Functions of the cardiovascular, hematological, immunological, and respiratory systems
- Functions of the digestive, metabolic, and endocrine systems
- Genitourinary and reproductive functions
- Neuromusculoskeletal and movement-related functions
- Functions of the skin and related structures
- Any other body functions

The International Classification of Functioning (ICF) splits each body functions in subfunctions; for example, sensory functions are divided into seeing, hearing, vestibular (incl. balance functions), and pain. Functions and sub-functions can be used as an outline for organizing the content of the final report, matching them with the journey phases. In each phase, some functions are more important than others; for example, in the information phase, mental and sensory functions are the most important; while getting on and off vessel requires neuromusculoskeletal and movement-related functions; and so forth. Data gathering will help to understand, for each phase, which the most critical aspects of accessibility are in relation to each function.

7.3.4 Disabled Population in Izmir

In Turkey, the elderly population, which is the population at 65 years of age and over, was 5.7 million in 2012 with a proportion of 7.5% (Turkish Statistical Institute 2013). By 2023, this segment is expected to grow to 8.6 million, with a proportion of 10.2%.

According to figures of the Turkish Statistical Institute, the country has about 8.5 million people with different physical and mental disabilities, which count for approximately 12.29% of the total population (Afacan 2008). Though official data are not available for the

Izmir municipality, a simple proportion with its population permits the estimation that disabled persons should be about 400,000.

7.4 Research Methodology

The authors adopted a mix of qualitative and quantitative methods in order to compare and validate the findings. In particular, the implemented methods were: (1) unstructured field observation conducted on all the different types of vessels, (2) survey of 233 passengers, and (3) key-informant interviewing of delegates of the service operator and of expert members of associations for persons with disabilities.

In the first phase (October 2014), unstructured field observation was used to highlight the main issues and problems. The observations were used to develop a survey which was then conducted on Izdeniz passengers (from October to December 2014). The survey, consisting of a questionnaire composed of a mix of closed-ended and open-ended questions, was developed on a random sample, respecting the criterion of administering it to passengers during their use of the service. Interviews of key informants—operators and representatives of associations for disabled people—were conducted in December 2014.

All the data resulting from the three methods were collected in an MS Access database and classified according to tags and journey phases. By using queries, it is possible to inspect the results according to multiple criteria, both in connection with the profile of passengers (i.e., age, sex, profession) and with journey phases (i.e., buying tickets, sailing, etc.).

7.4.1 Field Observation

In the discipline of design, field observation is employed to determine explicit and tacit user needs. The authors adopted this method in the early phase of the project, to have a first image of the problem that could help in the design of the survey. The observation of passengers during the different phases of the trip contributed to the identification of the most critical issues in regard to accessibility.

Covert, nonparticipant observation was chosen in order to avoid language barrier problems, though the method may cause confirmation bias. Observations were recorded in structured forms, organized according to the seven phases of the adopted journey model. Fifteen journeys, encompassing all the stopovers and the different types of vessels, were recorded.

7.4.2 Survey

The survey, allowing the gathering of direct data about the passengers and their opinions, had a primary role in the research. In accordance to the idea of Design for All, the hypothesis of this research is that accessibility problems are not related only to disabled people, but may be experienced by everyone. Thus, it was decided to develop the survey on a genuine random sample, with the only choice of administering it to passengers over the age of 14 and during their use of the transit service. The waiting phase resulted as the best moment to deliver the questionnaire, while doing it during the sailing phase resulted as not being efficient, mainly because of the short sailing times.

In order to increase the response rate, the survey was planned to require a relatively short time to be completed. Close-ended questions can provide precise answers to specific topics; on the opposite, open-ended questions encourage full, meaningful answers using the subject's own knowledge, and thus could provide unexpected insights on user needs and expectations. Thus, the balance between the two types of questions was an important issue in the design of the questionnaire. In its final version, the form was designed to have a filling time of 5–10 min and to fit in a single A4 page (see Figure. 7.1 for the English translation of the form).

Four project assistants, all native Turkish speakers, had the role of individually interviewing the passengers and filling the forms, after quickly explaining the purpose of the project. This approach was preferred to self-filling for speed, and because assistants could explain the meaning of questions, avoiding cultural or education barriers.

The first part of the form (questions 1–9) provides demographic data about the passenger and his/her route. The second part (questions 10–25) explores the passenger's opinions about all the phases of the journey.

In total, 233 surveys were collected in 12 sessions. The survey was administered at different times of the day, on different days of the week, and at different stopovers, in order to randomize the sample.

7.4.3 Key-Informant Interviewing

Key informants are valuable sources of information, having specialized knowledge about other people, processes, or happenings, from a particular or privileged point of view, they can help in understanding user needs and expectations in the specific field. Key informants joined a semi-structured interview. In the first part, a series of prearranged questions was asked. The second part consisted of open-ended questions, allowing the exploration of the topic and to discover unexpected information.

Two separate interviews were organized with

- Delegates of the operator, in the persons of Caner Pense, DPA assistant and ISM office responsible, Gökhan Atilgan, general deputy director, and Çağdaş Uzgur, master deck superintendent
- Mahmut Akkin, director of the unit of the municipality that deals with disabled people and is responsible for the project EngelsIzmir

7.5 Findings

The sample of passengers was composed by 51% women and 49% men. In proportion to the population structure of Turkey, passengers in the age range 20–29 were double; male passengers in the range 70–74 were represented thrice more; and female passengers in the age range 50–64 were only 20%–30%. All the other age ranges show a smaller difference, in the order of ±30%. The high number of young passengers corresponds to the use of the ferry by university students. The other differences may be due to the small sample size or may be due to cultural reasons (married women stay more at home); and more surveys are needed for the purpose of clarifying this point.

Forty percent of the passengers declared themselves to be employees, 27% students, 17% retired, and 9% homemakers.

Ferry Accessibility Survey

v. 1.1 – October 2014

1. Date (D-M-Y): _____

2. Journey departure (1) and destination (2):

___ Bostanlı   ___ Bayraklı   ___ Pasaport   ___ Göztepe

___ Karşiyaka   ___ Alsancak   ___ Konak   ___ Üçkuyular

3. Sex: ○ male   ○ female

4. How old are you? _____

5. Select your status:
- ❑ Student
- ❑ Employee
- ❑ Unemployed
- ❑ Freelance
- ❑ Homemaker
- ❑ Retired

6. How often do you use ferries?

○	○	○	○	○
Less than once a month	Once or twice a month	Once or twice a week	3–5 times a week	More than 6 times per week

7. Who are you usually travelling with?
- ❑ Alone
- ❑ Friends, colleagues, or relatives
- ❑ Child
- ❑ Baby
- ❑ Elderly - disabled

8. Why do you use ferries mostly?
- ❑ Arriving to work
- ❑ Shopping
- ❑ Tourism
- ❑ Going to school
- ❑ Other _____

9. Please check the boxes which are appropriate for you, if any:
- ❑ I use a wheelchiar
- ❑ I use a walking stick
- ❑ I am blind
- ❑ I have low vision
- ❑ I use hearing aid
- ❑ I use other mobility aids: _____

10. Which kind of ferry do you prefer?
○ Passenger ship   ○ Catamaran   ○ Car ferry

11. Why?
_____
_____
_____

12. Consider your sea journey. Which phases are the hardest for you?
- ❑ Finding information and buying passes
- ❑ Travel to ferry
- ❑ Wait for sailing
- ❑ Board ferry
- ❑ Sailing
- ❑ Get off ferry
- ❑ Travel to destination

13. Why?
_____
_____
_____

14. And which are the easiest phases?
- ❑ Finding information and buying passes
- ❑ Travel to vessel
- ❑ Wait for sailing
- ❑ Board vessel
- ❑ Sailing
- ❑ Get off vessel
- ❑ Travel to destination

15. Why?
_____
_____
_____
_____

16. Are you happy with the toilets?

○	○	○	○	○
Not at all	Slightly	Moderately	Very	Extremely

17. What services do you want to have during the journey? (playground for children, TV, newspapers, food…)
_____
_____
_____

18. What disappoints you the most about your current ferry journeys?
_____
_____
_____

19. Do you think signage within terminals is clearly understandable?

○	○	○	○	○
Not at all	Slightly	Moderately	Very	Extremely

20. And what about signage inside ferries?

○	○	○	○	○
Not at all	Slightly	Moderately	Very	Extremely

21. Would you like to have more information through loudspeakers?
- ❑ No, I'm fine with it
- ❑ Yes, I would like the name of the stopover to be announced
- ❑ I'd like to have more info _____

22. Please evaluate how difficult is for you to get on/off the vessel from the quay:

○	○	○	○	○
Not at all	Slightly	Moderately	Very	Extremely

23. Have you ever asked the help of Izdeniz staff to get onboard of the ferry?

○	○	○
Never	Once	More than once

24. In case you demanded for help to Izendiz staff, please describe shortly your experience:
_____
_____

25. Would you add anything to this interview?
_____
_____
_____

FIGURE 7.1
Template of the questionnaire used in the survey.

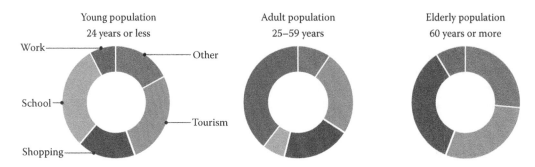

FIGURE 7.2
An example of the results of the survey: main reasons for traveling according to the age of passengers.

Figure 7.2 shows the main reason for traveling according to the age of the passenger: the young population uses ferries for going to school or for tourism, the adult population is mainly made up of commuters, and the most of the elderly population use ferries to go shopping or to meet family and friends.

According to gender, the male passengers use ferries mainly for commuting to work or tourism (59%), while the female population mainly for shopping or arriving to work (46%).

During covert observation, a small number of people using a wheelchair were spotted. Special electric platforms, which need to be operated by Izdeniz staff, allow them to get on and off passenger vessels, while car ferries can be accessed autonomously through their drawbridges. Current car ferry designs are not equipped with elevators, so these persons cannot seek shelter in the passenger areas, which are located on the upper deck. All the ships, except for the new catamarans, do not have accessible toilets. Passengers using a wheelchair were not present in the survey sample, but other passengers with mobility impairments were mostly disappointed by the process of getting on and off the ferry and traveling to the ferry terminal. All the passengers with mobility impairments declared they needed the help of Izdeniz staff at some points of the journey.

The majority of passengers with vision impairment declared that the most tiring phase of the journey is waiting for the ferry, as did the most of the passengers with good vision.

Almost all passengers agreed on the lack of signage both in terminals and in ferries, and on the need for more information through loudspeakers. Toilets received a very low evaluation too and surely need to be improved.

According to age groups, the most tiring phase was different. For passengers under 60 it is the waiting time, while for elderly people it is getting on and off board. The majority of elderly people declared that this phase is "very" tiring, while the rest of the passengers answered mainly "slightly" or "moderately." For all the passengers aged more than 75, sailing is the most relaxing phase, because they do not need additional help and they can enjoy the sights. This is the favorite journey phase for all passenger groups.

7.6 Conclusions

This chapter explains the methodology of the research and shows the first results that could be extracted from the collected data, but other interesting pieces of information could emerge querying the database with different criteria. Furthermore, the authors plan to expand the sample of the survey to obtain more reliable data.

All the findings are in the process of being reported in a guide, structured according to journey phases. Such a guide would be useful for designers, boat builders, and operators, because it may allow the easy checking of accessibility issues for the different components of short-route marine transit systems.

Acknowledgments

The authors would like to thank the staff of İzdeniz A.Ş.; in particular, Caner Pense, for his daily help in the development of the research, and Gökhan Atilgan and Çağdaş Uzgur, for having shared their experience through the interview. Thanks to Mahmut Akkin, for his interest in the project and for having been a valuable source of information.

The authors want to express their gratitude also to Laçin Aksoy, Yasemin Albayrak, Merve Bar, and Sinem Filis, students at Izmir Economy University, who worked in the project as assistants in the delivery of the field survey.

References

Afacan, Y. 2008. Designing for an ageing population: Residential preferences of the Turkish older people to age in place. In: *Designing inclusive futures,* eds. P. Langdon, J. Clarkson, P. Robinson. London: Springer.

Chapman, D. S. 2004. Ferry transit disabled access: New York–New Jersey Harbor Private Ferries. A case study. *Transportation Research Record: Journal of the Transportation Research Board* 1885: 111–120.

Chapman, D. S. 2006. Ferry vessel disabled access: Case study of toilet facilities. *Paper Presented at the Transportation Research Board 85th Annual Meeting*, Washington, DC, January 22–26.

Connell, B. R., M. Jones, R. L. Mace, J. Mueller, A. Mullick, E. Ostroff, J. Sanford, E. Steinfeld, M. Story, G. Vanderheiden. 1997. The principles of universal design. http://www.ncsu.edu/ncsu/design/cud/about_ud/udprinciplestext.htm (accessed September 1, 2015).

Department of Transport of Ireland, and NDA. 2012. *Guidelines for Accessible Maritime Passenger Transport.* http://www.dttas.ie/maritime/publications/english/guidelines-accessible-maritime-passenger-transport-0 (accessed September 1, 2015).

Ercoli, S., A. Ratti, and S. Piardi. 2014. Water-based public transport accessibility. A case study in the internal waters of Northern Italy. *Paper Presented at the 2014 AHFE International Conference,* Krakow, Poland, July 11–21, Vol. 9, pp. 11–21.

Fiedler, M. 2007. Older people and public transport. Challenges and chances of an ageing society. http://www.emta.com/IMG/pdf/Final_Report_Older_People_protec.pdf (accessed September 1, 2015).

International Maritime Organization. 1989. *Access to Marine Passenger Terminals for Elderly and Disabled Passengers.* London: IMO.

Minister of Public Works and Government Services Canada. 1999. *Code of Practice: Ferry Accessibility for Persons with Disabilities.* Ottawa: Canadian Transportation Agency.

Odeck, J., T. Hagen, and N. Fearnley. 2010. Economic appraisal of universal design in transport: Experiences from Norway. *Research in Transportation Economics* 29:304–311.

Story, F. M., J. L. Mueller, and R. L. Mace. 1998. *The Universal Design File: Designing for People of All Ages and Abilities.* Chapel Hill, North Carolina: Center of Universal Design at North Carolina State University.

Turkish Statistical Institute. 2013. Population Projections 2013–2075. http://www.turkstat.gov.tr/PreHaberBultenleri.do?id = 15844 (accessed September 1, 2015).

Waara, N. 2009. Older and disabled people's need and valuation of traveller information in public transport. *Paper presented at the European Transport Conference*, Leeuwenhorst, The Netherlands, October 5–7, pp. 1–21.

Waara, N. 2013. *Traveller Information in Support of the Mobility of Older People and People with Disabilities. User and Provider Perspectives.* Lund: Lunds universitet.

World Health Organization. 2001. *ICF: International Classification of Functioning, Disability and Health.* Geneva, Switzerland: WHO Press.

8

In Yacht Design, Contemporary Society Conditionings Require New Human Factors Solutions for Older Adults

Massimo Musio-Sale

CONTENTS

8.1 The Ergonomics of Yacht Design for Elderly Users: Issues Related to Accessibility of Recreational Boating Activities for Older Adults

We have reviewed criticalities as follows:

1. Boarding and water access systems
2. Transit on walkways to reach deck and sun areas
3. Docking maneuvers
4. Access to lower deck spaces
5. Access to guest cabins and use of facilities

8.1.1 Boarding a Yacht

Access onto and off a boat is one major criticality of boating. The relation of land and water is always variable because it engages two different elements. On the one hand, there is land, a fixed element by definition; on the other hand there is water, ocean water, impacted by waves and tidal movements that force a boat to follow them.

This condition must also consider the variables determined by the infinite possibility of changing levels as water levels rise and fall, in respect of both the yacht's landside connection and onboard access. These two bodies coexist but are never aligned except for no more than fractions of a second even in the best weather conditions.

While seemingly banal or negligible for younger adults, these aspects represent insuperable conditions for seniors aged 80 and over, such that they may jeopardize the possibility to start a cruise in the first place. To this purpose, we investigate solutions that optimize the reciprocal balance of boat and boarding pier. While we may not prevent a boat from floating on a body of water or the ocean from following tidal movements, the wisest thing to do is to fit in a buffer element that mediates between the boat's movement and the stability of land.

This solution is easily satisfied using a floating boarding pier. Floating docks ensure stability on water while replicating its movement, as boats do, at the same time the elevation of the pier surface remains unchanged above water level, unaffected by tidal movements as water levels rise and fall. Floating docks are normally connected to land by means of large, modular gangways that accommodate to the floating pier movements, with a hinged connection at their landside end. No matter what their health status, the ability of older adults to ambulate is no longer automatic or instinctive, nor is it a banal effort. Cautious, self-protective behaviors in old age are more than motivated by a combined series of at least three factors: decaying muscular readiness, reduced flexibility and mobility of lower and upper limbs, and, last but not least, a significant decrease in vision that is normally corrected by means of spherical, multifocal lenses.

These progressive lenses are corrective lenses used in eyeglasses to correct a multiplicity of vision disorders of older adults, at the same time that they do not facilitate peripheral vision downward, affected by the refracted direction through the part of the lens optimized for reading, with a near vision focus, that is ill suited for descending stairs or, more in general, to help you look where you are going. In these circumstances, older adults are particularly ill at ease with walking on uneven, irregular ground, or with the difficulty of access onto a boat. This is why to facilitate access to a floating pier from land by means of a gently sloped, stable gangway may be regarded as a good solution to begin with.

Our central focus is on the accessibility of short cruises on a motorboat to older adults. To facilitate access onto and off a motorboat is easy enough to handle. Motorboats normally sport twin-engine propulsion that enables agile maneuvers in reverse. Also, bow thrusters are installed that are standard equipment on board motor yachts to provide additional control during docking, also in the event that the boat may fall leeward, enabling easier stern in maneuvers.

For boats that use outboard or stern drive engines, aft spaces can be configured freely enough. Conversely, due to their volumes, outboards significantly affect access of elderly users, who may have to bypass the engines' volume while precariously transiting on slim platforms, if any, on the engines' sides.

From these observations we understand that the architecture of a boat designed to enhance accessibility to older adults combines badly with outboard propulsion. We go back to our earlier concept of free aft spaces to establish that optimized design solutions consist in seeking to minimize changes in level between the elevation of the pier surface and the elevation of the yacht's boarding level. Floating piers normally consist of a series of adjacent floating tanks, aligned to both sides of a rectangular boarding platform with grated flooring. This structure forms modules of 10 m in length and 2.5 m in width, for a walking surface normally at 40–60 cm above water level.

Based on this simple observation, there is significant useful information for yacht designers to enhance accessible recreational boating and its usability by older adults. Onboard access elevation that is 50 cm above water level ensures there is a good matching with the elevation of the boarding pier surface at water level. Proximity to a boarding pier is normally defined at a distance no greater than 50 cm. This enables reciprocal, noncolliding movements of boarding pier and yacht. To take a step forward and overcome this distance is no problem for younger adults. Conversely, older people may lack the momentum or the ability to take longer steps as required to ensure safe access on board. This instance represents a new inconvenience that is minor at the same time that it may become insuperable for older adults.

To approach docks or piers closer is theoretically possible, but poses risks of collision. Also, there is a risk that feet may become trapped in the gap, and this may be made more severe by pinching or crushing in the event that the boat is further pushed ashore. A day cruiser of average comfort is normally 10 m in length, with an estimated displacement of 5–6 tons. It is clear how the force induced by a wave onto the boat toward the dock cannot be stopped or hindered by human force. This said, it is also clear how boarding older adults in need of assistance onto and off the boat is certainly difficult and potentially dangerous.

To obviate this, as in the earlier discussion on how to facilitate access onto and off a boarding pier from land, it is also useful to fit in a buffer element that mediates between the boat's movement and the stability of land.

Gangways have long existed, serving this purpose. Gangways are normally pivoted to the boat's end, which enables a certain degree of freedom, suspended on one side to twin davits or, alternatively, to the mast head by means of a topping lift, as on board sailboats. This equipment is normally unsteady to walk on, not reassuring for elderly users in particular. There is a variety of electrohydraulically actuated gangway models that are currently on the market. Efficient and flexible, gangways are normally costly, heavy equipment. Stable and adjustable, gangways remain rigid once fixed in a position as established. Hydraulically activated gangways replicate the boat's movement exactly. Designed to enable access onto and off the boat, gangways provide no buffer between the reciprocal movements of both the boat and the floating dock. A dangerous blade-like effect may be caused in the event that the boat yaws unexpectedly. These structures would add 250 kg to a vessel sized 10 m and 5 tons, which equals one-twentieth of the boat's original displacement. A nonsensical choice. To this purpose, a simple handcrafted gangplank may do, providing greater efficiency, sized some 120 × 35 cm, of marine plywood, to locate on skid pads between the boat's edge and the boarding pier. The reciprocal movement of boat and pier is hence harmonized, enabling acceptable stability. Also, this equipment is easily movable after use that has a very short duration. However, older adults may need assistance to pass safely across.

8.1.2 Swim Platforms and Ladder Systems

Recreational vessels are normally equipped with ad hoc ladders that favor access to sea for recreation.

Safety norms require that yacht design provides for access on board of nonelderly, nondisabled adults without the use of any ad hoc movable accessory.

While the simplest solution consists in designing ad hoc transom stairs, it has become more common lately to integrate a bathing zone as part of the yacht's design, creating space level with the water to enhance a convivial atmosphere of guests, in and out of water. This solution consists in extending the upper deck toward the yacht's transom area, which remains in its lower part only, and obtaining an open terrace to facilitate boarding and water access, guest relaxation, and to house the yacht's tender during navigation.

Bathing platforms and beach deck spaces are normally at an elevation of between 30 and 50 cm above water level. Not too much to climb aboard in an emergency, but certainly more to exit the water in comfort and ease. Water access is generally provided by means of a simple metal tube ladder that has wooden steps to enable a broader tread base. Each half has two steps, at a distance no greater than 28 cm each. Now, four steps plus the platform height serve an elevation of 90 cm. Because swim platforms are normally 30 cm above water level, ladders extend for 60 cm underwater with two steps only, which is too modest to enable accessibility to older adults. To curl up and raise a foot to search for the lowest step on a swim ladder is virtually impossible for adults aged 80 or older due to their functional inability to perform actions as required to hook up to the lowest step underwater. It is necessary that a swim ladder be installed that is sufficiently extended to enable users of all abilities to utilize it in comfort.

Physiological necessities in this respect are to leverage a foothold that is deep enough underwater and requires no curling up. An ideal elevation should be measured in terms of a relation of a flexed leg no greater than 90° in respect of the bust. This measure corresponds to an elevation of 35 cm of the foot. If we consider that the distance of shoulder and foot is normally 130 cm in older adults, we observe that a swim ladder must extend no less than 90 cm below the surface to enable easy accessibility to older adults. Simply, one step more than the standard models in use.

Once on the swim ladder, one next criticality for older adults is that they need to hold on and pull themselves up to keep their balance in an upright position while climbing up the yacht's swim platform. In other words, while it may be no high priority for younger adults to emerge and climb the ladder to reach the boat's platform in a crouched position, and stand up subsequently, older adults may find this maneuver extremely difficult if not impossible to perform (see Figure 8.1).

Handrails above deck level may be suitable and convenient to serve this purpose and enable users of all abilities to maintain their balance as required in an upright position.

While we approximately calculate a reference to the 50th percentile, we observe that the handrails' elevation must activate at a level of no less than the level that enables a bust flexion forward of 45° approximately. We should also consider a simultaneous flexion of the

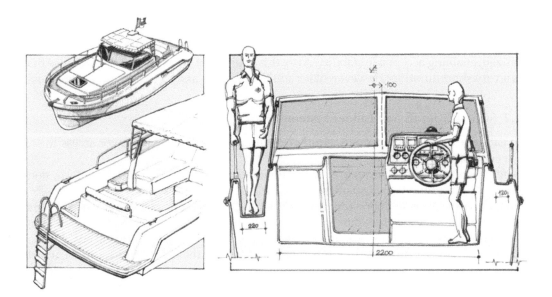

FIGURE 8.1
Left side, (top to down): handrail above the deck for sunbath area and a sketch of the ladder solution for the stern swimming platform. Right side: the cross section showing the main walkway on port-side so to increase the comfort for elderly users (see Section 8.1.3).

legs with an inner popliteal angle no greater than 90° while holding on with a vertically extended arm to mid-femur elevation. If we add these anthropometrical kinematic mechanisms, we clarify how handrails must be no less than 45 cm above deck level to serve this purpose. Lower handrails only force elderly users to use the ladder incorrectly, while ineffectively assisted in a way that can only be poorly supportive of their unsteady maneuvers.

For larger vessels than a 10 m day cruiser, electrohydraulically actuated equipment exists that, designed to be utilized as a tender lift, performs all functions as described above to facilitate boarding and water access of older adults and people with special needs. A product manufactured by Opacmare, Rivalta Scrivia, Italy, the Transformer hydraulically motions a platform served by automatically adjustable stairs for an extension of approximately 270°, thanks to a double pantograph hinge. The Transformer opens vertically to an elevation of 1 m, extends also 1 m onto the water surface and 50 cm below.

This is certainly the most complete, universal solution available today. Major criticalities of the Transformer are its weight, a significantly complicated utilization, and nonnegligible costs, which may substantially impact a yacht's sale price.

8.1.3 Transit on Side Walkways to Reach Foredeck Sun Areas

To access deck spaces and leave the confinement of the cockpit area means to enjoy boating fully. After navigation, when the yacht has reached the destination of choice for the day cruise, the yacht sits at anchor. Cast overboard from the most forward part of the yacht, near the bow stem, the anchor holds the boat placidly put, preventing it from falling leeward against a neighboring boat or ashore, or from drifting out to sea. At anchor, the yacht's bow is normally into the wind. Irrespective of where the wind blows, strong or weak, a boat at anchor is always with the bow against the direction of the wind and, generally, of the waves. This creates two different livable environments, upwind and downwind. Fore spaces, ahead of the yacht's windscreen are upwind. Upper deck areas downwind of the yacht's windscreen and lower deck spaces are downwind.

When the yacht is at anchor, on clear summer days, foredeck spaces are most agreeable to enjoy the sun and natural landscape while a gentle breeze blows. This experience is a major opportunity for livability, which we may not want to miss on a day cruise. Reflecting the yacht's aesthetics more than a functional utilization of lower deck spaces, access to the fore is very often made difficult by narrow side walkways that are sometimes totally absent.

Access to the fore requires equilibristic balance and proven agility for the difficulty of finding handrails as appropriate along the transit from cockpit to fore. For a yacht designed to meet the needs of older adults, this aspect should not be neglected. To enable the utilization of forward spaces, design concepts must incorporate features that address the issue of older adults who are unsteady on their feet and need to find a balance to motion and ambulate.

Onboard movement aft to fore must be adequately protected, wide, and comfortable. Steep stairs should be avoided, accessing the yacht's side walkways from the cockpit, not quite for the difficulty to climb them, but rather for the difficulty to descend them coming fore to aft. Also, handrails must adequately protect this passage on the outer side at least, better though if on both sides of the walkway. Older adults should be able to hold on to them and keep their balance, minimizing their uneasiness with the yacht's floating movement.

To design acceptable handrails for walkways on board a 10 m boat, we must initially determine a basic walkable surface in terms of two near parallel feet: 220 mm should be enough. To this we may add the width of a gunwale or a bulwark on the yacht's outer edge, and the deckhouse profile on the inside. On boats that are 90% fiberglass, construction gunwales are normally constituted by an inside wall, an upper part, which normally holds the handrails' stanchions, and an exterior part that reconnects to the junction between deck and hull.

These three elements project a width of no less than 150 mm. The yacht's deckhouse profile, which spreads outward top to bottom, has an aerial projection of approximately 100 mm. All things considered, on a small yacht designed to be accessible to elderly users side walkways should measure approximately 450 mm in width. A 10 m yacht has normally a beam no larger than 3.50 m. Maximum width is measured at approximately one-third of the yacht's length from fore. A boat's cockpit starts at mid-length approximately, where the hull beam's width starts diminishing gradually. Access to the yacht's side walkways is normally at two-third of its length from fore. Here the boat's width is at least 30 cm less than its beam. Usable width in this location is 3.20 m. If we take off both sides a width as required to create an accessible walkable surface, we must subtract no less than 900 mm, which leaves us with 2.30 m. From this we must subtract additional width for the cockpit's sides encircling the yacht's console and the windscreen base, which in a fiber glass construction are no less than 20 cm thick each. This means that the net cockpit area remains of 1.9 m width only. There is very little left to arrange the helm station, access to lower deck spaces and, possibly, additional place to host one guest or the yacht's skipper. This is the reason why side walkways on a yacht are normally narrow undersized spaces to independently and safely host the passage of older adults.

To tackle the issue of how to design accessible walkways on a 10 m yacht seems virtually unsolvable. In fact, it is geometrically impossible to have both accessible walkways and enough cockpit space, based on the traditional design concept of a perfect symmetry between the port and starboard sides of the yacht. But if we misalign the yacht's deck station profile by 100 mm only, we observe a prodigious redefinition of structural proportions and spaces. Subsequent to this choice, we may no longer have twin walkways versus one port side ergonomically satisfactory walkway of 450 mm and, at starboard, an auxiliary passageway where walkable surface is calculated at 120 versus 220 mm.

Also, the secret is this, to design a starboard walkway that is nonrecessed in respect of the deck's floor. This enables a design of 100 mm less than 150 mm as required for deckhouse and gunwale profiles, respectively. This reduced walkable surface of 300 mm enables a 2.20 m cockpit, which represents a very different size than 1.9 m. The yacht's control station features a dominating console, complete with steering wheel, throttles, navigational instrumentation, engine instrumentation, compass and facility controls, which project an 800 mm width. Seating to host—narrowly—the yacht's pilot plus one guest cannot be less than 1 m in width (see Figure 8.1).

While we consider this volume plus a possible enclave of the aft cabin that insists on cockpit volumes of no less than 600 mm in width, we obtain about a 600 mm width design to arrange for the access to lower deck spaces, which is acceptable to allow easy passage through a sliding hatch to lower deck quarters.

While the yacht sits at anchor, guests may just wish to enjoy sunbathing on its foredeck while a gentle breeze blows refreshingly. The functional mobility of older adults in the performance of actions including sitting, lying down, and rising again is impacted by the size of spaces and the equipment provided to serve their needs. As we earlier discussed in reference to access in and out of water of older adults, the functional mobility of lower limbs in older adults is limited to a range of motion of 90° of femur and knee. This mobility is assisted by modest, unsteady muscular tone. There is a need to design spaces that are large enough and equipped to enhance the independent mobility of older adults.

While current aesthetics define deck surfaces in terms of sleek aerodynamics, well-defined deckhouse volumes enable better usability of foredeck spaces to older adults, who may favor the comfort of sitting, subsequently allowing a rotation of the pelvis to lean back onto the sun pad instead of being forced to lie down on it at floor level. An elevation of 45 cm of the deckhouse in respect of the deck's level, as of a chair's seat, is most suited to meet this need. To avoid volumes that interfere with the yacht's aesthetics, recessed side walkways may be designed that help conceal the deckhouse profile, in part at least. This design concept provides for the accessibility of open deck spaces to older adults in the comfort required to meet their needs.

A design of foredeck spaces with a C-shaped handrail central to the sun lounge would ideally complement their accessibility to older adults. Installed longitudinally, aligned to the yacht's keel, handrails offer a nonindispensable but very useful support to older adults, while enabling their mobility from sitting to lying down onto the yacht's sun pad and, more importantly, to help them stand up, unassisted and in comfort. One key aspect at issue in this respect is that this layout splits a yacht's sun pad into two, which, under non-specific conditions, may constitute a functional and aesthetic constraint to a free utilization of these spaces. Handrails may also serve an ornamental purpose and be utilized to arrange accessorial instrumentation including horn, radio antenna, flags, and radar.

8.1.4 Docking Maneuvers

That older adults perform these operations may seem excessive. Because this implies the assumption that older adults operate a boat unassisted, while to dock a yacht is normally performed by younger adults who are in charge of both conducting the maneuver and agilely seizing the lines to secure them to mooring bitts, fore and aft. On board a 10 m yacht, stern mooring bitts are normally located on the structural bulwarks of the aft deck on both sides of the cockpit.

Mid-ship or forward, bitts are generally situated at deck level, at a higher elevation than the cockpit to enable larger lower deck volumes. In general, designers would find it senseless to sacrifice lower deck volumes to favor the accessibility of mooring bitts to elderly users. Notwithstanding, in the event that habitability of interior spaces should be no priority, port walkways may be recessed entirely to obtain a relative elevation of mooring bitts at approximately 400 mm.

8.1.5 Access to Lower Deck Spaces

Access to lower deck spaces is one major criticality for older adults on a 10 m yacht. Access to lower deck spaces is normally via a passage door from the yacht's cockpit to below deck, sheltered by the windscreen, near the helm. There is normally a 1 m difference in elevation between the cockpit's floor level and below deck. The door has vertical and horizontal top openings that uncover part of the deckhouse to facilitate access through the deck down the companionway, in nautical terms. A common feature is a sliding hatchway door that slides transversally into space below the yacht's console. A four-step stairs with vertical height between steps of 25 cm enable access below deck.

Requirements under applicable safety regulations are that there must be a raised threshold between exterior and interior spaces, to prevent water which may invade the cockpit area and cover the cockpit floor then passing through to lower deck spaces, thus compromising flotation of the craft.

Ergonomically, this threshold is a hindrance and poses potential trip risks. Because of this, there is a need for warning systems to alert users. Alternatively, to reposition the top step of the stairs to below deck at the same level as the threshold may help mitigate this hazard.

Due to space constraints, stairs to below deck have a significant slope, 60° approximately, which requires users to descend stairs in the backward direction. While younger adults may normally cover this 1 m downward distance in one jump, this elevation poses hazards to older users. There is a need to facilitate older adults by means of stable, comfortable side rails, which may not be standard equipment on board.

8.2 Conclusions

Yachts designed to enable accessibility of recreational boating to older adults are no special boats nor are they different than designs that already exist. Based on our earlier discussion, accessibility of recreational boating to older adults, safely and independently, only requires few essential considerations that yacht builders normally neglect.

The expected results of this research determine design indications that help define a product that meets the mobility and usability expectations of older adults as a potential user group. We have analyzed the relations between the functioning levels of older adults and basic onboard operations. Physical functionality is considered in terms of reduced articular and muscular agility, and in reference to the psychological effects of cautious, self-protective behaviors in old age.

Finally, there is a need for both designers and yacht skippers, to a lesser extent, to address issues as follows.

Design concepts that enhance accessibility of recreational boating for older adults focus on large, unobstructed aft spaces to eliminate trip hazards; four-step swim ladders that extend 1 m underwater, with handrails 500 mm above deck level; one large, protected side walkway to enable safe passage aft to fore, complete with handrails on both sides; foredeck sun lounges at an elevation of 450 mm above deck plus handrails central to the sun lounge 500 mm above floor level that allow older adults to rise from a seat to a standing position unassisted.

To have older adults actively involved in operational activities including mooring operations requires that mooring equipment is positioned to enable use without needing to stoop or bend.

Access to lower deck spaces should be via stairs with solid, safe handrails on both sides.

To enable mobility of older adults with reduced agility and an inability to perform torsional body movements, interior design must focus on avoiding steps and elevations to eliminate trip hazards.

Crew officers have responsibility for supervising access of older adults onto and off the boat, which is facilitated if the elevation of the pier surface is aligned with the yacht's aft platform and a short, stable, accessorized boarding gangplank is temporarily arranged on skid pads at both its ends with no wheels to ensure better stability.

Crew officers need sensibility and etiquette to assist older guests in the performance of actions requiring support including access to restroom facilities, exit from the water, and access to lower deck spaces, descending stairs preferably in the backward direction and holding on to handrails to ensure safety.

Finally, as discussed in our earlier introduction, old age in industrialized societies should no longer be regarded as a handicap as opposed to an opportunity for development and growth of the yachting industry that all builders should evaluate as strategic for their business.

Further Reading

Accolla, A., 2009. *Design for all. Il progetto per l'individuo reale.* Milano, FrancoAngeli.

Casiddu, N., 2008. *Anziani a casa propria. Linee guida per adeguare spazi e oggetti.* Milano, FrancoAngeli.

Cesabianchi, M., Cristini, C., 2002. *L'invecchiamento fra corporeità e creatività.* Enna, Oasi Editrice.

Dreyfuss, H., 1994. *Le misure dell'uomo e della donna.* Milano, Be-Ma.

Fantini, L., 2001. *Superare le barriere architettoniche, migliorando il confort e la sicurezza.* Santarcangelo di Romagna, Maggioli Editore.

Musio-Sale, M., 2009. *YACHT DESIGN. Dal concept alla rappresentazione.* Milano, Tecniche Nuove.

N.N.A. 2009. *L'assistenza agli anziani non autosufficienti in Italia.* Santarcangelo di Romagna, Maggioli Editore.

Ornati, A., 2000. *Architettura e barriere.* Milano, FrancoAngeli.

Spadolini, M. B., 2000. *La progettazione amichevole. L'evoluzione del concetto di* accessibilità *per l'utenza debole. Dall'habitat al design dell'oggetto.* Milano, Rima Editrice.

Stabilini, F., 2008. *Design e spazio terapeutico: il giardino Alzheimer.* Firenze, Alinea.

Urbani, G., 1991. *L'anziano attivo: proposte e riflessioni per la terza età e quarta età.* Torino, Fondazione G. Agnelli.

W.H.O., 2002. Active ageing: A policy framework, World Health Organization Report, Geneva.

http://www.lergonomica.it/index.php?option=com_content&view=article&id=7&Itemid=109

9

Human Diversity in the Maritime Design Domain: Social and Economic Opportunities of the Innovative Approaches of Design for Inclusion

Giuseppe Di Bucchianico

CONTENTS

9.1 Maritime Domain and Maritime Design Domain

The maritime domain is defined as all areas and things of, on, under, relating to, adjacent to, or bordering a sea, ocean, or other navigable waterway, including all maritime-related activities, infrastructure, people, cargo, vessels, and other conveyances.

This definition is strictly connected with the concept of maritime domain awareness (MDA), defined by the International Maritime Organization (IMO) as the effective understanding of anything associated with the maritime domain that could impact security, safety, economy, or environment.

So it is a very vast sector, which involves environments, products, and activities of man, that is all the proper elements of ergonomics: seas, rivers, lakes, protected areas, tourist stops, commercial ports, harbors, docks, piers, jetties, lighthouses, platforms, ships, boats, tugs, tankers, cargo, ferries, sailing yachts, motor boating, fishery, trade, transport, control of the coasts, sport, decks, cabins, cockpits, onboard equipment, etc.

Design, as a "foreshadowing of possible desirable" (Heidegger), obviously plays a central role in all of the changes and alterations made by the project in the maritime domain in view of human needs (Morris, 1881). Moreover, you can say that the multidimensional complexity of most of the design areas of the maritime domain, particularly the one referred to as the means of transport, goes through extremely complex systems of constraints that involve multiple disciplines and skills, sometimes highly specialized, from structural engineering to materials engineering, from fluid dynamics to marketing. To these they must be added also the prefigurative complexity of products to be used in inhospitable environments and sometimes in extremely adverse conditions, or just in spatial conditions of constant motion.

In relation to the human dimension of the project for the maritime domain, ergonomics has always played a central role during the last decades.

9.2 The Contributions of Traditional Ergonomics to the Maritime Design Domain

In the field of ergonomic design, the user-centered design places at the center of its identification process, by means of the users' demands, some ergonomic requirements required of the products. Ergonomic literature, on the other hand, already placed four different requirement levels referable to the users' demands in a close hierarchical relationship (Jordan, 2000).

In reference to these four levels, a similar articulation already been identified referred to the marine products (Di Bucchianico and Vallicelli, 2011). It is possible now to think about extending this articulation also to the entire universe of artifacts (environments, products, systems, and services) of the more complex maritime domain. Thus, the requirements concerning the safety of equipment, systems, infrastructures, and environment are identified, those regarding their functionality, ease, and practicality of their use, including the requirements that concern their pleasantness, the physical and mental "pleasure" that is experienced in interacting with them. For each one of them, ergonomics is able to offer, and indeed already offers, a precious contribution in relation to the project of environments, stations, equipment, and finishings, as well as, in a figurative sense, to establishing all those tasks and activities, even apparently secondary, which are carried out on the crafts, on terra firma on piers and docks, and even under the sea level.

In particular, it is known that regarding safety, ergonomics already provides all the most suitable instruments to analyze the characteristics of individuals and the limits of their psychophysical abilities, even in extreme environmental and postural conditions; on the functionality plane, ergonomic research is able to influence, through the project, with a significant impact on morphological and typological evolution of the equipment, systems, and infrastructures, in their entirety or with respect to their various parts, in order to allow a critical, objective, well-pondered reading of it; on the plane of usability of the equipment and postural comfort, ergonomic practice allows the identification of more innovative solutions also by means of direct observation and a critical analysis of organizations and structures of tasks on board or on terra firma; on the pleasantness plane, ergonomics has defined the most useful instruments and methods to assess the psychosensory interactions of individuals with components, equipment, and environment.

Furthermore, in all the cases the multidisciplinary approach of ergonomics to the design and the availability of methods, intervention procedures, and operative instruments which it offers, allows the study of the requirements of the user's well-being to be faced whether in relation to the single product/equipment, or to the task/position or to the environment/context in which he finds himself.

9.3 Limits of Traditional Ergonomics and New Concepts of Diversity

Traditional ergonomics too often relies on a mechanistic and deterministic view of man. Born of industrial culture, in order to ensure safety and well-being in work and use contexts closely related to both a production and consumption approach much tied to Taylorism and the derived Fordist mass production, ergonomics has however, over decades of development and maturation of the discipline, placed the center of its attention on the "standard" man, though with all possible percentile variations of the case. For decades, in fact, ergonomic design has dealt with the average individual (man or woman, young or adult), perfectly healthy and intact, fully alert, attentive and informed, with logical and rational behavior in most situations in life and work. The other "specific" cases were and still are largely studied and analyzed as "specific" categories, thus giving rise to design for "specific" users: the elderly, children, disabled, etc. Furthermore, the aggravation of the so-called "segmentation" of the users, from a design point of view, leads to the design of artifacts whose physical and cognitive "measures" and whose use are designed for groups of increasingly restricted individuals, to the realization of many "niche" products, even with evident inconsistency.

The complexity of contemporary society, however, can also be described through the diversity between individuals, expressed not only in terms of mental and physical ability, but also with respect to social and cultural differences. Let us, therefore, better define what could be defined as "diversity."

The concept of "diversity" has an ancient origin in the history of thought: the concept of "multiplicity" of Plato, that of the "otherness" of Aristotle, both with a meaning not necessarily bad, until reaching the philosophers of the last century as Lévinas, for which instead otherness not only is not a negative value, but is considered the higher ethical value. In common sense, the concept of diversity is ambivalent: on the one hand it arouses curiosity and interest, on the other hand it tends to strike fear in the sense of mystery that brings. Sometimes what is different is rejected, but other times in the difference between people and in the specificity of each one can be seen the wealth that can be derived from the comparison and mutual understanding. The "other" may be different for mental and physical abilities, but also for language, culture, customs, religion, and sensitivity.

Diversity, which relates directly to social "conventions" and "common" sensitivity, is therefore often seen in a negative way as a threat of their own identity, both from individuals and from social groups, whereas on the other hand not considering that regarding "diversity as an opportunity" could contribute to reaffirm, in a conscious and critical way, exactly their own identity.

Diversity can therefore be regarded as a limit or a resource both for the economic development of society and to its well-being in the broadest sense.

On the economic level, there is an added value deriving from the awareness that there is a market made up of "nonstandard" users (which in reality are always the vast majority), partially different from those who the design of environments, products, and services has so far addressed, and therefore the ability to propose new ones that are actively comparing the complexity of social articulation. This added value offers even the smallest company new competitive advantages (increase of potential users, customer retention, etc.).

At the social and collective well-being level, to base the development of society on the concept of diversity, promotes inclusion and the independent and active participation of all. Thus, by reducing the social tensions and costs (and the resulting fiscal pressure) and increasing productivity, substantially also the wealth and the overall level of wellness increases.

9.4 Design for Diversity

Over the past 30 years, different approaches and design methodologies for the inclusion of design-enhancing human diversity were developed, considering it as a resource rather than a constraint, thus obtaining solutions more content-rich, innovative, and therefore attractive.

It should, however, make a clarification on the issue of disability. Design for disability, which, especially in the last 70 years, has had a very rich and varied history and which was also the launching pad for the most recent and innovative approaches of design for inclusion, however, should be regarded as a separate area, because it refers to specific users. In fact, focusing its interest on specific disabilities with the aim of enabling people, through dedicated aids, to play out their daily life cycles peacefully, design for disability can be compared to other areas of "design for specific users," such as design for aging or children.

Another things, then, are those approaches and methodologies that cater to the extensive multiplicity of different groups of individuals: the most significant of them are Universal Design, Inclusive Design, and Design for All (DfA).

9.4.1 Universal Design: The Pragmatism of Checklists

Universal Design, created in the United States at North Carolina State University, and from where it spread throughout the world, is an expression of Anglo-Saxon pragmatism. It expresses a fundamental objective of good practice and theoretical project: to meet the needs of as many users as possible. This design methodology is a set of dimensional requirements, complying with codes, standards, or special features of specific users with disabilities, but it expresses the tension toward a higher objective: general design principles that are simple to be implemented and verified. Universal Design in fact does not only focus on people with disabilities, but for the first time defines the user in a wider way, suggesting the making of all products and spaces accessible and usable by people to the greatest extent possible. Not everything must be fully usable by all: the term "universal" refers more to a methodological attitude than to a rigid and absolute assumption.

The didascalic reduction of the approach to seven design principles, easy to be applied and then which rapidly spread throughout the world, however, tends to an extreme schematization of the project and especially not to take into account the complexity of the individual and the diversity and variability of mankind. Moreover, the definition of verification tools projects made from simple checklists, if on the one hand are useful to guide the design process and educate designers and consumers about the characteristics that products and environments should have, on the other hand requires an assessment and verification action "a posteriori" on existing or already developed projects, rather than constituting an inspiring concept.

9.4.2 Inclusive Design: Attention to Diversity

Inclusive Design has its main center of development in the United Kingdom and in the main countries of British influence. Unlike Universal Design, it does not lay out dogmatic design principles, but it defines a real careful approach to human diversity, based on the idea that no policy, principle, or guideline may be absolute but must always deal with the multiplicity of users, contexts, and objectives. Inclusive Design, in fact, considering the wider range of skills, languages, cultures, genders, ages, and all other possible forms of difference among users, bases its approach on three "dimensions": recognizing the diversity and the uniqueness among individuals, the inclusiveness of the tools and design methodologies, and the magnitude of the impact in terms of benefits.

9.4.3 DfA: The Participatory Process

DfA was briefly defined as "design for human diversity, social inclusion and equality." It represents the concrete application of a philosophical, social, and political assumption ("everyone should have equal opportunity to participate in all aspects of society"), and a scientific assumption ("good design enables, bad design disables," Paul Hogan). DfA, in fact, aims to improve the quality of life of individuals through the enhancement of their specificity and diversity: a holistic approach to design processes and methods referred to environments, equipment, and services, that must be experienced "in a condition of autonomy" by people with special needs and diversified skills.

A first innovative aspect of the DfA is therefore the passage from an interest to the "use" to an attention to the "experience" of environments, products, and services. The transition from the user-individual to the experiencer-individual seeks to achieve primarily the fulfillment of the aspirations and desires as well as needs. Not only, therefore, the search for simple "accessibility/usability" of environments and products (which is the starting "basic" condition), but also the search for the "well-being" for everyone through the active, comfortable, and pleasant enjoyment by the widest variety of subjects, different among them for social and cultural characters and for perceptual, motor, and cognitive skills.

The innovative aspect that most of all differentiates DfA from the two previous approaches, however, is the attention to "process," that is the path of the project definition. It shall involve, in different forms and at different times, all actors in the "supply chain project": decision makers, planners, producers/workers, users. This in the idea that to carry out really "inclusive" projects the skill and experience of the designer are not sufficient, and that the project cannot be exhausted in a simple comparison among client, designer, and experienced employees. On the other hand, the solution must come from a consciousness of widespread social participation, involving all other social, economic, and above all political decision makers in different ways.

9.5 Some Significant Design Experience

In different areas of the vast maritime domain and in the narrower one of maritime transportation, design approaches and design solutions for inclusion are spreading on a global scale, both due to the development of ever more demanding standards for accessibility and usability of environments and products, both for the diffusion of a growing sensitivity to the topics, and especially for the growing awareness of the social and economic benefits deriving from it. By now there are numerous examples of marinas, commercial ports, docks, piers, jetties, lighthouses, and infrastructure referred to maritime transportation addressing the issues of design for inclusion. Unfortunately, this often does not happen in an intelligent and proactive way, but only blindly applying rules and laws on accessibility.

Not as extensive numerically is the series referred to as the means of transport, probably due to the design complexity which was mentioned earlier. Surely, however, this is the area in which are recorded the most interesting examples of design for inclusion. Below are some thoughts on three different areas related to maritime transportation: passenger transportation, yachting, and coasting. Through the description of some good practices, it is clear that human diversity may represent an opportunity to innovate through an inclusive approach to a project.

9.5.1 Passenger Transportation (Ferries and Cruises)

Extremely interesting design for inclusion experiments were performed on ferries. For example, for the design of the new high-speed passenger ferries in Ryfylke (Norway) a specific call for tender was made in which one of the design criteria was that the ferry had to be designed considering multiusers. In this case, on the initiative of the same shipping company, a system of requirements was established, developed also through the auditions of representatives of sensitive groups of users (disabled, blind, etc.). The project has identified a number of solutions oriented to the self and pleasurable use of high-speed ferries, some of which are also extremely simple to implement. In particular, for example, the gangways were designed with two levels of handrails, nonslip surfaces, and without any stumbling element, hydraulically adjustable in height, length, and angle of inclination, so as to facilitate access for both wheelchair users, parents with prams, and trolleys used by service personnel on board. Moreover, inside the ferry the colors and surface finishes have been designed to facilitate the blind and visually impaired; lighting returns an overall feeling of naturalness and furnishings are coated with hypoallergenic materials and fabrics; amplification systems with magnetic induction promote understanding of messages even to people with hearing impairments and a system of monitors allows one to instantly know the location of the ferry and facilitates the knowledge of the sequence of stops even to those unfamiliar with the Norwegian language.

For cruise ships, nowadays most of the shipping companies are sensitive to design for inclusion, as they are aware of the potential revenues arising from the possibility of accommodating users with disabilities. They also know the importance that the multiusers who attend cruises (consisting of a large share of families with children and the elderly) retain a pleasant memory of life on board, which may also derive from friendly usability and accessibility of workplaces, equipment, services, and activities. So the ships of today have wide doors, often even sliding doors; escalators without stairs and large elevators; accessible swimming pools and whirlpools; theaters, casinos, dining halls, and other entertainment areas capable of accommodating all with different needs; tactile routes and signage systems; wireless or magnetic induction amplification systems. These design solutions are often flanked by additional services such as services for boarding and disembarkation for people with special difficulties, priority check-in and checkout, or menus for people with specific food intolerances. In the case of cruises, however, the biggest problems come from the integration with ground services enjoyed by guests especially during their excursions. It is necessary in this case to extend the principles of design for inclusion to the entire vacation system, involving therefore also the infrastructures and services of terra firma.

9.5.2 Yachting (Sailing Yachts, Motor Yachts)

The field in which probably in recent years particularly innovative design solutions of maritime transportation have been developed in respect to the issue of design for inclusion is yachting, which deals with the planning of pleasure boats, which have an internal living area and are for open-sea, sports, or cruiser navigation.

In this, the dimensions of the object have an important role in the definition of the relationships between interiors (referred to as the idea of "house") and exteriors (referred to as the idea of "vehicle"): in general, it can be retained that the bigger it is, the narrower does the relationship become with the anthropometric, proxemics, and figurative dimensions of the housing area inside (also as a consequence of the greater effective "stability" of the object) and, consequently, its relation to the sea becomes weaker (Di Bucchianico and Vallicelli, 2011). Furthermore, in contrast with other means of "habitable" locomotion, the crafts are forced to resolve the antinomy between movement and stop with a greater planning effort: "(…) A boat is represented as an object which moves even when it is still. When a camper van is still, it is static like a house; a craft is always in movement even when it is still" (Spadolini, 1987).

Among the projects for sailing yachts of greater interest in recent years, it should be noted the master's degree thesis "10 meters daysailer for all. Sustainable technological solutions for easy navigation" (Rossi et al., 2012), whose design was derived from the development and application of a specific tool for identifying limit users with regard to specific activities (Di Bucchianico et al., 2012) (Figure 9.1).

It is a project that offers highly innovative solutions for the living area, the steering position, and for going up/down on/from the sailing yacht (Figure 9.2).

It can also be worth mentioning the initiative of the three-masted British sailing ships, Lord Nelson and Tenacious, launched, respectively, in 1986 and 2000 thanks to the action of a group of volunteers, disabled or not. The purpose was to allow people with any physical, mental, and nautical ability to navigate side by side on equal terms. So far over 36,000 people have already sailed on the two ships since they were launched.

9.5.3 Coasting (Small Boats and Vessels)

Within the framework of smaller boats, the dimensions represent often a particularly strong constraint to develop solutions that are truly inclusive. This is also due to the fact that, especially in a sailboat, the technical, physical presence, and the importance of the riggings is so prevalent that it is difficult to think of simplifying them to the point of making them accessible to everyone. Therefore, the framework of small boats, although it has produced the first sailing models used by people with disabilities, so far could not produce products that do not appear specifically designed for the disabled. Among the boats specifically designed for people with disabilities, the most popular classes are: Access, Artemis 20, Challenger, Freedom Independence, GOS 16, and Martin 16. Added to these are the following paralympic classes: 2.4 mR (one-person keelboat), SKUD18 (two-person keelboat), and Sonar (three-person keelboat).

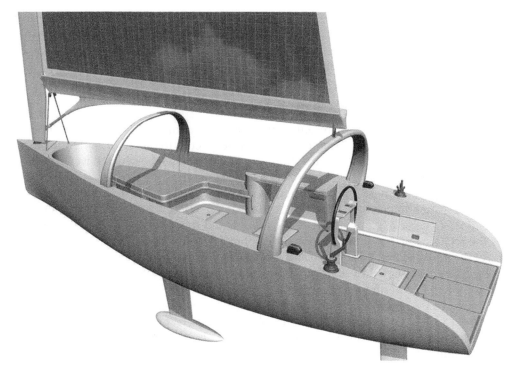

FIGURE 9.1
Overview of a 10 meters daysailer for All. (Adapted from Rossi, E., G. Di Bucchianico, M. Di Nicolantonio, 2012. 10 meters daysailer "for all." Sustainable technological solutions for easy navigation. In: Vink, P. (Ed.), *Advances in Social and Organizational Factors*. Boca Raton, Florida: CRC Press, Taylor & Francis Group.)

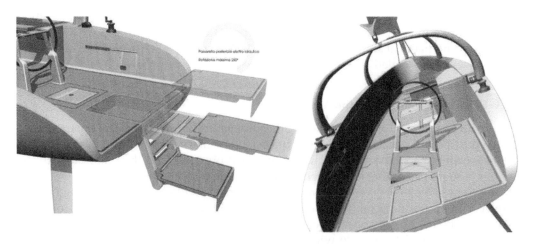

FIGURE 9.2
Innovative solutions for the steering position and the boarding/disembarking system. (Adapted from Rossi, E., G. Di Bucchianico, M. Di Nicolantonio, 2012. 10 meters daysailer "for all." Sustainable technological solutions for easy navigation. In: Vink, P. (Ed.), *Advances in Social and Organizational Factors*. Boca Raton, Florida: CRC Press, Taylor & Francis Group.)

Also some specific manuals were written for the adaptability of small boats (e.g., *U.S. Sailing Adaptive Sailing Resource Manual*): in fact, through the insertion of onboard seats, transfer benches, steering assists, and electronics, it is possible to share the experience of sailing with disabled people on small crafts.

9.6 Conclusions

The good practices that have been described confirm the usefulness and economic, social, and cultural opportunities and benefits resulting from the development of projects that are really "inclusive," aimed to the enhancement of human diversity. This is especially true in those areas and sectors that more than others tend to deal with multiple dimensions of users, such as the maritime domain.

These good practices, moreover, suggest how the recent, innovative, approaches of design for inclusion can really represent a major opportunity not only for business, but also for the entire social system and for collective well-being.

References

Di Bucchianico, G., M. Gregori, E. Rossi, 2012. Designing from activities. The "ability/difficulty table," a useful tool to detect the "limit users" in the Design for All Approach. *Work: A Journal of Prevention, Assessment and Rehabilitation, Vol. 41, Supplement 1/ 2012 Special Issue: IEA 2012: 18th World Congress on Ergonomics—Designing a Sustainable Future*. Delft: IOS Press.

Di Bucchianico, G., A. Vallicelli, 2011. User-centered approach for sailing yacht design. In: Karwowski, W., M. Soares, N. Stanton (Eds.), *Human Factors and Ergonomics in Consumer Product Design. Uses and Applications*. Boca Raton, Florida: CRC Press, Taylor & Francis Group.

Jordan, P. W., 2000. *Designing Pleasurable Products. An Introduction to the New Human Factors*. London, Taylor & Francis.

Morris, W., *Prospects of Architecture in Civilization*, 1881.

Rossi, E., G. Di Bucchianico, M. Di Nicolantonio, 2012. 10 meters daysailer "for all." Sustainable technological solutions for easy navigation. In: Vink, P. (Ed.), *Advances in Social and Organizational Factors*. Boca Raton, Florida: CRC Press, Taylor & Francis Group.

Spadolini, P.L., 1987. Le analisi delle funzioni d'uso e loro relazioni, gli spazi minimi ed i problemi dimensionali nelle imbarcazioni da diporto a vela e a motore. In: Vallicelli, A. (Ed.), *Architettura imbarcazioni da diporto*, Vol. 1. Firenze, Cesati.

Section II

Rail Domain

Introduction

The rail and road domain is by far the largest subsector of transportation, employing about nine out of 10 transport workers and also representing a significant percentage of the GNP (gross national product) of the entire world economy. Given its size, it is, however, also responsible for the most important problems connected with the entire transport sector: from environmental problems caused by harmful emissions to strictly ergonomic issues, and to the psychosocial stress induced on whole populations, groups, or individuals users of transport systems.

The 10 chapters within Sections II and III of the book are divided into two groups, respectively, related to the road transportation (Chapters 10 through 14) and rail transportation (Chapters 15 through 19). You will notice how, starting from fields of application that are often very specific, but which have references to wider issues, it is possible to face the broader issue of the pervasiveness of new technologies and its impact on the evolution of social and cultural aspects of contemporary society.

In particular, the research and experiences described in these chapters relate largely to the behavior of drivers, cyclists, pedestrians, users, and to the relationships established between them, and with respect to the means of transport, infrastructure design and driving assistance systems that are increasingly advanced and complex.

Neville Stanton

10

Leading Indicators of Operational Risk on the Railway: A Novel Use for Underutilized Data Recordings

Guy Walker and Ailsa Strathie

CONTENTS

10.1 Introduction

10.1.1 Data Recording

Data recording is the act of automatically logging information on system parameters over time. Data recording has become increasingly ubiquitous in rail transport operations. Entire national train fleets are now required to carry recorders which continuously extract data on how individual trains are being driven, at increasing rates, and across an increasing range of parameters. The outflow of data is extensive and growing, yet comparatively neglected. What could it be used for? In this study we argue it could be used to tackle the most important strategic risk issues currently faced by rail operators and authorities worldwide.

10.1.2 Brief History

The act of automatically recording data on system parameters over time is referred to as "data logging" or "data recording." In the aviation industry the generic term data logging falls under the specific heading of Flight Data Recording, which itself comprises several individual procedures and devices. The most prominent of these is what is termed colloquially as the "black box," which represents the combination of a flight data recorder (FDR) and a cockpit voice recorder (CVR). Other systems under the heading of flight data monitoring (FDM) include various Aircraft Condition Monitoring Systems (ACMS), such as engine health monitoring (e.g., the Rolls-Royce EHM programme) and the wide range of parameters available from modern avionics (e.g., ARINC 573) via so-called "quick access recorders" (QARs).

Data recording can trace its origins back to the allied fields of metrology, instrumentation, telemetry, predictive maintenance, and condition monitoring. The Wright Brother's 1903 "Wright Flyer 1," one of the world's first powered aircraft, was equipped with "instruments to record air velocity, engine revolutions and time while in the air" (Ford, 2012) and herein lie the very early antecedents for the sophisticated Flight Data Recording and Monitoring that exist today. The rail sector, however, can lay claim to even earlier and more sophisticated examples of instrumentation. Stephenson's 1829 Rocket, for example, had instrumentation for boiler pressure and water level, and in 1838 the Great Western Railway in the United Kingdom constructed the first "dynamometer car," using equipment designed by Charles Babbage to integrate various readings into an accurate representation of train performance.

The use of data logging as a tool in safety science is a postwar development. It evolved amid a wider context that included a marked increase in postwar air travel, the development of new jet airliners, and accidents in which passenger aircraft "crashed without trace" leaving investigators perplexed as to the root cause. Most notable among these were the De Havilland Comet crashes of 1953 and 1954. During the subsequent investigations it was noted that "anything which provides a record of flight conditions, pilot reactions, etc. for the few moments preceding the crash is of inestimable value" (Warren, 1954). The prototype "Flight Memory Unit" (as the black box was then referred) was manufactured from early magnetic audio recording technologies and a primitive crash survivable enclosure. The device could superimpose signals from some of the aircraft's primary controls onto approximately 30 ft of metal wire at a rate of approximately eight signals per second. The device was configured so that the metal wire looped continuously, storing 4 h of voice and data, continually overwriting itself.

In 1958 the UK Air Registration Board became aware of the Flight Memory Unit and due to the national importance of the jet aviation industry and the potential safety barrier that the comet crashes represented to continued foreign sales, the concept was considered important enough to warrant further development. A British clock-making company called S. Davall and Sons was able to acquire production rights and develop the first commercial "black box," or Davall Type 1050 "Red Egg," as it was then called. Improvements now enabled readings to be captured at a rate of 24 per second, greater accuracy in the data collected from aircraft instruments and controls, and the flexibility to record voice, data, or both. To do this, up to 40 miles of stainless steel wire was needed as a recording medium.

An unexplained air crash in Queensland in 1960 led to the mandatory fitment of cockpit voice recorders (CVRs) like these in Australia. Regulations also appeared in the United States as early as 1958, and legislation also followed in 1960 (Morcom, 1970). In Britain, changes were made to the Air Navigation Order as early as 1960 although a lengthy period of consultation and evaluation ensued, meaning that it did not become mandatory to carry an FDR until 1965. With legislation imminent, however, the supply and fitment of recorders was well underway prior to this. Indeed, the first crash investigation to make substantial use of the data provided by an FDR occurred in 1965 when a BEA Vanguard fitted with a Davall Type 1050 "Red Egg" crashed in poor weather at London's Heathrow airport.

Early data recorders were relatively stand-alone devices. The recorder carried its own sensors and, apart from an electrical supply, operated relatively independently of the host aircraft (Campbell, 2007). Calibration proved to be a problem, with the actual state of the aircraft systems not necessarily being identical to those indicated by sensors in the recorder, or even sometimes the same as those displayed to the pilots on their cockpit displays. This "system architecture" was to change with the advent of avionics. Avionics is the collective term given to aircraft electrical systems. The Boeing 787 and Airbus A380 represent the current state of the art and an expression of what is sometimes referred to as "electronic aircraft." Here, air, mechanical, and hydraulically operated systems are replaced by electrical systems, all of which reside on a communications network that can be interrogated by various aircraft systems, including flight data recorders (FDRs). Rather than a stand-alone device, data recorders are now part of a comprehensive data acquisition architecture that relies on the integration of data from myriad sources via a Flight Data Acquisition Unit (FDAU), common communications protocols (ARINC 573, 717, and 767), and the use of QARs as well as crash survivable "black boxes." Modern FDRs are solid-state devices with the ability to continuously record over 2000 separate parameters for in excess of 30 days. The separation between a crash survivable data and voice recorder, mandated by law and used for accident investigation, and a quick access recorder (QAR), not mandated but used for operational and safety purposes by airlines and regulators, occurred in the 1970s. It arose from a growing recognition that easy access to flight data, both routine and abnormal, was of value.

While the aviation sector has a long history with on-vehicle data recording devices for the purposes of safety and crash investigation, these are a much more recent innovation in the rail sector. Experience in the United Kingdom is quite typical. Here, fitment has only been mandatory since 2002 but has been the subject of discussion within the industry for many years (Uff, 2000), indeed, a Railway Group Standard (GO/OTS203) was issued in 1993 in recognition of the fact that the technology existed and was beginning to be fitted in isolated cases. The situation the industry faced was one in which costs (in terms of installation and operation) of fitting data recorders were estimated at £13,000 per unit, with savings due to investigations and repairs estimated at only £3,200. In simple terms, this required investment of £75 million and would need to prevent at least two equivalent fatalities each year to show positive financial benefits (Uff, 2000, p. 177). On this basis, widespread fitment of data recorders could not be justified. Privatization of the rail industry in 1994 and a number of coincident high profile crashes (Southall in 1997 and Ladbroke Grove in 1999) served to accelerate the adoption of data recorders. In the Southall Inquiry report it is noted that "In my view, the cost-benefit figures produced and the conclusions that they suggest amply demonstrate the shortcomings of CBA [Cost Benefit Analysis] as a decision-making tool [...] I believe that the general fitting of data recorders is long overdue and that this view is shared by the great majority of the industry." (Uff, 2000, p. 178).

By 2002, Railway Group Standard GM/RT 2472 made data recorders mandatory in all new UK trains from December 7, 2003 onwards, and required that existing trains be fitted with them by December 31, 2005.

Modern trains share some conceptual similarities with aircraft in that they too make extensive use of electrical actuation (the brakes are "electro-pneumatic" for example), rely on communications between disparate devices and systems, data buses (i.e., the Train Data Bus), and various forms of standardized communications protocol. In other words, they possess a roughly equivalent form of "avionics" and a data bus (or "buses") through which an onboard recorder can acquire information. There is not the same degree of conformity as in comparable avionics systems. Critical differences between the rail and aviation data acquisition architectures are, first, that the functions of a "Flight Data Acquisition Unit" (FDAU) are incorporated within the on train data recorder (OTDR) device itself. Likewise, so are some of the functions of a QAR, and as a result the data must be downloaded manually via serial cable, USB (universal serial bus), or other memory device. At the present time there is not a standard "data frame" for OTDR data, with each device manufacturer using a proprietary version and associated analysis software. At the present time the emphasis is on individual data download and analysis for the purposes of driver training and assessment (as per the Southall inquiry recommendations) or else incident investigation, rather than large-scale data storage and industry-wide analysis of "normal" operations. Some modern rolling stock is able to wirelessly download diagnostic information for the purposes of condition monitoring but at the time of writing this is the exception rather than the rule.

10.1.3 Pushing the Envelope

Regardless of measure, whether it takes into account exposure by distance or time, the risk to the traveling public and workforce of using and operating the railway is exceedingly low. In Europe the probability of a fatality is approximately 0.57 per billion miles (Evans, 2011), or two fatalities per 100 million person travel hours (EU, 2003). This figure arises despite the fact that exposure in time and distance have increased dramatically in some countries. In the United Kingdom, for example, between 1995 and 2012 the risk exposure by passenger distance rose by 25 billion kilometres or 58% (DfT, 2011). At the same time estimated mean fatal train accidents per billion train kilometres have fallen by approximately 9.1% annually (Evans, 2011). Risk exposure is accompanied by an increase in the overall intensity of operations. The UK railway system currently supports 1.3 billion passenger journeys (ORR, 2012) with 16% more trains timetabled in 2013 compared to 1995, most of which are running at higher passenger occupancy levels. This equates to 296.2 million train miles traveled in 2010/2011 (ORR, 2012), all of which have at one point or another been recorded on an OTDR device.

The UK rail network is the fifth busiest in the world behind China (1.86 billion journeys), Germany (1.95 billion), India (8.03 billion), and Japan (22.67 billion). Relative to population size the United Kingdom has the third highest rail usage in the world at approximately 21 journeys per head of population. The network is currently loaded with 1.5 billion passenger journeys per year, which compares to an historical peak of 1.43 billion which occurred in 1946. At this time the network comprised approximately 30,400 route kilometres compared to a 2013 network of 15,777 route kilometres. To support even greater numbers of journeys to those achieved in 1946 on a network 48% smaller means, quite simply, that more trains are using less track at higher speeds, or in other words, the system is being constantly pushed "back to the edge of the performance envelope" (Woods and Cook, 2002, p. 141). The "performance envelope" is defined by Rasmussen (1997) as a set of specific boundaries within which transport systems reside. There is a boundary of economic failure: these are the financial constraints on a system that influence behavior toward greater cost and operational efficiencies. Then there is a boundary of unacceptable workload: these are the pressures experienced by people and equipment in the system as they try to meet economic and financial objectives. The boundary of economic failure creates a pressure toward greater efficiency, which works in opposition to a similar pressure against excessive workload. Because transport systems involve human as well as technical elements, and because humans are able to adapt situations to suit their own needs and preferences, these pressures introduce variations in behavior (Clegg, 2000; Qureshi, 2007). "Over a period of time, [human] adaptive behaviour causes [the system] to cross the boundary of safe work regulations and leads to a systematic migration toward the boundary of functionally acceptable behaviour. This may lead to an accident if control is lost at the boundary" (Qureshi, 2007, p. 6). Systems that exist at these "limits of controllability" represent a significant challenge for the human operators within them (Lupu et al., 2013).

10.1.4 The Black Box Paradox

There are three key paradoxes inherent in this wider picture. First, because so few major rail accidents occur there is no longer enough data to construct reliable forward looking estimates (e.g., Evans, 2011). When safety performance data reaches the level of that achieved in the rail sector it instead starts to become characterized by unpredictable periodicities, cycles, or discrete events. This is becoming evident in EU rail safety data, with one large-scale rail accident occurring on average every 6 years (EU, 2003). In other words, the safety data is "leveling off" with a persistent class of human/machine–systems accident now elevated to the status of a key strategic risk (RSSB, 2009).

Second, "there is widespread concern within the industry that the background indicators—rather than the headline grabbing ones—have remained worryingly stable" (Wolmar, 2012). An example of this is UK data on signals passed at danger (SPAD) incidents. In the period since the introduction of a countermeasure called the train protection and warning system (TPWS), after which there were initial improvements, there has been comparatively little variation in the overall SPAD rate. For example, the rate for quarter 4, 2012 is the same as quarter 3, and indeed the same (or very nearly the same) as on seven previous reporting periods since 2005 (e.g., ORR, 2012).

Third, and most paradoxically, is that the opportunities to use black box data for their original purpose (i.e., postaccident analysis) are diminishing at the same time as the technical capabilities of data recorders are increasing. What this means is that enormous quantities of nonaccident data are being collected day in and day out, but not currently used.

10.1.5 Flight (Rail) Data Monitoring

This study is premised on best-practice safety science approaches developed within the aviation domain, specifically a process called flight data monitoring (FDM). This is "a systematic method of accessing, analyzing, and acting upon information obtained from digital flight data records of routine flight operations to improve safety. It is the proactive and timely use of flight data to identify and address operational risks before they can lead to incidents and accidents" (CAA, 2003). FDM is mandatory for operators of airplanes of a certified takeoff mass in excess of 27,000 kg. In effect, it is a way of using data collected from routine operations to detect trends which, if allowed to continue, might eventually lead to an accident. Changes are made to address issues, and the changes themselves are monitored for their possible effects on other parts of the system. The traditional approach to FDM is focused on exceedence or event detection. Events are defined as: "deviations from flight manual limits, standard operating procedures and good airmanship" (CAP 739, p. 16). Computer software is used to automatically scan FDR data for instances of these deviations, and a set of core events that cover the main areas of interest are quite standard across operators. Event detection is commonly based on simple statistical techniques and automatic algorithms that detect different phases of flight and events therein. FDM is a highly successful way of making use of black box data in a proactive manner, but it too is challenged by the emerging class of human–systems problems that is the focus of this study. Indeed, while having systems that automatically detect events, it is still incumbent on so-called FDM analysts to manually interpret the lower-order "trace plots" that data recorders produce in order to derive meaning from them. As such, there is considerable value in being able to robustly transform these trace plots into higher-order representations, to detect psychologically meaningful patterns therein, and to automatically derive human performance metrics that can help to assess the risk.

10.1.6 Human Factors Leading Indicators

Leading indicators are measurable precursors to major events such as an accident (Reiman and Pietikainen, 2012). The indication of a precursor "leads," or comes before, the actual event itself. Lagging indicators are the opposite. These are so-called "loss metrics" that can only become apparent after an event (Rogers et al., 2009). Leading indicators are said to be "proactive" because they enable steps to be taken to avoid seriously adverse consequences. Lagging indicators are said to be "reactive" in that a seriously adverse event needs to occur before it can be learnt from (Hinze et al., 2013). For this reason, leading indicators are also sometimes referred to as "positive performance indicators" and lagging indicators as "negative performance indicators." The concept of leading and lagging indicators originally derives from the field of economics and the need to understand "business cycles" (i.e., growth, recession, investment, divestiture, etc.) and to predict when one phase of a "cyclical process" such as this will change to another (Mitchell and Burns, 1938). The terms have been appropriated more recently by the safety science community (e.g., Grabowski et al., 2007; Reiman and Pietikainen, 2012; Hinze et al., 2013), particularly in view of developments in safety management systems (SMS) since the 1990s. Leading indicators, in a safety management context, can be defined as "something that provides information that helps the user respond to changing circumstances and take actions to achieve desired outcomes or avoid unwanted outcomes" (Step Change, 2003, p. 3).

The basic "research problem" can be stated thus: despite considerable improvements in safety performance in both the rail and aviation sectors, a persistent class of accident/near accident continues to occur. This class of accident/near accident resides at the interface of people and machine systems. What is required is a means to detect the presence/emergence of such problems before they manifest themselves as a serious operational incident. This study describes how black box data can be coupled to existing human factors methods to provide leading indicators of trends residing at this interface. Specifically, it examines

OTDR data on driver responses to an in-cab warning to reveal the types of errors that may be more likely to arise if the discovered trends continue.

10.2 Method

10.2.1 Data File and Parameters

The OTDR data file is a continuous download for a single traction unit. The recording started at 05:34:57 on July 6, 2012 and ceased at 21:36.32 on July 11, 2012. This is a period of 136 h, 1 min, and 35 s during which the train made 107 journeys and traveled 1638 miles. The raw data takes the form of a comma separated values (CSV) file containing a data matrix 191,021 time samples (rows) deep by 72 parameters (columns) wide: a total of 13,753,512 data points. The mean sampling rate is 2.56 s. The logger itself scans the parameters for changes at a rate of 20 mS but, in the present system, to economize on memory requirements data are only logged when one of the 72 parameters changes. The OTDR device itself was a UK Railway Group Standards compliant Arrowvale unit which recorded 72 parameters, 25 of which are in addition to those mandated. In terms of data classification, four of the parameters, time, distance, and two speed signals derived from a driven and non-driven axle, are continuous ratio data. The remaining 68 are nominal/binomial (i.e., on or off).

10.2.2 Rolling Stock

The sample of OTDR data was obtained from a Class 153 "super sprinter," unit number 153 306. This is a single-unit diesel powered railcar built between 1987 and 1988. Class 153s are 23.2 m in length and have an unladen weight of 41.2 tons. They seat 72 passengers, comprise a riveted aluminum body shell affixed to a steel under-frame, and are equipped with four electrically powered single-leaf Bode doors. The prime mover is an underslung turbocharged six cylinder Cummins NT855 diesel engine producing 285 bhp. A Cummins-Voith T211r hydraulic transmission drives both axles of the leading BT38 bogie via a Gmeinder final drive. The unit's maximum operating speed is 75 mph. It is fitted with electro-pneumatic clasp brakes, with cast iron brake pads acting directly on the tread of the wheel(s) via compressed air actuation. Air suspension is provided for additional passenger comfort and refinement. Tight-lock compact BSI auto-couplers mean that Class 153's can work flexibly in unison with several other DMU classes. The present unit worked solo for the duration of the data collection period and an example of the class is shown in Figure 10.1.

10.2.3 Journeys and Routes

Data collection took place on the Great Eastern (Route 7) and West Anglia (Route 5) regions of the UK's rail network. The strategic "backbone" of the Great Eastern region is the Great Eastern Main Line (GMEL) originating from London Liverpool Street and traveling north east to Norwich. There are numerous branch lines attached to the GMEL providing services to commuter areas and freight hubs. Data for the present analysis was derived from three journeys between the towns of Ipswich and Felixstowe (Figure 10.1), a distance of 25 km, over which there are 14 automatic warning system (AWS) sites.

Mondays to Fridays

								1	**1**			
London Liverpool St. ⊖... d						0600	0700	2000	21b00			
Harwich Int ⚓ 🖵 .... d							0750		2138			
Ipswich 🖵 .... d	0504	0604	0620		0714	0735	0825	2117	2217	2228		
Westerfield .... d	0510	0610	0626		0720	0741	0831			2234		
Derby Road................ d	0515	0615			0725		0836			2239		
Trimley .... d	0524	0624			0734		0845			2248		
Felixstowe .... a	0530	0630			0740		0851			2257		

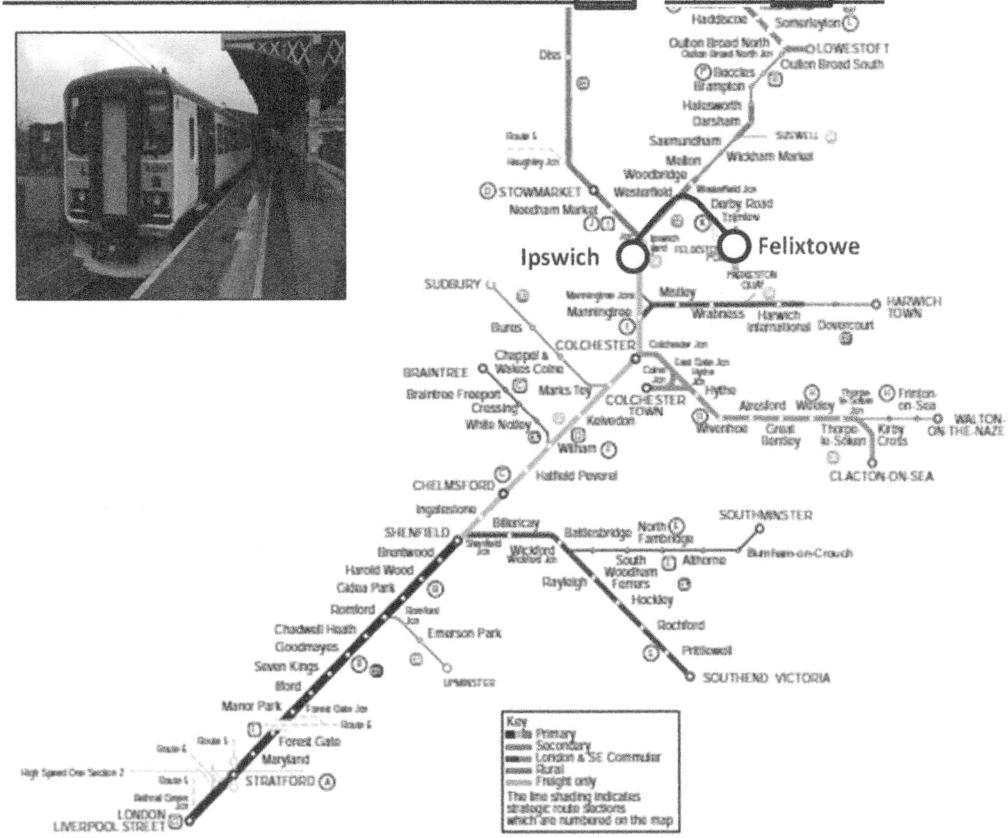

FIGURE 10.1
Rolling stock type, working diagram/timetable, and route map.

10.2.4 Automatic Warning System

The purpose of the AWS is described thus (McLeod et al., 2005):

> AWS serves two functions. The first function is to provide an audible alert to direct the driver's attention to an imminent event (such as a signal or a sign). The second function, linked to the first, is to provide an on-going visual reminder to the driver about the last warning. [AWS] is there to help provide advance notice about the nature of the route ahead, and thus communicate to the driver the need to slow down or stop. (McLeod et al., 2005, p. 4)

AWS alerts and reminders are triggered by an electromagnetic device placed between the tracks approximately 200 yards prior to the signal, sign, or other event to which it refers (although this distance can vary according to local circumstances). Sensors underneath the train detect the presence of magnetic fields and activate AWS accordingly. AWS has three system states:

State 1: No additional action required. If the event to which an AWS activation is caused requires no action by the driver (such as a signal showing a green aspect), then a bell or simulated chime (at 1200 Hz) will sound briefly. The visual display will also remain inactivated. The driver behavior in this case is to proceed; there is no requirement to cancel AWS using the cancelation button nor is there any specific need to enact driving behaviors in addition to those that are current or planned.

State 2: Attention is required toward some imminent signal, sign, or event. If the event to which an AWS activation is caused requires (or potentially requires) the driver to slow down or stop, then a steady alarm or horn sound (at 800 Hz) will sound. The visual display will activate (turning from an all-black display to yellow and black, known as the "sunflower").

State 3: Acknowledge (and continue to be reminded of) the imminent or previous signal, sign, or event. The act of canceling AWS (by pressing a button) stops the horn sound, and the sunflower continues to be displayed. Failure to cancel AWS within approximately 2 s leads to an immediate emergency brake application which cannot be canceled (Railway Safety, 2001). This level of braking may cause some discomfort to passengers and the event will be logged on the on train monitoring recorder (OTMR) equipment. Repeated failure to cancel the AWS within the time allowed is likely to lead to an investigation followed by some remedial training and coaching of the driver.

AWS is a legacy system that was originally conceived as a means to prevent signals passed at danger (SPADs). Several major accidents saw the use of AWS, and the number of events it now refers to, being extended. AWS now provides warnings in six circumstances:

1. (Certain types of) permanent speed restriction
2. All temporary speed restrictions
3. (Some) level crossings
4. SPAD indicators
5. Canceling boards
6. And other locations (such as unsuppressed track magnets, depot test magnets, etc.)

Unfortunately, the simple two state warning (bell/horn) and reminder (black/yellow) is unable to discriminate between these six different events. There are approximately 29,000 AWS sites around the UK railway network, which equates to a mean of 1.6 AWS indications received in the train cab every 1.6 route miles, or 2.7 activations (either a bell or horn sound) every minute when traveling at 100 mph. Many warning indications require no action from the driver, simply a press of the cancelation button. Many other warnings occur in situations when the correct behavior at that time is to accelerate the train (it is moving at slow speed or departing from a station, for example). The task of the driver, therefore, is not a simple one of hearing the warning and pressing a button. It requires them to interpret the source of the warning and the context in which it is occurring, and then decide on the correct course of action. The confusion that this could cause for drivers has been cited in several accident inquiry reports (e.g., Cullen, 2001).

10.3 Results and Discussion

10.3.1 Behavioral Clusters

Process charting techniques are used to represent complex real-world activity in an easy to read graphical format using standardized symbols and layout. The process chart methodology has a long application history, with early examples dating from the 1920s (e.g., Gilbreth and Gilbreth, 1921). It has been used extensively in military and high hazard domains as a way of understanding the interaction between people and systems, particularly in terms of identifying human error potential. The method has been used in both rail and aviation settings before. In this application, process charts offer a novel way of converting raw "trace plots" derived from data recorders into an alternative representation, one that makes it easier to

- Discern how larger journey phases break down into smaller component activities
- The order and timing that component activities occur in
- Who is performing what activity
- The presence of distinct activity clusters

The 72 parameters extracted from the data recorder were classified into:

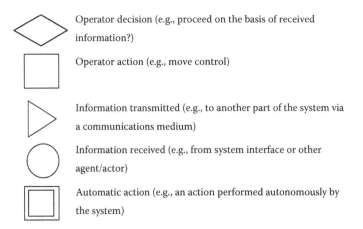

Operator decision (e.g., proceed on the basis of received information?)

Operator action (e.g., move control)

Information transmitted (e.g., to another part of the system via a communications medium)

Information received (e.g., from system interface or other agent/actor)

Automatic action (e.g., an action performed autonomously by the system)

Once classified the process chart itself was constructed. This involved creating a timeline and columns for each "agent" in the system. In the case of the railway example six such agents/columns have been used (Figure 10.2). As different recorder channels become active, the corresponding process chart symbol is inserted into the relevant column at the correct point on the timeline. The sequence of activities and their dependence on each other defines when these symbols are linked. Thus an activity/symbol that occurs *after* another activity/symbol becomes linked "vertically." Activities that are performed *concurrently* are linked "horizontally." Figure 10.2 shows the "normative" structure of operations associated with receiving and canceling an AWS alert.

Having the ability to detect behavioral clusters grants the opportunity to assess whether such structures are typical or atypical. Indeed, whether they are one of a number of different behavioral responses within a wider repertoire, and whether one cluster of behavior is implicated in risk outcomes more than another, and under what circumstances. From the collected data three different "clusters" were detected. The first cluster is the normative "perceive–decide–act" sequence. Here the infrastructure on the track triggers an in-cab warning horn. The driver perceives (hears) this, has enough time to classify it (0.89 s) and respond by pressing the cancelation button. The second cluster is the "predictive cancel" sequence. In this case the infrastructure on track triggers the in-cab warning horn but the driver responds so quickly that it is not possible to have perceived, classified, and responded to the warning. Instead, the driver has seen the track infrastructure and has anticipated the in-cab warning and timed their response to coincide with it starting. The third cluster is the "multiple predictive cancel" sequence. As in cluster two, the driver can

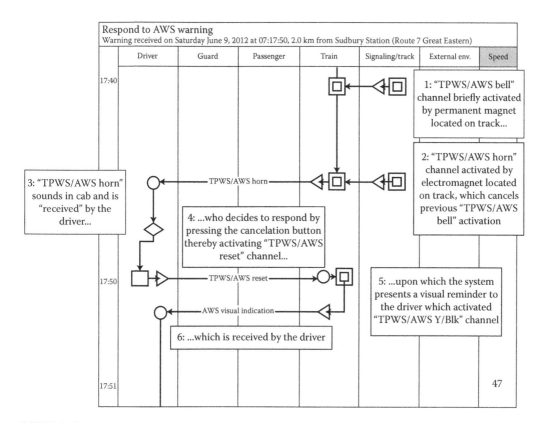

FIGURE 10.2
Annotated process chart showing the type and sequence of operations required to correctly respond to an AWS warning.

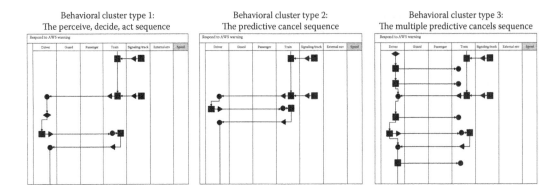

FIGURE 10.3
Three clusters of behavior associated with canceling an AWS alert were detected.

see the track infrastructure ahead and is pressing the cancelation button numerous times before hearing the in-cab warning horn, and several times after the warning has sounded and been canceled (Figure 10.3).

10.3.2 Response Bias

Based on the analysis above it is clear, first, that canceling an AWS warning is not merely a perceptual one of hearing and seeing the different system alerts and indications, it is also cognitive: driver's not only have to discriminate a "stimulus" from within a "noisy" environment, but correctly classify it and respond. Second, there are different strategies that drivers employ to perform this apparently simple task. Signal detection theory (SDT) helps to untangle these different aspects by separating out a person's sensitivity to stimuli (how easy it is to detect something) and their response bias (their preference for responding one way or another to the stimuli). SDT helps us to understand why a particular "stimulus," which might be very loud, visible, or unambiguous, is not always responded to in the ways we expect (or vice versa). SDT classifies human responses to stimuli in the environment in four ways. The responses that drivers made within each of these categories are shown in Table 10.1.

The ability to accurately detect stimuli in the environment and correctly classify them is the desired outcome. Under the SDT paradigm, this requires a high number of hits and a low number of false alarms. For example, if the reset button was pressed in response to ANY warning indication this will ensure a 100% hit rate but will also increase the rate of false alarms. Accuracy in this case is low. If, on the other hand, the driver is trying to do the opposite, to avoid false alarms and instead maximize correct rejections, they would not respond to ambiguous "signals," they would instead "play it safe." This would increase the number of correct rejections but it would also increase the number of "misses." Accuracy in this case is also low. SDT enables us to separate sensitivity (d') from decision bias (C). Sensitivity is a measure of accuracy and tells us how easy it is to distinguish a particular environmental stimuli (e.g., an in-cab warning). Decision bias tells us whether one response is more probable than another.

Human responses to certain stimuli vary because of incentives embodied in the environment. In train driving, for example, there is a strong incentive in normal operations to prevent unnecessary applications of the emergency brake because an AWS warning

TABLE 10.1

Driver Responses to the AWS Can Be Characterized According to the SDT Paradigm as Follows

Status of AWS/TPWS System	Categorization	Number	%
AWS horn followed by reset	Hit	21	30
AWS horn followed by no response	Miss	0	0
AWS bell followed by reset or no activation followed by reset	False alarm	22	31
AWS bell followed by no response	Correct rejection	28	39
Totals		71	100

was missed. This is because an unwanted brake application cannot be canceled and is highly inconvenient. On the other hand, the consequences of false alarms are low because nothing happens when the cancelation button is pressed in error. Because of this it would be expected that train drivers might adopt a "liberal" decision bias and be willing to prioritize false alarms (redundant presses of the cancelation button) over correct rejections (not pressing the cancel button unnecessarily). Added to this is the discriminability of the "stimulus" people are responding to, a stimulus that occurs in a "noisy" real-world environment. By noise we refer to other stimuli, competing demands and distractions in the environment as well as the background noise inherent in human perceptual and cognitive processes. Decision bias and sensitivity interact with these "noisy" transport environments to make some future responses more likely than others, in ways that are not always immediately apparent. For example, highly visible warnings that are apparently "missed," or control actions that are at odds with the situation.

10.3.2.1 Sensitivity

Sensitivity to a stimulus is given by the metric d', which was calculated as follows:

$$d' = z(FA) - z(H)$$

where $z(H)$ is the number of hits expressed as a z-value subtracted from the same Z-transformed false alarm rate. The results obtained are shown in Table 10.2.

The d' figure measures the strength of the stimulus, which in this case is an AWS warning. A value of 3.03 indicates that drivers are highly sensitive to it: in this situation it is unambiguous and easy to discriminate from the wider background of noise, distractions, other contextual factors, etc. Expressed more formally, the response drivers are providing when an AWS warning is overlaid on top of the "contextual noise" is 3.03 standard deviations "different" from the responses they give when the signal is absent (and only

TABLE 10.2

Driver's Sensitivity to the AWS Alerts

Journey	Hits	Misses	Correct Rejections	False Alarms	d'
1	5	0	11	0	4.65
2	9	0	6	7	2.30
3	7	0	11	15	2.13
				Mean	3.03

the "contextual noise" is present). Sensitivity provides an important leading indicator concerning the discriminability of information needed for drivers to develop accurate situational awareness. The same "stimuli" may yield different levels of sensitivity depending on external/contextual factors. A warning that was not expected, ambiguous, not fully understood, or masked may lower sensitivity.

10.3.2.2 Decision Criterion

Decision bias/criterion is given by the metric c, which was calculated as follows:

$$c = -\frac{Z(H) + Z(FA)}{2}$$

The results obtained are shown in Table 10.3.

Decision bias is independent of sensitivity and relates not to the discriminability of the "signal" but to the payoffs involved in making one response in favor of another. Thus, regardless of how easy it is to discriminate a stimulus, a counter intuitive response may still be favored. This is because the consequences of false alarms, misses, and correct rejections vary with the context. Psychological research shows that decision bias is more unstable and situationally dependent than sensitivity and, therefore, a potentially valuable leading indicator.

The mean decision bias value across the three sampled drivers was $c = -1.24$ which indicates a liberal bias. Drivers make more responses that indicate the AWS signal is present than it is absent. In other words, they are prioritizing false alarms over correct rejections which, in turn, provides a clue as to the sorts of error that may be more likely to occur in future (i.e., warnings that are canceled incorrectly). Assuming that drivers' "internal responses" to the AWS warning are normally distributed (as per SDT), it is possible to plot individual driver decision bias into the chart below which provides an important diagnostic tool in defining risky psychological/decision-making states.

According to Figure 10.4, driver 1 shows no systematic bias in their responses to the AWS warning. They respond correctly to the AWS warning on every occasion and his/her false alarm rate is zero. Drivers 2 and 3 are different. They are exhibiting a strong "liberal response bias" meaning that they are much more inclined to exhibit "false alarm" responses (and behavioral clusters 2 and 3). With the ability to detect these changes in decision bias comes the possibility to analyze (a) the extent to which different biases interact with accident/incident rates (i.e., is a liberal bias of this magnitude associated with particular types of risk) and (b) how the context influences human decision making (and therefore how that context can be modified to "unbias" human responses).

TABLE 10.3

Results of Decision Bias (c)

Journey	Hits	Misses	Correct Rejections	False Alarms	Criterion
1	5	0	11	0	0
2	9	0	6	7	−1.21
3	7	0	11	15	−1.26
				Mean	−1.24

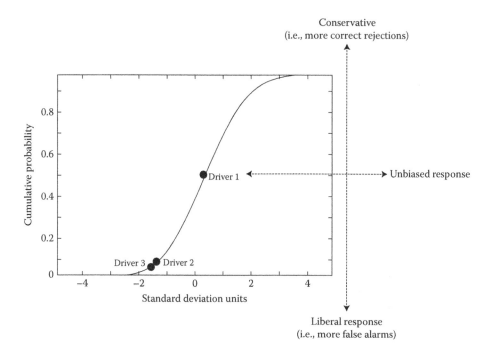

FIGURE 10.4
Cumulative probability plot showing the extent and direction of response biases exhibited by the drivers.

10.4 Conclusions

This study has described how the outputs from transport data recorders can be coupled to existing human factors methods to provide advanced indication of strategically important human factors risks. The "black box paradox" is that the opportunities to use these devices for their original post-accident purpose are diminishing at the same time as their technical capability and data richness are increasing. In addition, the types of risks are changing, with progress in technical areas of reliability and performance exposing issues around human–machine system interaction. To continue to ensure safety in the face of increased risk exposure and operational intensity, better use of this data needs to be made. This study demonstrates how black box data from the rail sector can be turned into useful "information" in the form of human factors leading indicators of risks associated with the use of an in-cab warning device. Sensitivity provides a measure of how much useful information there is in the environment and the extent to which drivers can discriminate it from the background of contextual noise. Warnings, stimuli, and so forth may, in an engineering sense, appear to be unambiguous, yet they may be considerably less so cognitively. Sensitivity provides a measure of this which can, in turn, be associated with changing risk. Decision bias reveals the likelihood that one type of driver response will be favored and how this interacts with risk. In a wider application it would be possible to examine decision bias in a systematic way looking at differences between drivers and between particular routes. This could provide insight into driving styles and indicate whether particular aspects of a route result in a shift in decision bias. For example, a specific AWS signal on a particular route may result in a high level of predictive pressing (high false alarms) relative to most others, identifying this as a more risky section of journey. Relationships such as these would need to be established based on large-scale future research but even on a smaller sample of data the method was able to detect potentially important differences between drivers, with some adopting a much more liberal response bias than others. The principle, however, is a much more important one. We have demonstrated that human factors methods like these can accept recorder data as an input, are amenable to the kind of software implementation that would be required in a full-scale application, and point the way toward leading indicators of strategic importance more generally in safety science.

Acknowledgments

This work was funded by the UK Economic and Physical Sciences Research Council (EPSRC) under grant reference: EPSRC EP/I036222/1. The involvement and support of the UK Civil Aviation Authority, the Rail Safety and Standards Board (RSSB), the Association of Train Operating Companies (ATOC), Greater Anglia Trains and Aerobytes Ltd. is also gratefully acknowledged.

References

CAA. 2003. *CAP 739: Flight Data Monitoring. A Guide to Good Practice*. UK: Civil Aviation Authority.

Campbell, N. 2007. The evolution of flight data analysis. In: *ASASI 2007: Proceedings of Australian Society of Air Safety Investigators Conference*, Civic Square, Australia. Available online at: http://asasi.org/papers/2007/The_Evolution_of_Flight_Data_Analysis_Neil_Campbell.pdf.

Clegg, C. W. 2000. Sociotechnical principles for system design. *Applied Ergonomics*, 31, 463–477.

Cullen, W. D. 2001. *The Ladbroke Grove Rail Inquiry: Parts 1 and 2 Report*. London, UK: HSE Books.

Department for Transport (DfT). 2011. Department for transport statistics. Retrieved July 12, 2012, from: http://www.dft.gov.uk/statistics.

European Transport Safety Council. 2003. *Transport Safety Performance in the EU: A Statistical Overview*. Brussels: European Transport Safety Council. Retrieved July 17, 2012, from: http://www.etsc.be/documents/statoverv.pdf.

Evans, A. W. 2011. Fatal train accidents on Europe's railways: 1980–2009. *Accident Analysis and Prevention*, 43(1), 391–401.

Ford, M. 2012. FDM technology and data recovery. *Presented at FDM Short Course*, Cranfield University, UK.

Gilbreth, F. B. and Gilbreth, L. M. 1921. *Industrial Efficiency*. New York: American Society of Mechanical Engineers.

Grabowski, M., Ayyalasomayajula, P., Merrick, J., Harrald, J. R., and Roberts, K. 2007. Leading indicators of safety in virtual organizations. *Safety Science*, 45(10), 1013–1043.

Hinze, J., Thurman, S., and Wehle, A. 2013. Leading indicators of construction safety performance. *Safety Science*, 51(1), 23–28.

Lupu, M. F., Sun, M., Xia, R., and Mao, Z-H. 2013. Rate of information transmission in human manual control of an unstable system. *IEEE Transactions on Human-Machine Systems*, 43(2), 259–263.

McLeod, R. W., Walker, G. H., and Moray, N. 2005. Analysing and modelling train driver performance. *Applied Ergonomics*, 36, 671–680.

Mitchell, W. C. and Burns, A. F. 1938. Statistical indicators of cyclical revivals, NBER Bulletin 69, New York; reprinted as Chapter 6. In G.H. Moore (ed.) *Business Cycle Indicators*, Princeton, New Jersey, Princeton University Press, 1961.

Morcom, A. R. 1970. Flight data recording systems: A brief survey of the past developments, current status and future trends in flight recording for accident investigation and operational purposes. *Aircraft Engineering and Aerospace Technology*, 42(7), 12–16.

Office of Rail Regulation (ORR). 2012. *National Rail Trends Yearbook, 2010–2011*. London: ORR.

Qureshi, Z. H. 2007. A review of accident modelling approaches for complex socio-technical systems. In: *Proceedings of the Twelfth Australian Workshop on Safety Critical Systems and Software and Safety-Related Programmable Systems*, Vol. 86, pp. 47–60. Adelaide, Australia: Australian Computer Society.

Rail Safety and Standards Board (RSSB). 2009. *The Railway Strategic Safety Plan 2009–2014*. London: Rail Safety and Standards Board.

Railway Safety. 2001. *Automatic Warning System (AWS) GE/RT8035*. London: Railway Safety.

Rasmussen, J. 1997. Risk management in a dynamic society: A modelling problem. *Safety Science*, 27, 183–213.

Reiman, T. and Pietikainen, E. 2012. Leading indicators of system safety—Monitoring and driving the organizational safety potential. *Safety Science*, 50(10), 1993–2000.

Rogers, A., Evans, R., and Wright, M. 2009. *Leading Indicators for Assessing Reduction in Risk of Long Latency Diseases*. Reading, UK: Greenstreet Berman Limited for Health and Safety Executive.

Step Change. 2003. *Leading Performance Indicators: Guidance for Effective Use*. Aberdeen: Step Change in Safety.

Uff, J. 2000. *The Southall Rail Accident Inquiry Report*. Sudbury: HSE Books.

Warren, D. R., 1954. A device for assisting investigation into aircraft accidents. ARL-MECH-ENG-TECH-MEMO-142.

Wolmar 2012, May *Rail Magazine*, 697: Safety is the hidden success story http://www.christianwolmar.co.uk/2012/05/rail-697-safety-is-the-hidden-success-story

Woods, D. D. and Cook, R. I. 2002. Nine steps to move forward from error. *Cognition, Technology, and Work*, 4(2), 137–44.

11

To Beep or Not to Beep: Developing a Non-Fail-Safe Warning System in a Fail-Safe Train Protection Environment

Richard van der Weide, Kirsten Schreibers, and Colete Weeda

CONTENTS

11.1 Introduction

11.1.1 Train Protection Systems in the Netherlands

The most deployed automatic train protection system (ATB-EG) in the Netherlands leaves a "gap" below 40 km/h, leading to 112 signals passed at danger (SPADs) in 2014 (ILT, 2015). Although the Dutch railway system is relatively safe, and SPAD figures are diminishing, some recent incidents and accidents reinforced the need for extra measures to reduce the number as well as the effect of SPADs. A significant example was a train to train collision in Amsterdam in 2012 involving 1 fatality and 188 people injured (Dutch Safety Board, 2012). One of the extra measures is a new warning system (with the acronym "ORBIT") to fill the "ATB-EG gap" to a large extent by warning the driver when approaching a stop signal at (too) high speed. In short, ORBIT continuously checks train position, speed, and brake characteristics against the distance to the next stop signal, and generates a warning

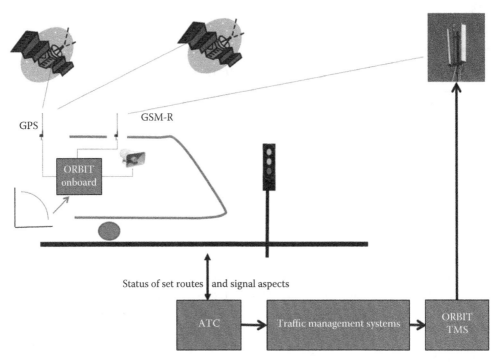

FIGURE 11.1
Schematic overview of ORBIT.

when a critical speed–distance curve ("alarm curve") is passed. A schematic overview of the system is shown in Figure 11.1.

After passing a yellow signal aspect, train speed must be reduced to 40 km/h and the driver must be prepared for a red signal aspect (stop) at the next signal. ATB-EG checks whether the required speed of maximum 40 km/h has been reached or if the brake has been activated to a predefined minimum. If one of these conditions has not been met an acoustic warning ("bell") will sound, and eventually the train will automatically be stopped by ATB-EG. However, the system allows passing a signal at danger with a train speed below 40 km/h or even with a speed over 40 km/h when a minimal brake has been activated.

11.1.2 ORBIT in Addition to Train Protection Systems

To overcome the period until nationwide implementation of the European train control system (ERTMS/ETCS)—with full brake curve supervision—a relatively simple warning system has been developed: ORBIT. ORBIT is based on knowledge from train traffic management systems (TMSs) about the end of set routes (red signal aspects) combined with GPS (global positioning system) coordinates of this lineside signal, and actual position, speed, and acceleration of the train based on GPS. This information is evaluated against the brake characteristics of the specific rolling stock. If a red signal is approached with a too high speed, ORBIT presents a warning to the driver.

ORBIT is not fail-safe as it is based on GSM-R (global system for mobile communications-railway) communication and GPS, both of which can fail or be unavailable. As a consequence of the operational concept design, ORBIT can only guard signals controlled by the train TMS, and not automatic signals. In practice this means that about 6,000 out of total 11,700 signals are covered; it must be stressed that controlled signals are located at most hazardous infrastructural situations, like points or switches. So ORBIT is planned to be in effect at the most risky locations. In fact, it was analyzed that in theory ORBIT would have been helpful in 84% of the SPADs during 2004–2008, as well as in two major accidents in 2009 and 2012. From a cost/feasibility perspective, ORBIT is advantageous compared to systems that need to be integrated into rail infrastructure. Basically, ORBIT uses information already present in TMSs, combined with—new to most of the Dutch fleet but commercially available—GPS information and a relatively simple algorithm. The ORBIT onboard unit has no controls, other than a secured power switch that may only be operated with permission. Sounds run via dedicated speakers and entry of train number—necessary for identification of the train—happens automatically via existing GSM-R (this functionality was not yet implemented during this study, and was done by a remote test facilitator during the test).

11.1.3 Human Factors Issues

The human factors specialist joined the team of (IT) systems designers to cover safety and effectiveness aspects related to human factors. In terms of system performance, ORBIT must enhance the driver's perception, assessment, and decision to brake in time when approaching a red signal. In Figure 11.2 (Wickens and McCarley, 2008) this is visualized: ORBIT adds the Imperfect Automated Diagnosis path to the situation where at speeds below 40 km/h performance only relies on human perception, assessment, and decision. More specific, ORBIT should help prevent a miss by human decision. From this signal detection theory, the following issues can be identified:

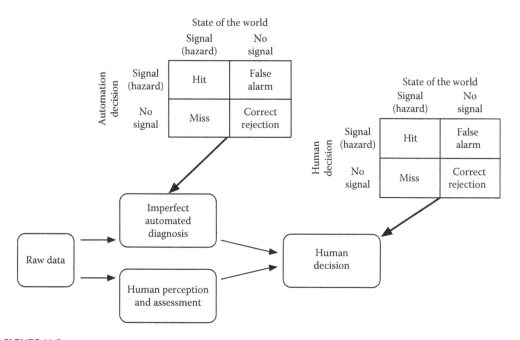

FIGURE 11.2
Signal detection theory applied to parallel human and automation alerting system. (Adapted from Wickens, C.D. and J.S. McCarley. 2008. *Applied Attention Theory*. Boca Raton, FL: CRC Press.)

- Hits by ORBIT must be conveyed to the driver in a perceivable and understandable way (alert salience) to result in the correct action by the driver.
- False alarms may be detrimental to driver's trust in ORBIT, and thus potentially result in ignoring the warning. The rate and nature of false alarms are discussed below in terms of reliability.
- Misses by ORBIT influence efficacy of ORBIT, but may not have a large effect on trust because SPADs (misses by human decision) are quite rare.

For an optimal alert salience in a train cab, auditory warnings seem to be preferred as the primary task is highly visual; visual attention must not be diverged or overloaded (Woodson, 1981; Sanders and McCormick, 1993; ISO 7731, 2003; Noyes et al., 2006; Peryer et al., 2010). The choice for an auditory warning for ORBIT was confirmed in preliminary tests with drivers, although some of them would prefer a combination of auditory and visual warnings. For practical reasons this combination was not further tested. Obviously, auditory warnings should be distinctive from existing warnings, discernable above environmental sounds, and fit with the "message" (Lees and Lee, 2007).

Introducing a non-fail-safe system in a fail-safe train protection environment raised questions about required reliability and the amount of warnings a train driver should encounter. Required reliability of ORBIT in terms of true and false alarms and misses could not easily be determined from the existing literature. First, terminology is quite diffuse: alarm and alert are not well defined. For example, EEMUA (2013) states:

- Alarm: Audible or visible means of indicating to the operator about an equipment or process malfunction or abnormal condition.
- Alert: Is used to attract the attention of the operator to changes that may require assessment or action when time allows.

However, in many studies (e.g., Wickens et al., 2009), alarm and alert are used as synonyms which makes it difficult to select required reliability specification just on the priority of the warning. Second, according to EEMUA (Engineering Equipment and Materials Users Association) definitions ORBIT would qualify as an alarm by elimination: the consequence of ignoring or missing the warning can be very serious. Still, ORBIT would be of lower priority than warnings from the ATB train protection system. It has been suggested that an automated attention guidance system (which ORBIT in essence is) will generally assist human performance so long as the reliability is above about 0.80 (Wickens and McCarley, 2008). A "naturalistic" study of Wickens et al. (2009) however showed that air traffic controllers accepted a 47% rate of false alerts of an automated conflict alerts system. The authors concluded that this relatively high rate was accepted because false alerts were considered "forgivable" as a result of a conservative detection threshold, resulting in more false alerts and less misses. Also, the alert did not distract controller's attention from their primary task as they are both addressing the same high priority task, and the alert was not using sound which would be more intrusive and possibly annoying when false.

During ORBIT development three types of false alerts were identified: (1) inadvertent/overdue alerts, when in fact the red signal aspect has changed to yellow or green, but that change has not reached the train because of long propagation times in TMS or GSM-R systems, (2) unnecessary alerts, when the train driver is in full control of the braking movement but an alert is produced anyway, and (3) random alerts, when a warning sounds while no stop signal is being approached, possibly because of incorrect GPS positioning of train or GPS coordinates of the signal. The first two types may be acceptable to drivers, as long as they understand what triggered ORBIT in the actual situation. However, too many of these alerts may cause nuisance. Random alerts are not comprehensible and may cause shock reactions, and undermine trust in the system (Lees and Lee, 2007). Eventually, high rates of false alerts may cause ignoring ORBIT warnings at all. Safety is not at stake directly, because ORBIT does not impede perception of outside signals, but a "cry wolf" effect (slow or no reaction to warnings) may be expected (Wickens and McCarley, 2008). The efficacy of ORBIT will then become low.

Very high reliability of ORBIT may cause overreliance issues: the train driver may be tempted to wait for the ORBIT warning before starting to brake. Anecdotally, some drivers rely on the ATB system (elsewhere also reported by Naweed et al., 2015), and with more recent train protection systems like ETCS, drivers are taught to follow indicated speed-brake profiles. This behavior is unwanted and hazardous with ORBIT, because ORBIT is not fail-safe and only in effect with controlled signals. In this perspective, a certain rate and type of false alerts may not be detrimental to calibrate trust (Parasuraman et al., 2000). Also, this calls for an accurate introduction of ORBIT to drivers to make them understand the nature of choices made in development of this imperfect diagnostic automation (Wickens and McCarley, 2008; Hoff and Bashir, 2015). Finally, there are indications that people tend to reduce their reliance on automation when greater risk is involved (Hoff and Bashir, 2015).

11.1.4 Design of ORBIT

Originally, the operational concept of ORBIT only consisted of an alarm to be given when it was certain that a red signal aspect was approached at too high speed, not allowing for false alerts. Consequently, the expected frequency of a warning would be approximately equal to the average frequency of SPADs, which is about once per 10–15 year per driver, and bears the risk to be too infrequent to be recognized by the driver (Noyes et al., 2006). Also, in this operational concept it could not be guaranteed that the train would come to a halt before the stop signal because of possible propagation time delays. Therefore, bearing in mind the discussion above about false alerts, it was decided to incorporate uncertainty in the nature of the warning in accordance with the concept of likelihood alarms (Wickens and McCarley, 2008). We introduced a pre-alarm warning ("attention" warning), which is meant to direct attention of the driver to the next—supposedly red—signal aspect. The attention warning is triggered a fixed number of seconds before the alarm curve would be crossed; in fact an attention curve is defined in addition to the alarm curve (see Figure 11.3). This will cause the warning to be triggered more frequently to improve driver experience with ORBIT, and allowing the driver some extra time to assess the situation and act appropriately. The algorithm for both curves takes into account whether the driver is already braking, by delaying the warning with the time it takes for the brakes to couple.

In addition to these curves an algorithm based on positive acceleration and speed within a distance of 20 m of the red signal is used to warn at unauthorized departure.

In this research we focus on alert salience by determining specifications for the (auditory) ORBIT warning, and on the balance between reliability and number of (true and false) alerts taking into account risks of compliance and reliance. The human factors part

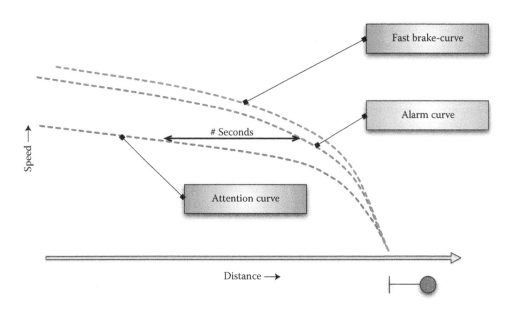

FIGURE 11.3
Schematic display of ORBIT-warning triggering curves (*x*-axis: distance to red signal; *y*-axis: train speed).

of all research done in the development of ORBIT consisted of the following parts, following an initiating HAZID (hazard identification) session with stakeholders and a literature review (see Schreibers et al., 2014):

1. Simulation I: Explorative of nature to gain driver acceptance and fine-tuning of parameters
2. Simulation II: Validating driver acceptance and system efficacy
3. Field test: Validating system efficacy and driver acceptance during 20 days in regular service on The Hague–Venlo route

11.2 Simulation I

11.2.1 Method

Three experienced passenger train drivers from one company were involved in 5 half-day sessions using the ProRail MATRICS simulator extended with an ORBIT simulation module. This simulator consists of two laptop computers, one of which is controlled by the test leader, the other connected with a RailDriver™ console for actuation of brake, traction, doors, and deadman placed on a driver's console. The simulated outside view was projected with a beamer (see Figure 11.4). One route (Breukelen-Geldermalsen) of about 45 min was used for evaluation of several ORBIT settings. The test route considered a stop train service in a narrow timetable, in order to meet as many red signals as possible, including crossing a complex yard (often prone to SPADs). The main goal of simulation I was to narrow down the number of experimental conditions for

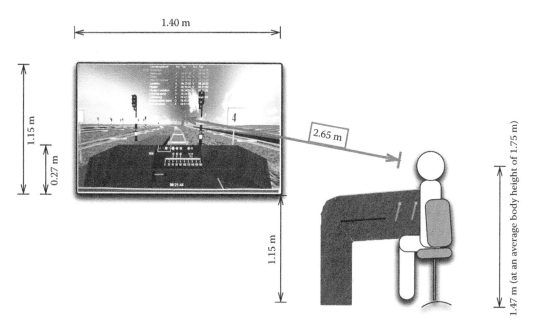

FIGURE 11.4
Simulator setup.

simulation II, and to improve the simulation scenarios. The following conditions were tested:

- Auditory warning: Tone (attention: single tone 700 Hz, 0.2 s, 1/1 s; alarm: three-tone: 700/930/700 Hz, 0.7 s, 1/0.85 s) or a woman's voice (attention: "Attention, SAD;" alarm: "Brake now! SAD!," translated from Dutch).* In accordance with Noyes et al. (2006), a female voice was chosen because it is more unusual than a male's in a train cab environment and this would result in it attracting more interest and attention. Also, the alarm text was spoken with an emotion of urgency to fit the situation. Drivers were instructed to check the signal aspect first when an *attention* warning was given, and then act/brake if required, and to brake immediately when an *alarm* warning was given and check the signal aspect subsequently. Volume of all ORBIT warnings was about 10 dB(A) higher than simulated ambient noise of 60 dB(A).
- Distance between alarm curve and attention curve: Adjustable between 3 and 9 s (see Figure 11.3).

Drivers were allowed to be present during other drivers' tests to get maximum experience with all test conditions. Train drivers are used to the presence of colleagues in the cab for, for example, educational purposes. In order to distract drivers, as often happens during SPADs, a secondary visual and cognitive task was added. Within the primary field of view every 15 s a new sum was to be solved (±7 or 13 to/from a two-digit

---

* In Dutch the technical/professional term "STS" is used, which is an abbreviation of "Stop Tonend Sein." Translation of this is "signal at danger" (SAD).

number between 30 and 70). Evaluation was done by questionnaires and debriefing semi-structured interviews.

11.2.2 Results

All three drivers assessed the simulator as being sufficiently real to life for this purpose. They were all positive about the concept of ORBIT (8–9 on 10-point scale).

The drivers had a slight preference for the spoken warnings compared to the tone warnings, mainly because the tone had to compete with about 10 other tones in the train cab, and the voice directly gives meaning to origin and required action. The fact that a spoken warning may also be audible and comprehensible to passengers in some cabs was mentioned as a disadvantage.

Contrary to the instruction it was observed that all drivers actuated the brake following an attention warning instead of checking the signal aspect first. They declared that this was their natural reaction to auditory warnings in general. Furthermore, the attention warning was too long in their experience: they suggested that "Attention: SAD" could easily be replaced by "SAD," as the auditory warning is an attention in itself.

Because of the urgent character of the warnings, the drivers prefer the ORBIT warnings to be sparse, leaving just enough time to react appropriately. In the simulation this meant that the shorter conditions between alarm and attention curve (3–5 s) were preferred to longer distances (6–9 s). It was concluded that the shorter distances were suited for further testing and validation. As the preference for voice warnings did encounter some practical dilemmas, it was decided to use both speech and nonspeech warnings in the validating simulation.

Some adjustments were done in the simulator due to technical instability of the software, like a shortened test track. The scenario was adapted to better fit actual driving along this track.

11.3 Simulation II

11.3.1 Method

Due to practical constraints—a limited amount of train drivers with limited time available—a proper counterbalanced design was not feasible. Fourteen experienced passenger train drivers (average 52 years old, 21 years of train driving experience, all male) from four different companies were involved in half-day tests in groups of three (one group of two). All drivers were allowed a 15-min test drive on a part of the experimental route to get accustomed to the MATRICS simulator. After that, each train driver drove Breukelen–Houten Castellum (30 min). They could witness the colleague drivers from their test group drive during their practice and test runs to get maximal experience during the half-day. Drivers were not allowed to talk to each other until the final debriefing interview. Each driver drove the same experimental scenario once, but with different settings. These settings were:

- Distance between alarm and attention curve 3 s + spoken warnings
- Distance between alarm and attention curve 4 s + spoken warnings
- Distance between alarm and attention curve 5 s + tone warnings

For practical reasons (time constraints) tone was coupled with the longer distance between alarm and attention curves, because the longer distance potentially triggers more ORBIT warnings and voice was argued to be too intrusive for frequent alerting (Noyes et al., 2006). These settings were not communicated to the drivers in advance. The spoken and tone warnings were demonstrated during briefing before the test runs. The last signal on the route was set as a technical fault causing the signal to suddenly show the red aspect during approach. An ORBIT warning was thus inevitable for the driver. This was to make sure that every driver was confronted with an ORBIT warning at least once. The secondary task from simulation I was used to distract vision and load cognition.

ORBIT loggings were analyzed for the number and nature of ORBIT warnings, and reaction times (time between start of ORBIT warning and brake application) with a sample frequency of 1/300 ms. Unfortunately, it appeared technically impossible to simulate system propagation delays. So in fact, simulation tests were done with an "ideal" ORBIT system in terms of GPS accuracy, almost zero propagation delay, etc.

11.3.2 Results

Due to the small number of test persons and the fact that ORBIT only triggers a warning in exceptional cases, it is hard to make quantitative statistical analyses. In Table 11.1, the number of ORBIT warning per experimental condition is shown. Figure 11.5 shows the distribution of reaction times to ORBIT warnings—defined as the time difference between warning occurrence and brake actuation. As loggings were taken at 300 ms intervals, the distribution is shown in categories of 300 ms: median 1.2 s and mode 0.9 s. Adding an average 1.5 s of brake coupling time (for passenger trains with electromagnetic brakes), this would mean that average 3 s and maximum 4 s would be present between the ORBIT warning and start of train deceleration.

Scores for the secondary task did not differ. From Table 11.1, it might be concluded that the 3 s condition causes less warnings. However, from observations and from the literature, we argue that individual driving styles are of major influence. As all drivers were able to witness the other drivers in their group (with other conditions), responses to questionnaires seem to be more valuable. From these questionnaires seven drivers preferred the 3 s condition, four drivers the 4 s condition, and one driver the 5 s condition. Two drivers did not notice the difference. Written clarifications were quite consistent with simulation I: ORBIT should be (1) not disturbing during "controlled" braking and (2) an ultimate warning leaving just enough time to act. Supporters of the 4 s condition thought the 3 s condition too "tight," and preferred a second of extra time to act.

All 14 drivers preferred the voice warning above the tone warning, and 13 were in favor of nationwide implementation of ORBIT. The one driver who voted against this thought

TABLE 11.1

Mean (Min–Max) Number of Attention and Alarm
Warnings per Experimental Condition

3 s ($n = 5$)		4 s ($n = 5$)		5 s ($n = 5$)	
Attention	Alarm	Attention	Alarm	Attention	Alarm
1 (0–3)	0 (0–1)	4 (3–6)	0 (0–1)	3 (0–7)	0 (0–0)

Note: To obtain five observations per condition one driver performed two test drives; the inevitable ORBIT warning at the last—red—signal was not counted.

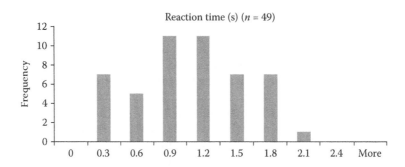

FIGURE 11.5
Frequency distribution of reaction times to ORBIT warnings.

ORBIT superfluous: a well-trained driver should not need this assistance/warning device, and the non-fail-safe nature of the device was not appreciated. Average overall appreciation of ORBIT was 8.2 on a 10-point scale (differing from 7 (3x), 8 (6x), 9 (4x) to 10 (1x)).

The vast majority thought volume, distinctiveness, and appropriateness of both attention and alarm (spoken and tone) were good. So the design of the warnings cannot explain preference for a timing condition. It was concluded that the 3 or 4 s distance between alarm and attention curve in combination with the voice warning was best to be implemented in the field test. Taking into account results from reaction time (see above), the 4 s distance was chosen for the field test.

11.4 Field Test

11.4.1 Method

Ten intercity trains (type ICR) were equipped with ORBIT onboard units in one cabin (BDS). During 20 days in regular service on The Hague–Venlo route, ORBIT loggings were analyzed. Loggings contained information about ORBIT unit number, train number, next red signal guarded by ORBIT, distance to that signal, speed, acceleration, type of ORBIT warning (attention/alarm), and type of ORBIT algorithm that triggered the warning (braking/non-braking/departure). In real time, ORBIT was activated by a remote test facilitator when an instructed driver was present in the ORBIT equipped cabin. Drivers from home stations along the route were instructed about the presence and working principles of ORBIT. The instruction was done by information flyers, information on their handheld device, and personal instruction by their manager (who were instructed by our team).

Every ORBIT warning was followed up by a structured evaluation interview by telephone. Human factors specialists analyzed all objective loggings information and interview results, and classified ORBIT warnings to be necessary, inadvertent/overdue, unnecessary or random/false.

11.4.2 Results

During the test period, 218 valid train journeys on (parts of) this route and 5800 red signal approaches guarded by ORBIT were recorded. Analysis of loggings showed 78 ORBIT

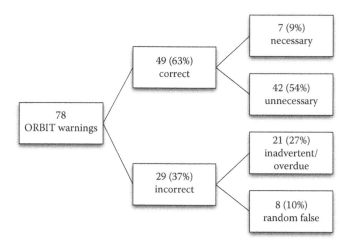

FIGURE 11.6
Classification of ORBIT warnings during the field test.

warnings were triggered (see Figure 11.6). About 70% were only attention warnings and 30% attention followed by alarm warnings. Further analysis of loggings showed that 29 out of 79 warnings appeared (technically) incorrect. Eight of those (10%) seemed to be associated with drift in GPS positioning due to dense and high buildings along the track or a large station roof (mainly Rotterdam). In 21 other incorrect warnings (27%), the warning was inadvertent/overdue: 12 took place due to an error in TMSs (not safety-related) and the rest was caused by excessive signal propagation times (often more than 60 s, where 2 s is a nominal value) most probably caused by GPS fix or GSM-R connection problems in and after leaving a tunnel. Although hard to establish, no automation misses could be identified.

Out of 49 technical correct warnings 42 (54% of total) were evaluated by both train drivers and human factors experts as unnecessary: the speed profile showed that the train was in control. Often, when nearing a standstill, drivers release the brake which may cause a temporary positive acceleration. If this happens within 20 m of the red signal the "departure algorithm" causes an ORBIT warning. Most of these ORBIT warnings occurred at one signal; post hoc analysis revealed that its GPS coordinates had not been updated after the signal was moved about 5 m during infrastructural works. This stresses the importance of infrastructural data management.

Seven ORBIT warnings (9% of total) were considered necessary by the human factors experts, based on visual comparison of the corresponding speed profiles to those of other trains that did not generate a warning. The braking was at least uncomfortable, although some drivers declared that they had braked in a "normal" way and the ORBIT warning was in their eyes unnecessary. In one case, in which an attention warning was followed by an alarm, the train driver admitted ORBIT was useful urging him to brake in time, and thus preventing a potential SPAD.

The subjective evaluations with 37 drivers—equaling 57 ORBIT warnings—resulted in 28 completed interview forms. A few drivers experienced ORBIT more than once. For nine drivers only summarized interview results were delivered by their manager, which were not suited for further analysis. We did not succeed in obtaining information from the train drivers regarding 21 other ORBIT warnings.

Some train drivers could not recall an ORBIT warning on the trains. Most probably the speakers of two ORBIT units were defective during (part of) the field test. In other cases about half of the drivers thought the sound volume was too low. A sample verification measurement in two trains confirmed this feedback: the attention warning had about the same volume as the ambient driving noise at 40 km/h (~55–60 dB(A)), which was 10–15 dB(A) less than specified. These results also indicated the need for an integrated test of speaker functionality in the train's onboard unit.

Despite incorrect or unnecessary warnings or low volumes, 80% of the interviewees support the nationwide implementation of ORBIT. The average appreciation figure is 7 on a 10-point scale, ranging from 4 to 9. Three drivers scored lower than 6, two of them arguing that a well-trained driver should not need a warning device, and one driver was really shocked by a technically incorrect warning. All train drivers judged incorrect warnings as not acceptable. The unnecessary signals were however acceptable to most drivers as they understood that ORBIT could generate a warning in that specific situation. The timeliness and duration of the warnings were considered adequate on average.

11.5 Discussion

The concept of ORBIT is a promising measure to reduce the number of SPADs by capturing and guiding the train driver's attention, and improving his situation awareness. Also, as expectations about set routes may drive an incorrect situation awareness, ORBIT may help to correct the driver's mental picture of the situation. This is important because expectations are a major contributing factor to SPADs in the Netherlands (ILT, 2015).

However, during the field test some technical issues arose that generated incorrect ORBIT warnings. These warnings—where drivers cannot relate the warning to a potentially hazardous situation—are not acceptable because they can be detrimental to automation (ORBIT) trust (Wickens et al., 2015). After the field test, technical measures were taken to overcome errors and latencies originating from the TMS. Moreover, at locations with unreliable GPS or GSM-R signals, it was decided to exclude signals from ORBIT coverage. By implementing other safety systems at those locations, the risk of SPADs is to be mitigated.

The relatively high percentage of unnecessary warnings can be overcome by changing some parameters for determining whether a train is braking or accelerating. Post hoc analysis shows that over 85% of the unnecessary warnings would then have disappeared without missing one of the necessary warnings. Still, in accordance with other sources (e.g., Wickens and McCarley, 2008), it was confirmed that "unnecessary" warnings can be acceptable to a certain extent: unnecessary warnings can be interpreted as an understandable or forgivable form of false alerts. They confirm presence of the warning system and can serve to reinforce the driver's assessment of the outside environment.

Contrary to the ATC conflict alert system (Wickens et al., 2009), ORBIT uses auditory warnings. Wickens et al. suggest that sound is intrusive and thus potentially annoying when false. Our results suggest that sound is in the favor of and preferred by train drivers because the auditory channel is relatively free compared to the primary visual channel. A visual warning in the train driver's console would redirect the driver's outside visual focus, which we consider undesirable (see also Noyes et al., 2006). An alternative would be to project the visual warning as a heads up display. This solution was not considered from a cost/feasibility perspective. It might have been an adequate alternative or add-on although a potential "competition" between the visual warning and the lineside signal must be taken into account. Furthermore it must be noted that the auditory pre-alarm attention warning in ORBIT is distinctive but "gentle."

The preference of drivers for a warning that is infrequent (not many warnings which are relatively reliable, but leave little time to react appropriately) in combination with a preference for a spoken warning is in accordance with theory (e.g., Noyes et al., 2006). It certainly prevents issues that were identified with extended AWS (automatic warning system) in the United Kingdom where the same warnings for different states, ambiguous warnings, and warnings regarding a variety of causes were applied (McLeod et al., 2005). The spoken warning directly reveals to the driver which system triggered the warning, what the identified hazard is ("SAD"), and how to act (alarm: "SAD! Brake now!").

It was calculated that—when the proposed measures on technology and algorithms are taken and effective—a train driver would be confronted with an ORBIT warning once every 2 weeks on average. For "conservative" drivers this frequency may well be (much) lower.

Still two risks of ignoring or overreliance on the ORBIT warning were identified. By implementing the mentioned measures the number of incorrect and unnecessary signals are estimated to drop to 0% and 10%, respectively. This is expected to mitigate the risk of ignoring the warning. Obviously, proof of efficacy of the measures has still to be delivered. Further risks of overreliance will be mitigated by a thorough introduction of ORBIT to train drivers, emphasizing the not fail-safe and imperfect nature of ORBIT. Besides, it is believed that unnecessary warnings—in the expected low rate—contribute to the understanding of the imperfect nature. In addition, as automatic signals (about 50% of all signals) are not guarded by ORBIT, it would require a deliberate decision by the train driver to allow himself to rely on ORBIT in guarded areas. This seems not very likely to happen. Finally, it is suggested by Hoff and Bashir (2015) that under high-risk conditions, operators may have a tendency to reduce their reliance on complex automation. However, the risk cannot be excluded beforehand. By monitoring the number of ORBIT warnings and the average distance (in seconds) to the attention and alarm curves, it is expected to be able to identify drifts in driving behavior at an early stage in future. The monitoring system will also be used to obtain real-time information about ORBIT warnings. This will help to identify faults like incorrect GPS coordinates or (temporary) GPS signal quality problems, to allow for immediate action to be taken.

Reactions of train drivers to the attention warning were not compliant to instruction: they braked immediately instead of checking the signal aspect first. Although drivers indicated that the nature of the attention warning ("SAD," spoken in a neutral tone of voice) fitted the original instruction, they naturally actuated the brake handle. In respect of their preference for a relatively late warning, it was decided to change the instruction to allow for natural braking behavior: brake and check signal aspect. For calculation of efficacy of the ORBIT system, this is advantageous, because estimated driver reaction times could be lowered. Theoretically, with these measures and recalculations into effect 57% of all SPADs in the Netherlands can be prevented and in 63% the danger point will not be reached. These figures need to be evaluated in practice, but a reduction of several tens of percent is realistic.

Recently, we started studying the effectiveness of ORBIT with freight trains. These trains are less predictable in their braking capabilities and behavior, because of varying composition, load, and brake settings. Thus the rolling stock specific ORBIT attention and alarm curves cannot be implemented. It seems feasible to use a universal attention and alarm curve for freight trains, from heavy coal trains to locomotives without wagons. Obviously, as a locomotive has better braking capacities, speed during a stop signal approach may be higher, and it may encounter an ORBIT warning more frequently. There are two reasons why this may not be a problem: there are operational rules into effect that prohibit relatively high speeds even with locomotives, and because drivers operate freight trains with different characteristics it may not be detrimental to encounter an (unnecessary) ORBIT warning once in a while, as argued above.

Finally, special procedures like passing a signal at danger with permission of the train signaler—which will cause ORBIT to trigger warnings—will be a subject of study. Moreover, most probably ORBIT will be extended with static stop signs, like stop boards on shunting yards and buffer stops.

References

Dutch Safety Board. 2012. Train collision Amsterdam Westerpark. http://onderzoeksraad.nl/uploads/items-docs/1841/Rapport_Westerpark_EN_web.pdf

EEMUA. 2013. *191 Alarm Systems—A Guide to Design, Management and Procurement.* London: The Engineering Equipment and Materials Users' Association.

Hoff, K.A. and M. Bashir. 2015. Trust in automation: Integrating empirical evidence on factors that influence trust. *Human Factors* 57(3):407–434.

ILT. 2015. *SPAD Analysis 2014 (In Dutch: STS-passages 2014).* The Hague: Human Environment and Transport Inspectorate (ILT).

ISO 7731. 2003. *Ergonomics. Danger Signals for Public and Work Areas. Auditory Danger Signals.* Geneva: International Organization for Standardization.

Lees, M.N. and J.D. Lee. 2007. The influence of distraction and driving context on driver response to imperfect collision warning systems. *Ergonomics* 50(8):1264–1286.

McLeod, R.W., G.H. Walker, and A. Mills. 2005. Assessing the human factors risks in extending the use of AWS. In J.R. Wilson, B. Norris, T. Clarke, and A. Mills (Eds.), *Rail Human Factors: Supporting the Integrated Railway* (pp. 109–119). Aldershot: Ashgate Publishing.

Naweed, A., S. Rainbird, and J. Chapman. 2015. Investigating the formal countermeasures and informal strategies used to mitigate SPAD risk in train driving. *Ergonomics* 58(6):883–896.

Noyes, J.M., E. Hellier, and J. Edworthy. 2006. Speech warnings: A review. *Theoretical Issues in Ergonomics Science* 7(6):551–571.

Parasuraman, R., T.B. Sheridan, and C.D. Wickens. 2000. A model for types and levels of human interaction with automation. *IEEE Transactions on Systems, Man, and Cybernetics—Part A: Systems and Humans* 30(3):286–297.

Peryer, G., J.M. Noyes, C.W. Pleydell-Pearce, and N.J.A. Lieven. 2010. Auditory alerts: An intelligent alert presentation system for civil aircraft. *International Journal of Aviation Psychology* 20(2):183–196.

Sanders, M.S. and E.J. McCormick. 1993. *Human Factors in Engineering and Design.* New York: McGraw-Hill.

Schreibers, K.B.J., R. van der Weide, and C.E. Weeda. 2014. Human factors specialists to the rescue for (other) systems designers—Case study in railway. *Paper Presented at the XI Symposium on Human Factors in Organizational Design and Management,* Copenhagen, pp. 65–70.

Wickens, C.D., B.A. Clegg, A.Z. Vieane, and A.L. Sebok. 2015. Complacency and automation bias in the use of imperfect automation. *Human Factors* 57(5):728–739.

Wickens, C.D. and J.S. McCarley. 2008. *Applied Attention Theory.* Boca Raton, FL: CRC Press.

Wickens, C.D., S. Rice, D. Keller, S. Hutchins, J. Hughes, and K. Clayton. 2009. False alerts in air traffic control conflict alerting system: Is there a "cry wolf" effect? *Human Factors* 51(4):446–462.

Woodson, W.E. 1981. *Human Factors Design Handbook.* New York: McGraw-Hill.

12

Evaluating Design Hypotheses for Pedestrian Behavior at Rail Level Crossings

Gemma J. M. Read, Paul M. Salmon, Michael G. Lenné, and Elizabeth M. Grey

CONTENTS

12.1 Introduction

Collisions at rail level crossings (RLXs) involving pedestrians represent a significant public safety concern in Australia and internationally. The most recent statistics available show that between 2002 and 2011, 92 pedestrians were struck by trains at RLXs in Australia (Australian Transport Safety Bureau, 2012). In the United Kingdom, over a similar time frame, 72 pedestrians were killed at level crossings (Rail Safety and Standards Board, 2015). The European Railway Agency has reported 373 fatalities associated with collisions at RLXs in Europe in 2012 alone, with approximately 40% of those killed being pedestrians (European Railway Agency, 2014). It has been noted that Australia and the United States have achieved reductions in the numbers of motor vehicle–train collisions, but not in pedestrian–train collisions (e.g., Australian Transport Safety Bureau, 2012; Metaxatos and Sriraj, 2013). To make gains in improving pedestrian safety at RLXs, a new approach is required. Such an approach recognizes that RLXs are complex sociotechnical systems. Taking this perspective, safety at RLXs is the outcome of interactions between social and technical components such as road users, vehicles (road and rail), equipment, and infrastructure. The interactions can be diverse and random, particularly due to the openness of the system with no barriers to system entry in place for many road users including pedestrians and cyclists (i.e., there is no licensing, training, or significant supervision of these users).

To understand and improve the performance of complex systems, such as RLXs, the application of systems analysis and design methods is required. Cognitive work analysis (CWA) is a framework of methods that can be used to understand complex systems (Vicente, 1999). In contrast to most other tools for understanding human behavior which specify how behavior should be (normative approaches) or how behavior is (descriptive approaches), CWA takes a formative approach by specifying the constraints of the domain within which behavior can occur. In this study, the design of RLXs will be examined using the first phase of CWA, work domain analysis (WDA).

While safety education and enforcement of rules is often promoted as a means to improve safety at RLXs, changes to improve user interaction with the physical design of the crossing, through a consideration of the affordances available to users and the constraints imposed on behavior, have the potential to provide more effective safety gains in this context. The WDA phase is well placed to explore the affordances at RLXs.

Woods (1998) explains that designs represent designers' hypotheses about the relationship between technology and human cognition. The hypotheses underlying the design of RLXs are based on a normative approach. That is, the current designs are based upon a series of predefined tasks required for pedestrians to cross safely. For example, at RLXs where warnings are provided, pedestrians are expected to search for and detect warnings, to stop in a particular place when warnings are detected, to wait until warnings are deactivated, and then to complete crossing. These norms are reinforced by legislation regulations which prohibit certain behaviors and are enforced through the application of sanctions.

This chapter will focus on identifying the affordances and constraints that enable and influence pedestrian behavior at RLXs. We also identify the hypotheses underpinning current RLX (rail level crossing) design and use naturalistic observations of pedestrians to gain insight into the extent to which the hypotheses are supported by actual user behavior. We begin by identifying affordances and constraints using the WDA phase of CWA.

12.2 The RLX Domain

In Melbourne, Australia, where this work was focused, RLXs are designed to operate in one of three ways. The first type of design which is common in urban environments provides warnings that a train is approaching (e.g., bells, flashing lights) as well as physical barriers (automatic pedestrian gate) to stop pedestrians from entering the track area when a train is approaching. The second type provides warnings but no barrier (the warnings are provided for an adjacent road RLX but no pedestrian gate is provided) and the third type provides signage to inform the pedestrian that a crossing is present but relies on users to identify whether a train is approaching and to make the decision to stop or go accordingly. The former type of design incorporating the pedestrian gate is generally considered the most effective in minimizing collisions, at least for road vehicles (e.g., Wigglesworth and Uber, 1991). However, even with the widespread use of physical barriers, collisions still occur. Photographs of example RLXs with pedestrian gates are provided in Figure 12.1.

FIGURE 12.1
Examples of RLX sites in Melbourne, Australia.

12.2.1 WDA Representation

To better understand the RLX system, an abstraction hierarchy from the WDA phase of CWA was developed (Figure 12.2). An abstraction hierarchy provides a functional view of a sociotechnical system, encompassing five levels of abstraction, with means-ends links between nodes at adjacent levels. It describes the constraints of the work domain within which behavior is possible. The representation identifies the physical resources available within the system (e.g., flashing lights), the processes afforded by

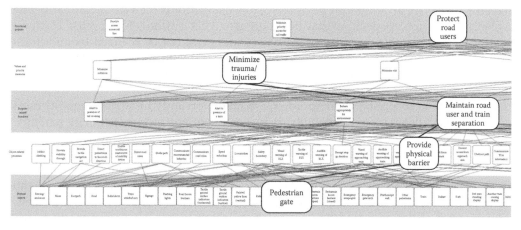

FIGURE 12.2
Extract from WDA.

those resources (e.g., provide visual warning of approaching train), the functions supported by the processes (e.g., alert road user to presence of train), the values and priorities that are measured and monitored within the system (e.g., minimize collisions), and finally, the overall purposes of the goal-directed work domain (e.g., to protect road and rail users).

The abstraction hierarchy presented considers RLXs from the perspective of pedestrian behavior. However, the analysis also includes aspects of the system that may be designed for other road users (i.e., motorists) such as boom barriers, which can also influence pedestrian behavior, even if this is not an intended outcome of their installation.

Within Figure 12.2, one chain of means-ends links showing relationships within the abstraction hierarchy is highlighted. Starting from the bottom level of the hierarchy, the physical object highlighted is the *pedestrian gate*. This object provides a *physical barrier*, which supports the system function of *maintain road user and train separation*. The ability of the system to achieve this function can be measured through its success in *minimizing trauma and injuries*, which relates to a functional purpose of the system, to *protect road users*.

The relationships in the abstraction hierarchy were identified through documentation review (i.e., engineering standards and legislation relating to pedestrian behavior at RLXs) as well as input from subject matter experts. Thus, the relationships documented are those intended by system designers.

12.2.2 Affordances and Constraints at RLXs

Of particular interest to the present discussion is the means-end link between the bottom two levels of the abstraction hierarchy (the physical objects and object-related processes). A number of authors describing the object-related processes level of abstraction suggest that it identifies the affordances of the physical objects present in the work domain (e.g., Naikar et al., 2005; Jenkins et al., 2008; Stanton and Bessell, 2014). It could be interpreted, in applying WDA in an intentional domain at least, that object-related processes can be described as what an object in the work domain offers to actors "… what it provides or furnishes, either for good or ill" (Gibson, 1979, p. 127). Gibson also discussed affordances as properties of the environment that provide goal-relevant opportunities for action by an actor.

It should be noted that affordances are actor-dependent meaning that what can be done depends on the actor's (i.e., pedestrian's) capabilities such as their height or strength. What an object affords for an adult may be different to what it affords for a child. Similarly, what an object affords for a pedestrian may be different to what it affords for a cyclist. However, an affordance is a stable property of the object, and need not be realized by an actor to exist. An affordance exists independently of the actor and their motivation at that point in time. It is also important to note that affordances do not cause behavior; they just provide a means for it to occur (Withagen et al., 2012).

Gibson's (1979) conception of an affordance as an opportunity for action suggests that it supports human behavior and flexibility. However, in our abstraction hierarchy, a number of physical objects were present within the system specifically to constrain human behavior (i.e., to obstruct a person's path through the environment) and thus do not easily fit the definition of an affordance as providing an "opportunity for action." For example, an object-related process associated with the physical object of a pedestrian gate was identified as providing a physical barrier. These relationships in the abstraction hierarchy reflect the normative philosophy underpinning design in this domain which aims to restrict pedestrian behavior so as to enforce the task of stopping at the crossing when a train is approaching.

In developing the WDA representation, it was identified that the intended effect of a number of the physical objects within the RLX domain is to impose restrictions on the behavior of pedestrians. This may reflect the unique nature of the RLX domain, which trades off the goals of road users (i.e., to traverse the RLX efficiency) with safety purposes of the overall domain (i.e., to protect road and rail users and to maximize priority access over the RLX for the train). Thus, in relation to the human actors in the system, this provided a constraint rather than an affordance (removing an opportunity for action rather than providing one). Given this finding, which may be limited to certain types of domains where there are conflicting goals and purposes, we propose categories of object-related processes that could be used to provide further insight into how behavior may be influenced by the physical objects available to actors in the work domain.

The categories of object-related processes have drawn upon the work of Don Norman (1988, 1999), who discussed the role of affordances and constraints in design. A set of proposed categories and examples from the WDA (relationship between the physical objects and object-related processes) are outlined in Table 12.1. It should be noted that multiple categories may be associated with a single physical object and that multiple affordances and constraints may interact to shape behavior; it suggested that it is useful to consider how these different processes are more or less present in a design.

Some types of affordances are likely to be stronger in influencing behavior than others. For example, a physical constraint will have a greater influence on behavior than a cultural constraint as it forces behavior (Norman, 1999). Therefore, such categorizations of object-related processes in intentional systems can provide insights into the flexibility for action

TABLE 12.1

Categorization of Object-Related Processes

Category	Description	RLX Examples
Affordance	A property of the world that is actionable by an actor; an opportunity for action. Includes both affordances that are readily perceived and those that are available even if they are not perceived by an actor/s	• The footpath surface affords locomotion • The pedestrian gate can afford jumping/leaping over • The emergency exit gate affords pulling open to gain access to the RLX when the pedestrian gate is closed
Physical constraint	A barrier or obstacle that constrains behavior	• The fencing obstructs locomotion and restricts the pedestrian's path • The pedestrian maze restricts the pedestrian's path • The pedestrian gate obstructs locomotion
Cultural constraint	A convention shared by a cultural group that limits actions/behavior	• Most pedestrians would not open an emergency exit gate to enter the track while the warnings are operating, even though this is possible in most locations • Pedestrians tend to avoid coming into close proximity with one another and have a culturally influenced amount of physical space that is kept from others
Logical constraint	A constraint that involves the use of reasoning to determine the available options	• Pedestrians walk forwards through the crossing starting at one gate, following a path and exiting at another gate
Semantic constraint	A constraint that suggests its meaning through the situation or context—relying on our knowledge of the situation and of the world	• The train whistle/horn provides an audible warning of an approaching train • The bells provide an audible warning of an approaching train

available within the domain (based on the number of affordance relationships identified) as well as the degree and strength of constraints on behavior imposed by physical objects in the work domain. Reviewing the means-ends links in complete version of the WDA (of which an extract is provided in Figure 12.2), it is clear that there are many more constraints on behavior imposed by the physical objects at RLXs than there are affordances in the form of opportunities for actions.

12.2.3 Design Hypotheses

Designers are comparable to experimenters. During the design process, they develop hypotheses (explicitly or implicitly) about what the impact of the designed object will be on human behavior and the system. These hypotheses are explored and evaluated during the design process. Once implemented, however, the designed objects become part of the system and change the system which itself is dynamic and evolves over time in response to various pressures. Accordingly, it is valuable to periodically reevaluate the extent to which the hypothesized effect remains valid over time, particularly in complex systems such as RLXs.

The hypotheses underlying the design of infrastructure and warnings intending to influence pedestrian behavior at RLXs were identified based on the affordances and constraints defined in the abstraction hierarchy as well as a further review of the road rules and design standards and documentation. The standards reviewed included the Australian Standard for Traffic Control Devices at Railway Level Crossings (AS1742.7) and the Victorian Rail Industry Group Criteria for Infrastructure at Railway Level Crossings (VRIOGS 003-2-2006). The legislation outlining the offences applicable to behavior at RLXs included the *Road Safety Road Rules 2009* (Vic) and the *Transport (Conduct) Regulations 2005* (Vic).

Design hypotheses for each physical object were determined. For example, for the automatic gate (when closed) the design hypothesis was determined to be that: *pedestrians will stop at the gate when it is closing, closed, or opening.* This reflects both the physical constraint that a gate provides (stopping pedestrians when it is closed) as well as the legislated rules for crossing which make it an offence for a pedestrian to cross an RLX if *"a gate, boom or barrier at the crossing is closed or is opening or closing"* (*Road Safety Road Rules 2009* (Vic) Section 235(2)(b)). The hypotheses represent the normative behavior expected by the designers of the technology and the wider system (e.g., the legislature).

12.3 Observations of Pedestrian Behavior

12.3.1 Site Selection

Seven RLX sites in metropolitan Melbourne were selected for the conduct of covert observations of pedestrian behavior. The sites were selected based on the features of the crossing (e.g., infrastructure, equipment, warnings present) as well as incident history. The features of each site are described in Table 12.2. The site selection process ensured that a range of crossing features were represented across the sites including automatic gates, automatic gates with locked emergency gates, pedestrian boom barriers, pedestrian mazes, crossings adjacent to stations, and crossings adjacent to road level crossings (exposing pedestrians to features such as flashing lights and road boom barriers, etc.). Some crossings incorporated tactile ground surface indicators (TGSIs) to define the edges of the pedestrian path (areas

TABLE 12.2

RLX Observation Sites and Times

Site	Features	Incident History (2005–2013)	Observation Day/Times
Site 1	• Automatic gates • TGSIs • Adjacent to train station • Adjacent to road level crossing	• 2 collisions • 54 near misses • Ranked 1 of 20	Friday • 7:00 to 10:00 a.m. • 2:00 to 4:00 p.m.
Site 2	• Automatic gates • Painted yellow lines • Adjacent to train station • Adjacent to road crossing • Independent gate operation	• 3 collisions • 51 near misses • Ranked 2 of 20	Friday • 2:00 to 4:00 p.m.
Site 3	• Automatic gates • TGSIs • Emergency gate latch • Red man standing display • Another train coming display • Adjacent to train station • Adjacent to road level crossing • Independent gate operation	• 1 collision • 20 near misses • Ranked 4 of 20	Thursday • 7:00 to 10:00 a.m. • 2:00 to 4:00 p.m.
Site 4	• Pedestrian boom barriers • Painted yellow lines	• 1 collision • 12 near misses • Ranked 8 of 20	Thursday • 7:00 to 10:00 a.m. • 2:00 to 4:00 p.m.
Site 5	• Pedestrian maze • TGSIs (south crossing) • Painted yellow lines (north crossing) • Adjacent to road level crossing	• No collisions • 10 near misses • Ranked 14 of 20	Friday • 7:00 to 10:00 a.m. • 2:00 to 4:00 p m.
Site 6	• Automatic gates • TGSIs • Adjacent to train station • Adjacent to road level crossing • Independent gate operation	• No collisions • 10 near misses • Ranked 15 of 20	Wednesday • 7:30 a.m. to 12:00 p.m.
Site 7	• Automatic gates • Painted yellow lines • Adjacent to road level crossing	• No collisions • 8 near misses • Ranked 20 of 20	Wednesday • 7:00 to 10:00 a.m. • 2:00 to 4:00 p.m.

of raised studs or bars used to provide a tactile cue to pedestrians with visual impairments). Other crossings had painted yellow lines to define the edges of the path.

Where the site is documented in Table 12.2 as incorporating the feature of independent gate operation, this means that the RLX comprises two independently operating sets of pedestrian gates with one located on each side of an adjacent train station with an island or center platform. This gate design enables pedestrians to access the island platform when a train is approaching from the far track (i.e., a track that they need not cross to reach the train station). Pedestrians who wish to traverse the whole crossing will be able to cross the first track/s, but will then wait in the center of the tracks at a closed gate until the train on the far side has departed. At these locations, which are all adjacent to road crossings with flashing lights, boom barriers, etc., the gate remaining open is the only indication pedestrians receive that the train is not approaching on that track (i.e., the general warnings of bells, flashing lights, road boom barriers, etc. will operate as usual).

The RLX at site 3 has additional countermeasures implemented including a latch on the emergency gate to restrict pedestrians from being able to open the gate from the approach side of the crossing, a red man standing display (similar to a road pedestrian signal, however, instead of showing green it extinguishes when no train is approaching), and an another train coming display (to inform waiting pedestrians that the gates remain closed because another train is approaching). Previous investigations have indicated that the another train coming display and red man standing display may not provide additional benefits in influencing behavior where a locked emergency gate is provided (Warwick, 2009).

As well as identifying the features of each site, Table 12.2 also displays the recent incident history for each site (taken from chart titled *Top 20 Crossings by Pedestrian-only Incidents 2005 to 2013*, provided by G. Sheppard, personal communication, May 10, 2013). All sites were within the top 20 list which are ranked according to the total number of incidents (collisions and near misses).

Observations were held on weekdays and were planned to occur in the mornings and early afternoon to cover peak travel periods. At some locations the planned observations were unable to be undertaken due to operational requirements restricting access to some rail signal boxes and other unforeseen events. The actual observation times are provided in Table 12.2.

12.3.2 Observation Protocol

The observations were conducted in a covert manner to avoid influencing the behavior of crossing users. Observations were undertaken from signal boxes with windows overlooking crossings, or from a vehicle parked close to the crossing. Pedestrians were selected for observation using a convenience sampling method. Due to the unpredictable and sometimes very low flow of pedestrians through the crossing, this strategy was used to ensure as many pedestrians as possible were able to be observed.

The protocol required that the pedestrian should be identified and selected when they were approaching the crossing, but not yet on the crossing. The person was then observed while they crossed and until they exited the crossing and moved away from the area. Where a group of pedestrians were approaching the crossing, one person in the group was selected to be observed, with the effect of other pedestrians on their behavior documented. In addition to pedestrians, cyclists who chose to use designated pedestrian crossing were observed however, these results are not reported here unless the cyclist dismounted before or during traversing the RLX and used the RLX as a pedestrian.

A structured form was used to record the observed behavior. The paper-based form was completed for each pedestrian observed. The following items were recorded:

- Date and time of the observation
- The system state encountered by the pedestrian (e.g., check box for: warnings not activated, warnings activated as the user approached, warnings activated as traversing crossing, etc.)
- The behavior of the pedestrian in relation to each physical object present (e.g., for fencing/enclosure—check box if the user: looked through, jumped over, leaned on, walked within, walked around, other [with free text to specify the behavior])
- A description of the path taken by the pedestrian and their behavior, including information about the person if it affected their behavior, such as a mobility impairment (free text description)
- A representation of the pedestrian's path through the RLX, including the starting point and destination, overlaid on an aerial view of the crossing

12.4 Results

12.4.1 Inter-Rater Reliability

After receiving training on the observation protocol and data collection form an independent rater concurrently documented the behavior of pedestrians using the RLX during 3 h (~10% of total observation time) at the first observation site. Ratings of 28 pedestrians were gathered during that period. Inter-rater reliability calculations were performed on two aspects of the observations for each of the 28 pedestrians observed: the classification of the system state and the classification of behavior in relation to each physical object present. Between the raters there were 1264 agreements (e.g., both raters recorded that the pedestrian *walked within* the *fencing/enclosure* or both raters did not check the box that the pedestrian *walked within the fencing/enclosure*) and 93 disagreements (e.g., one rater recorded that the pedestrian *walked within* the *fencing/enclosure* however the other rater did not check this box). The calculations took into account where the physical object was not present during the observation providing no opportunity for behavior in relation to the object. Where an object was not present it was excluded from the analysis (i.e., was not counted as an agreement nor disagreement). The percentage agreement score was 93.15. Once this satisfactory level of inter-rater agreement was obtained, the remaining observations were conducted by a single rater alone.

12.4.2 System State during Observations

In total, 337 pedestrians (including pedestrians using mobility devices such as motorized scooters) were observed. The majority of pedestrians observed (175) approached and traversed the RLX with the warnings not activated at any point. The term "warnings" included any technology at the crossing designed to inform users of the presence of a train and included bells, gates, flashing lights, and boom barriers (even where these were only present at an adjacent road crossing). In 64 cases, the warnings activated as the user approached meaning that these pedestrians needed to make a decision about whether to stop or to proceed across the RLX. For 71 pedestrians observed, the warnings were activated during the whole time of approach, meaning that they had information to make a decision very early in the approach. During 21 observations, the warnings activated while the user was traversing the RLX and in three cases the warnings activated just after the user had exited the crossing. In two cases the warnings stopped just as the user reached the crossing and in one case multiple system states were relevant as the pedestrian had approached and traversed the RLX multiple times within a short period of time with the warnings being active at different points in time.

12.4.3 Evaluation of Design Hypotheses

The key design hypotheses for selected physical objects are displayed in Table 12.3. Those selected include those with the most interesting findings and where the hypothesis related to behavior was conducive to observation.

An interesting finding that did not align with the design hypothesis was that the onset of the bells resulted in a number of pedestrians increasing their speed, rather than stopping. This is the opposite of the design hypothesis that the user will stop when the auditory warning is present. The bells are an initial warning and there are a number of seconds between their onset and the closing of the pedestrian gates. Pedestrians may determine

TABLE 12.3

Evaluation of the Design Hypotheses for Key Physical Objects, Based upon Observed Pedestrian Behavior

Physical Object	Object-Related Processes and Design Hypothesis	Observed Behavior and Discussion
Fencing/enclosure	*Object-related processes* • Guide path (physical constraint) • Provide physical barrier (physical constraint) *Relevant road rule* • A pedestrian must not cross a railway line at a level crossing unless there is a pedestrian facility at the crossing and the pedestrian uses this facility (Road Safety Road Rules 2009 (Vic) Section 235(1)(a)) *Design hypothesis 1* • Pedestrians will walk within the fencing	*Design hypothesis 1: Supported* Overall, the majority of pedestrians walked within the fencing provided at the RLX • 324 pedestrians walked within the fencing at least during some part of the crossing traversal • 26 pedestrians walked around the fencing at least during some part of the RLX traversal • 4 pedestrians walked outside of the fencing during the entire crossing traversal *Additional observations* • 6 pedestrians were observed to lean on the fencing while waiting for trains to pass and gates to open
Bells/alarm	*Object-related processes* • Audible warning of approaching train (cultural constraint) • Prompt stop/go decision (cultural constraint) *Relevant road rules* • A pedestrian must not cross a railway line at a level crossing if warning bells are ringing (Road Safety Road Rules 2009 (Vic) Section 235(2)(a)) • If warning bells start ringing after a pedestrian has started to cross a railway line at a level crossing, he or she must finish crossing the line or tracks without delay (Road Safety Road Rules 2009 (Vic) Section 235(2A)) *Design hypothesis 2* • Pedestrians approaching the RLX will stop when the bells begin to sound *Design Hypothesis 3* • Pedestrians on the RLX when the bells begin will increase their speed to complete crossing without delay	*Design hypothesis 2: Not supported* More pedestrians approaching the RLX when the bells activated increased their speed than decreased their speed or stopped • 42 pedestrians increased their speed on approach to the RLX in response to the bells sounding • 17 pedestrians decreased their speed and eventually stopped (where gates remained closed) in response to the bells sounding • 2 pedestrians stopped immediately when the bells activated *Design hypothesis 3: Somewhat supported* • The majority of pedestrians who were on the crossing when the bells activated maintained their speed when the bells begin. Some increased speed and a small number stopped and returned to the entry point. • 12 pedestrians increased their speed to cross more quickly • 2 elderly pedestrians stopped immediately. One stopped and walked back to wait behind the closing gate. The other stopped and walked back to wait on the track side of the closed gate. Once the train had passed this pedestrian walked behind it while the warnings remained activated. *Additional observations* • 46 pedestrians looked up the railway tracks when the bells sounded suggesting they were looking for an approaching train

(Continued)

TABLE 12.3 (Continued)

Evaluation of the Design Hypotheses for Key Physical Objects, Based upon Observed Pedestrian Behavior

Physical Object	Object-Related Processes and Design Hypothesis	Observed Behavior and Discussion
TGSIs (Defining the Edge of the Path)	*Object-related process* • Guide path (cultural constraint) *Relevant road rule* • A pedestrian must not cross a railway line at a level crossing unless there is a pedestrian facility at the crossing and the pedestrian uses this facility (Road Safety Road Rules 2009 (Vic) Section 235(1)(a)) *Design hypothesis 4* Pedestrians will walk across the RLX within the boundaries set by the TGSIs	*Design hypothesis 4: Supported* The majority of pedestrians walked within the boundary of the path marked by the TGSIs • 217 pedestrians walked within the TGSIs • 27 pedestrians walked around the TGSIs (at least during some part of their traversal of the RLX) • 4 pedestrians walked on the TGSIs while traversing the RLX *Additional observations* • 1 pedestrian using a wheelchair appeared to have difficulty keeping the wheelchair from turning toward the tracks due to the cross fall gradient of the footpath. The tactile indicators stopped the wheelchair from going onto the tracks
Pedestrian Gate (When Closed)	*Object-related process* • Provide physical barrier (physical constraint) *Relevant road rules* • A pedestrian must not cross a railway line at a level crossing if a gate or barrier at the crossing is closed or is opening or closing (Road Safety Road Rules 2009 (Vic) Section 235(2)(b)) *Design hypothesis 5* Pedestrians will stop at the gate when it is closing, closed or opening	*Design hypothesis 5: Somewhat supported* The majority of pedestrians stopped at the gate, but a considerable number walked around the gate, either on approach to the RLX or when caught on the RLX when the gate closed or began to close • 51 pedestrians stopped at the closed or closing gate • 31 pedestrians walked around the closed or closing gate *Additional observations* • 3 pedestrians pushed or attempted to open a closed or closing gate • 1 pedestrian pushed the gate as it began to open • 1 pedestrian leaned on the gate while it was closed • 1 pedestrian's walking stick became caught in the gate while it was closing *(Continued)*

TABLE 12.3 (Continued)

Evaluation of the Design Hypotheses for Key Physical Objects, Based upon Observed Pedestrian Behavior

Physical Object	Object-Related Processes and Design Hypotheses	Observed Behavior and Discussion
Emergency Gate	*Object-related processes* • Pushing open to achieve exit from track (affordance) • Pulling open to achieve entry to track (affordance) *Relevant road rules* • A pedestrian must not cross a railway line at a level crossing if a gate, boom, or barrier at the crossing is closed, or is opening, or closing (Road Safety Road Rules 2009 (Vic) Section 235(2)(b)) • A person must not open or interfere with any locked gate or door on rail premises (Transport (Conduct) Regulations Section 46(c)) *Design hypothesis 6* Pedestrians on the crossing when the gates are closing or closed will use the emergency gate to exit the crossing *Design hypothesis 7* Pedestrians will not access the emergency gate from the approach side of the crossing	*Design hypothesis 6: Supported* The majority of pedestrians caught on the RLX when the warnings activated used the emergency gate to exit the crossing • 25 pedestrians pushed the emergency gate open to exit the RLX after they were caught on the tracks • 2 pedestrians waited in the refuge area rather than using the emergency gate to exit the track completely • 1 user walked out onto the road to exit the RLX, rather than using the emergency gate *Design hypothesis 7: Somewhat supported* While the majority of pedestrians stopped at the RLX did not use the emergency gate as a means to enter the RLX when the pedestrian gates were either closed or in the process of closing, some did do so • 14 pedestrians pulled the emergency gate open from the approach side of the RLX and used it to enter the track area • 2 pedestrians held the emergency gate open to allow others to enter the RLX
Other Pedestrians	*Object-related process* • Obstruct path (physical constraint) *Relevant road rule* • A pedestrian must not cross or attempt to cross railway tracks at a place provided for crossing by pedestrians if the crossing or the path beyond the crossing is blocked (Transport (Conduct) Regulations (Vic) Section 39(1)(d)) *Design hypothesis 8* Pedestrians will stop if the crossing or the exit to the crossing is blocked (e.g., by other pedestrians)	*Design hypothesis 8: Mixed findings* Only a small number of pedestrians faced a situation where the path past the RLX was congested with other pedestrians • 1 pedestrian walking with a pram stopped prior to the RLX to enable a pedestrian coming from the other direction to finish crossing before they entered the RLX • 1 pedestrian failed to stop before entering the RLX and found themselves stuck on the track side of the crossing while a large crowd of pedestrians entered the RLX coming from the other direction. The pedestrian was caught on the crossing for some time while the crowd dissipated *Additional observations* In general, a number of pedestrians (41) were observed to actively avoid being in the path of other pedestrians when near or on the RLX

that if they increase their speed they will be able to cross before the gates close. Further, at train stations, the onset of the bells may suggest to user that their train is approaching, motivating them to increase their speed and reach the station before the train. Interestingly, in some of the situations where pedestrians increased their speed after the bells activated, they were seen to stop just after the RLX, in some cases to catch their breath after running across, suggesting that they were not necessarily in a hurry but for some reason felt compelled to rush across. The bells are quite loud and their onset is sudden. Potentially their onset activates the physiological fight or flight response in pedestrians with a natural response being to rush or run across the RLX. Potentially a more gradual and nuanced transition from the state of no warning to full warnings (i.e., bells, flashing lights, gates closed) could lead to more desirable pedestrian behavior.

While most users stopped at closed gates or pedestrian boom barriers, there were users who avoided them either to cross at the road or to access the crossing through an emergency gate. The gates and barriers are intended to provide a safety boundary or safety zone for users and using the emergency gate or road to access the crossing would be considered most undesirable behavior from the perspective of the designers of the technology. The gate latch added to the emergency gate at site 3 is intended to prevent access to the crossing from the approach side of the emergency gate. No users were observed using the emergency gate to access the crossing at that location. However, three pedestrians were observed to avoid the pedestrian facilities and cross at the road crossing with the warnings beginning to activate while they were on the road, indicating that even with a locking gate installed pedestrians can continue to access the RLX when a train is approaching.

In addition to identifying behavior relating to the design hypotheses, behaviors were observed that uncovered additional affordances. For example, fences and gates afforded leaning on (i.e., pedestrians resting their arms on the gate or fencing while watching the train approach and pass). This behavior may indicate a preference for comfort during the waiting period which can be quite lengthy depending on the time of day and types of trains traversing the crossing. Aspects of RLX design such as user comfort and user experience appear to have been traditionally overlooked. Potentially, RLX design could explore improving comfort and engagement while pedestrians are waiting at the RLX which could in turn promote desirable behaviors such as not rushing through the RLX when the warnings are activated.

Related to the user experience at RLXs, there were instances of pedestrians attempting to exert some control of the infrastructure at the RLX. For example, some pedestrians attempted to push open a gate that was closing or opened the emergency gate to enter the track area. The need for agency may also explain why some users choose to cross by accessing the road, rather than waiting for a barrier to open. Potentially, RLX design could be improved by increasing the control that pedestrians have over the environment, or perhaps increasing a sense of control to avoid feelings of frustration and actions to overcome the environment that is seen as being restrictive and full of constraints. Providing more affordances for RLX users to act upon may be a fruitful design strategy.

12.5 Conclusions

The aim of this study was to consider the intended or designed-for affordances of RLXs from the perspective of pedestrians using the crossing and the related hypotheses held by designers of crossings, as inferred by reference to design standards. Evaluating the hypotheses using data gathered from actual behavior at rail crossings has shown that they are not always supported in practice and that other affordances and constraints, both positive and negative with respect to safety, are created. These finding have implications for RLX design.

While the observations were based on a convenience sample and observation times were not equal across all sites, the intention of the research was to record the range of behavior and gather insights into the design hypotheses rather than formally evaluating the hypotheses. However, this could be achieved in future research perhaps comparing behavior against the hypotheses taking into account different site features, situational factors, etc. The numbers provided in the results only intended to provide an indication of the extent of the behavior within the sample of pedestrians observed.

This work has shown that the design philosophy underlying current RLX design is normative and does not acknowledge the normal performance variability occurring within the domain. It is assumed that users will be influenced by the affordances and constraints intended by designers and will follow the rules set by the legislature. However, the observations show that there is significant variability in human behavior, regardless of constraints and rules. Modern safety science has moved away from rule-based systems and now acknowledges the need to recognize emergence and performance variability and to support flexibility in the means for attaining goals. It is suggested that safety at RLXs could be better achieved through recognizing that humans are the glue that hold complex sociotechnical systems together. Accordingly, RLX design should explore attempts to support the adaptive capacity of humans through appropriate performance variability and emergence. This could be achieved by moving from a domain dominated by physical and cultural constraints to one where humans are provided with affordances that assist them to meet their goals.

Acknowledgments

This research was funded through an Australian Research Council Linkage Grant (ARC, LP100200387) to the University of Sunshine Coast, Monash University, and the University of Southampton, in partnership with the following organizations: the Victorian Rail Track Corporation, Transport Safety Victoria, Public Transport Victoria, Transport Accident Commission, Roads Corporation (VicRoads) and V/Line Passenger Pty Ltd. Professor Paul Salmon's contribution to this chapter was funded through his Australian Research Council Future Fellowship (FT140100681).

References

Australian Transport Safety Bureau. 2012. *Australian Rail Safety Occurrence Data: 1 July 2002 to 30 June 2012*. Canberra: ATSB.
European Railway Agency. 2014. *Railway Safety Performance in the European Union*. Valenciennes, France: European Railway Agency.

Gibson, J. J. 1979. *The Ecological Approach to Visual Perception*. Hillsdale, New Jersey: Erlbaum.

Jenkins, D. P., Stanton, N. A., Salmon, P. M., Walker, G. H., and Young, M. S. 2008. Using cognitive work analysis to explore activity allocation within military domains. *Ergonomics*, 56, 798–815.

Metaxatos, P. and Sriraj, P. S. 2013. *Pedestrian/Bicyclist Warning Devices and Signs at Highway-Rail and Pathway-Rail Grade Crossings*. Chicago: Illinois Center for Transportation.

Naikar, N., Hopcroft, R., and Moylan, A. 2005. *Work Domain Analysis: Theoretical Concepts and Methodology*. Melbourne, Victoria: Defence, Science and Technology Organisation.

Norman, D. A. 1988. *The Design of Everyday Things*. New York: Doubleday.

Norman, D. A. 1999. Affordance, conventions, and design. *Interactions*, 6, 38–43.

Rail Safety and Standards Board. 2015. *Annual Safety Performance Report: A Reference Guide to Safety Trends on GB Railways 2014/15*. London: Rail Safety and Standards Board.

Stanton, N. and Bessell, K. 2014. How a submarine returns to periscope depth: Analysing complex socio-technical systems with cognitive work analysis. *Applied Ergonomics*, 45, 110–125.

Vicente, K. J. 1999. *Cognitive Work Analysis: Toward Safe, Productive, and Healthy Computer-Based Work*. Mahwah, New Jersey: Lawrence Erlbaum Associates.

Warwick, J. 2009. Active pedestrian crossings controls and technology. *Paper Presented at the IRSE Technical Convention*, Melbourne.

Wigglesworth, E. C. and Uber, C. B. 1991. An evaluation of the railway level crossing boom barrier program in Victoria, Australia. *Journal of Safety Research*, 22(3), 133–140.

Withagen, R., de Poel, H. J., Araújo, D., and Pepping, G.-J. 2012. Affordances can invite behavior: Reconsidering the relationship between affordances and agency. *New Ideas in Psychology*, 30(2), 250–258.

Woods, D. D. 1998. Designs are hypotheses about how artifacts shape cognition and collaboration. *Ergonomics*, 41, 168–173.

13

Drivers' Visual Scanning and Head Check Behavior on Approach to Urban Rail Level Crossings

Kristie L. Young, Michael G. Lenné, Vanessa Beanland,
Paul M. Salmon, and Neville A. Stanton

CONTENTS

13.1 Introduction

Crashes at rail level crossings (RLCs) constitute a major safety concern, as they are often catastrophic, involving multiple fatalities and traumatic injuries. In 2011, 49 RLC (rail level crossing) collisions were recorded in Australia, leading to 33 fatalities (ATSB, 2012). The costs associated with RLC crashes in Australia have been estimated at approximately AUD $24 million per year. In the European Union (EU), RLC collisions and fatalities represent more than one-quarter of all railway crashes occurring on the EU railway system, with 604 fatal and serious injury casualties recorded during 2011 (European Railway Agency, 2013). Figures from the United States are similar, with 247 fatalities and 705 injuries at RLCs in 2009 (US Department of Transportation, 2014). Given the high levels of trauma and disruption to rail and road networks associated with RLC crashes, their prevention represents a key priority area for rail and road organizations around the world.

Factors contributing to RLC crashes are poorly understood; however, driver behavior is believed to play a key role (Davey et al., 2007; Lenné et al., 2011). Direct causal factors relating to driver behavior typically fall into two broad categories (Lenné et al., 2011). The first involves intentional noncompliance with crossing signals, whereby drivers detect the train and/or the activation of crossing warnings and fully understand the meaning of the warnings, but will nevertheless intentionally cross. A propensity to engage in risk taking or sensation seeking and a low perception of risk have both been found to contribute to intentional noncompliance at RLCs (Davey et al., 2008; Witte and Donohue, 2000). The second, particularly prevalent, category is unintentional noncompliance where drivers, for a range of reasons, fail to detect the crossing signals, fail to comprehend the signals' meaning, or fail to detect the train itself, and will enter the crossing as a train approaches. Indeed, it has been estimated that unintentional noncompliance accounts for almost half of all RLC crashes in Australia (ATSB, 2002). Diminished situation awareness, distraction, and inattention are likely to be key contributors to unintentional noncompliance at RLCs (Caird et al., 2002; Salmon et al., 2013); however, the reasons why situation awareness is degraded, or why inattention occurs, are less clear from previous research.

Studies examining driver behavior in the RLC context have been typically observational, employing on-site observers or video analysis (e.g., Meeker et al., 1997; Tenkink and Van der Horst, 1990; Tey et al., 2011). The primary measure derived from such observational studies is driver noncompliance with crossing signals, which has been estimated to be between 14% and 38% for active crossings with flashing lights and boom barriers (Meeker et al., 1997; Witte and Donohue, 2000). Observational studies do not, however, allow for an in-depth examination of driver behavior in terms of the factors underpinning compliance and noncompliance; particularly, situation awareness, workload, attention, and the system-wide factors underlying each of these. Developments in vehicle instrumentation now make it possible to examine driver behavior at RLCs in greater depth in on-road settings using a suite of onboard driver and vehicle monitoring equipment coupled with human factors methods.

This study focuses on RLCs in urban environments. In such areas, one factor that is likely to shape driver behavior on approach to RLCs and contribute to unintentional noncompliance is the location of crossings in high workload segments of the road network. Urban RLCs are often surrounded by busy shopping strips with high levels of pedestrian, vehicle, and cyclist traffic and a high level of visual clutter (objects unrelated to driving). Complex road environments that contain dense traffic and visual clutter have been shown to increase driver workload and the potential for distraction by removing drivers' eyes off the road or impairing their visual scanning patterns (Horberry, 1998; Jahn et al., 2005; Patten et al., 2006). Thus, the complex traffic environment in which many urban RLCs reside could induce high levels of driver workload and distraction, which in turn may lead to drivers paying less attention to the RLC due to their attention being diverted elsewhere. Pickett and Grayson (1996) identified three types of drivers who are likely to be involved in a crash at RLCs, one of which involved those drivers who are unaware of the signals due to inattention or distraction. Further, an analysis of Canadian RLC crashes over a 19-year period found that a number of crashes involved driver distraction as a factor contributing to drivers failing to see the signals/train at all, or in time to stop (Caird et al., 2002). Driver attention being diverted from the RLC can lead to a failure to safely negotiate the crossing. When the crossing is currently active and drivers fail to detect the signals in a timely manner, the results can be catastrophic. However, distraction and inattention may also affect crossing behavior when no train is immediately present, such as when drivers fail to detect traffic banking up on the far side of the crossing and are forced to queue on the crossing itself.

Little is known about where drivers direct their attention on approach to RLCs and the influence of a high workload environment on driver attention and behavior in relation to the crossing. This on-road study aimed to examine where drivers direct their visual attention on approach to urban RLCs situated in high workload areas—shopping strips. Drivers' eye glance data were examined for the 150 m approach to RLCs to identify what aspects of the road environment drivers focus their visual attention on and how much attention is focused on the crossing itself.

13.2 Method

13.2.1 Participants

Twenty drivers (11 males, 9 females) aged 18–53 years ($M = 26.8$, $SD = 9.2$) participated in the study. All participants held a current Victorian car driver's license, drove regularly in urban areas, and spoke English as their first language. Eight participants held a valid full car license while the remaining 12 held a valid P2 (second year provisional) license. Participants had held their car license for 8.5 years on average ($SD = 9.2$) and drove an average of 7.8 h ($SD = 5.5$) per week. Participants were recruited through a Monash University newsletter and were compensated for their time and travel expenses. Ethics approval was granted by the Monash University Human Research Ethics Committee.

13.2.2 On-Road Test Vehicle

The on-road test vehicle (ORTeV) is an instrumented vehicle equipped to collect vehicle and video data. Vehicle CAN-bus and video data were acquired using a Racelogic Video VBOX Pro system, which combines a GPS (global positioning system) logger, multiple cameras, and a 32-channel CAN (controller area network) interface. Vehicle data collected included: trip time and distance, GPS location, vehicle speed, brake pressure, and vehicle heading. Video data were derived from seven unobtrusive cameras which recorded forward and peripheral views spanning 90° each, respectively, as well as the driver, vehicle cockpit, and the rear of the vehicle. In the current study, the video data were used to manually code drivers' visual scanning behavior.

13.2.3 Driving Route

The driving route comprised an 11 km urban route in the southeastern suburbs of Melbourne. The route included arterial roads (80, 70, and 60 km/h) and shopping strips (40 and 50 km/h) and contained six RLCs, all with active controls (flashing lights with boom barriers and bells). The route took 20–25 min to complete. To ensure that participants would experience similar traffic conditions, all drives were completed on weekdays at 10 a.m. and 2 p.m. Direction instructions were provided prior to the drive and participants also carried a route map.

13.2.4 Procedure

Demographic details (age, gender, license type, driving history) were collected prior to the study. Participants were then seated in the ORTeV and the data collection systems were initiated. Participants completed the driving route while driving alone in the vehicle.

Two observers followed the participant at a distance in another vehicle to ensure they could redirect participants back on-course in the event that they took a wrong turn. Participants provided verbal protocols throughout the drive. After completion of the drive, drivers were taken back to the university where they completed an interview about their experiences during the drive.

13.2.5 Data Reduction

Five of the six RLCs encountered were examined in this study. One RLC was excluded from analysis as it was not located in a shopping strip area. Drivers' visual scanning behavior on the 150 m approach to each RLC (to the point where the vehicle cleared the train tracks) was manually coded using the onboard videos. The driver and forward-facing camera views were used to determine the location of each glance while the vehicle was moving. The locations of drivers' glances were coded across eight different areas including various segments of the forward and side roadway, mirrors, and inside the vehicle (Figure 13.1). The number and duration (ms) of glances to each of the eight areas, the mean distance from the RLC at which drivers glanced to off-road areas, and the percentage of time spent with eyes off the forward roadway was coded for the approach to each RLC. A glance was defined as an uninterrupted fixation to the area of interest. The video was recorded at 10 Hz, thus fixations were examined in 100 ms intervals by moving through the video frame by frame and recording in which area the driver's gaze was directed. Only glances where vehicle speed was above 0 km/h were coded.

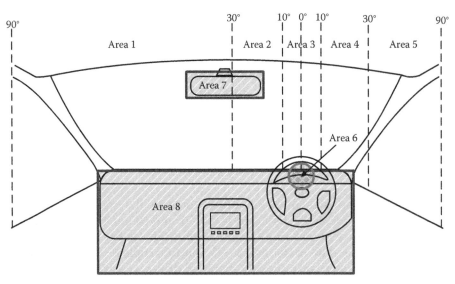

*Area 1: 30–90° left
Area 2: 10–30° left
Area 3: road ahead (0° ± 10°)
Area 4: 10–30° right

*Area 5: 30–90° right
Area 6: speedometer
Area 7: rearview mirror
Area 8: other area in vehicle

FIGURE 13.1
Areas used for coding driver eye glances on approach to urban RLCs. *Only coded from start of approach to 30 m before RLC. Within 30 m, glances to these areas were assumed to be toward the RLC itself, rather than off-road.

Head checks directed toward the RLC were also examined. Drivers were coded as having executed a head check if, within 30 m prior to the crossing, their head direction and gaze fixation deviated in excess of ±30° horizontally (where 0° indicates straight ahead). Outside of the 30 m approach, glances in excess of ±30° were coded as being directed toward the footpath and shops.

Eye glance behavior was coded by a trained coder. A sample of approximately 10% of the RLC approach videos were independently coded by a second coder and inter-rater reliability was examined using Pearson's r. The reliability between the two raters was excellent for the number of glances made to each area ($r = 0.94$, $p < 0.001$) and the duration of glances ($r = 0.95$, $p < 0.001$).

13.3 Results

Eye glance data were not captured for two drivers due to video recording issues. As drivers' eye glance behavior is likely to be affected by the presence of a train or activated crossing signals, crossing events were coded as to whether the driver encountered a train or not and crossings where a train was present were directly compared with those where no train was present. Eye glance behavior was pooled across the five RLCs and examined across the eight glance areas when a train was present and not present in a series of generalized estimating equations (GEE). GEE is an extension of the generalized linear model and is useful for analyses such as these because it factors in correlations due to the repeated measurements. The models to examine the mean duration of glances and total percentage of time fixated on each area were specified with a normal error distribution, an identity link function, and the correlation matrix was specified as exchangeable due to convergence problems. The model to examine the mean distance from the crossing when off-road glances were made was specified with a normal error distribution, an identity link function, and the correlation matrix was specified as unstructured. Finally, the model to examine the frequency of glances to each area was specified with a Poisson error distribution and a log link function as it was count data and the correlation matrix was specified as exchangeable due to convergence issues.

13.3.1 Frequency and Duration of Glances

The mean number and duration of glances taken to each of the eight areas on the 150 m approach to urban RLCs is displayed in Table 13.1, separately for when a train was present and when no train was present. As shown, the on-road areas had the highest number of mean glances and glances to these areas were also of longer duration than glances to off-road areas. A GEE model was fitted to examine if the number of fixations made to each area differed across RLCs with a train present versus no train present. A significant train presence by glance area interaction was found ($p < 0.001$), whereby, when there was a train, drivers made a greater number of glances to all areas except the speedometer, which they glanced at more frequently when no train was present.

The GEE model for the mean duration of glances to each area revealed a similar pattern of results, with a significant train presence by glance area interaction ($p < 0.001$) indicating that, on approach the crossings, drivers glanced to all areas longer when a train was present, apart from the speedometer and rearview mirror, which they glanced at for longer

TABLE 13.1

Mean (*SD*) Frequency and Duration of Glances Made to Each Area
on Approach to Urban RLCs as a Function of Train Status

	Mean Frequency		Mean Duration (s)	
Area	Train	No Train	Train	No Train
1: 30–90° left (footpath and shops)[a]	0.5 (0.8)	0.2 (0.5)	0.8 (0.2)	0.6 (0.1)
2: Roadside to 30° left	2.2 (1.7)	1.0 (0.8)	0.7 (0.5)	0.5 (0.2)
3: Road ahead (0 ± 10°)	9.3 (4.3)	6.7 (2.9)	2.7 (1.5)	2.7 (1.8)
4: Roadside to 30° right	5.0 (2.6)	3.8 (2.6)	1.0 (0.5)	0.8 (0.6)
5: 30–90° right (footpath and shops)[a]	0.9 (1.2)	0.6 (0.9)	1.1 (0.3)	0.8 (0.2)
6: Speedometer[a]	0.5 (0.6)	0.7 (0.9)	0.2 (0.1)	0.4 (0.3)
7: Rearview mirror[a]	0.6 (0.7)	0.6 (0.8)	0.4 (0.1)	0.6 (0.2)
8: Other area in vehicle[a]	0.6 (0.6)	0.5 (1.0)	1.2 (0.4)	0.9 (0.3)

[a] Defined as off-road glances.

when no train was present. Taken together, these results reveal that drivers made longer and more frequent glances to the forward roadway on approach to urban level crossings when a train was present compared to when no train was present. However, the findings also reveal that drivers made longer, more frequent glances to a number of off-road areas when a train was present, including to footpaths and shops on either side and to other areas inside the vehicle.

13.3.2 Percentage of Time Fixated on Off-Road Areas

Given that the individual glance duration and frequency data are substantially affected by travel speed and the overall duration of the approach period, the percentage of time spent fixated on a particular area was also examined as it controls for the total time spent on approach. Table 13.2 displays the percentage of time spent fixated on each of the eight areas on approach to the urban RLCs. The data show that drivers spent just under 10% of time on approach to the urban crossings with their visual attention *off* the forward roadway (9.5% when train present and 8.3% when no train present). The GEE

TABLE 13.2

Percentage of Time Fixated on Each Area
on Approach as a Function of Train Status

	% of Time	
Area	Train	No Train
1: 30–90° left (footpath and shops)[a]	2.7	0.7
2: Roadside to 30° left	4.6	2.5
3: Road ahead (0 ± 10°)	69.2	73.1
4: Roadside to 30° right	16.7	16.1
5: 30–90° right (footpath and shops)[a]	3.2	2.4
6: Speedometer[a]	0.5	1.5
7: Rearview mirror[a]	0.8	1.5
8: Other area in vehicle[a]	2.3	2.2

[a] Defined as off-road glances.

model for the percentage of time drivers spent looking at each area revealed a significant train presence by glance area interaction ($p = 0.044$). The majority of the time was spent glancing at the road ahead, regardless of whether a train was present or not, but when a train was present drivers spent slightly less time glancing at the road ahead and more time glancing at roadside areas, particularly to the right. In contrast, when no train was present drivers predominantly looked at the road ahead but spent a higher proportion of time glancing at the speedometer and rearview mirror. The proportion of time glancing at other in-vehicle areas did not vary between train present and no-train crossings.

13.3.3 Distance from the RLC When Off-Road Glances Made

Examining the distance drivers were from the crossing when they glanced to off-road areas can provide insights into drivers' visual scanning strategies and how they may regulate their off-road glances in relation to the crossing. Drivers were defined as reaching the crossing when the front of the vehicle was level with the first rail of the train tracks. Table 13.3 shows the mean distance (in meters) drivers were from the crossings when they glanced to each of the five off-road areas on approach to each RLC. As displayed, the drivers were quite far from the RLCs when they made their glances to the off-road areas, particularly to the off-road area that was unrelated to driving—"other area inside the vehicle," suggesting a fairly conservative off-road scanning strategy. Results of the GEE model revealed a significant train presence by glance area interaction ($p = 0.002$). Drivers glanced to all off-road areas a longer distance from the crossings when a train was present, compared to when no train was present, apart from the rearview mirror, which drivers were presumably using to monitor vehicles behind them as they came to a stop for the train.

13.3.4 Head Checks toward the RLCs

Driver head checks within the 30 m immediately prior to entering the crossing were examined. Table 13.4 details the mean number of head checks made and the distance from the crossing when the first and final head checks were made. Data are reported descriptively due to the limited number of head checks made by drivers (the median number of head checks made was 0, regardless of whether a train was present). Drivers made only a small number of head checks of the RLCs, with many drivers making no head checks. Drivers made slightly more head checks overall when a train was present, although head checks were typically performed earlier when no train was present at the crossing.

TABLE 13.3

Mean (*SD*) Distance (m) from RLCs When Glances Were Made to Each Off-Road Area as a Function of Train Status

Area	Train	No Train
1: 30–90° left (footpath and shops)	91.7 (3.5)	71.5 (24.6)
5: 30–90° right (footpath and shops)	97.5 (19.0)	84.4 (25.7)
6: Speedometer	103.3 (34.8)	80.6 (36.6)
7: Rearview mirror	53.9 (18.2)	86.5 (39.3)
8: Other area in vehicle	92.9 (29.9)	91.9 (37.8)

TABLE 13.4

Mean (*SD*) Number of Head Checks and Distance
(Meters) from RLC When First and Final Head
Checks Made as a Function of Train Status

	Train	No Train
Mean Number of Head Checks		
Left	0.14 (0.14)	0.14 (0.14)
Right	0.42 (0.37)	0.23 (0.25)
Mean Distance from Crossing: First Head Check		
Left	6.9 (12.2)	15.0 (11.2)
Right	12.1 (8.0)	15.4 (9.2)
Mean Distance from Crossing: Final Head Check		
Left	−1.3 (11.6)	6.4 (10.6)
Right	8.9 (8.1)	11.2 (8.1)

13.4 Discussion

Eye glance and head check behavior were examined to determine to what areas, and for how long, drivers direct their visual attention when approaching RLCs situated in urban shopping strips. Results revealed that drivers spent the majority of their time on approach to urban crossings with their visual attention focused *on* the forward roadway (over 90%). However, the findings also show that drivers do direct their visual attention off the forward roadway to a range of areas inside and outside the vehicle when approaching urban RLCs. Within the 150 m approach, drivers spent around 10% of the time fixated on off-road areas.

The presence of a train at the urban RLCs did influence drivers' visual scanning behavior. Drivers took longer and more frequent glances to the forward roadway on approach to urban RLCs when a train was present compared to when no train was present. Drivers also glanced at the speedometer and rearview mirror for longer periods on approach when no train was present. However, results also revealed that drivers took longer and more frequent glances to a number of off-road areas when a train was present, including the footpaths and shops on either side and to other areas inside the vehicle. One explanation why drivers spent more time looking at these off-road areas when a train was present is that they were travelling slower in these situations due to the need to come to a stop at the crossing. That is, drivers may have felt more confident looking at areas unrelated to driving when travelling at the slower speeds associated with the presence of trains. Glances to these off-road areas were also typically short (<1.5 s) and were made when drivers were a fair distance from the crossings (>50 m), suggesting that drivers better regulated their off-road glances on immediate approach to the crossings. Further, due to the slower speeds drivers also spent more time on approach to the crossing, which explains the fact that overall glance durations and frequencies were greater when a train was present compared to when no train was present. When controlling for this time difference and instead comparing the percentage of time spent looking in each area, drivers still spent longer looking at off-road areas in the presence of a train, compared to when no train was present.

Driver head check behavior within the 30 m immediately prior to entering the crossing revealed that, on average, drivers made a very small number of head checks at the level crossings, with many drivers not making any head checks at all on approach. Drivers made a slightly higher number of head checks when a train was present, but these were performed later in the approach than when no train was present. These findings indicate that drivers rarely actively check for trains at urban crossings, particularly when the crossing signals are not activated on approach. In an earlier study of driver behavior at level crossings in rural areas, Lenné et al. (2013) found that drivers completed a higher number of head checks (5–6) at passive crossings (stop and give way) compared to actively controlled (boom barrier) crossings (1–2 checks). Taken together, the results of both studies suggest that, at urban RLCs located in high workload areas, drivers have become heavily reliant on the crossing signals to alert them to the presence of a train. It is also possible that drivers may be restricted in their ability to perform effective head checks in built-up urban environments, such as the ones examined in this study, as their sightlines are restricted by buildings and other infrastructure. Nevertheless, drivers failing to scan the crossing for trains, regardless of the reason, could be problematic in a number of instances; for example, if the RLCs warning infrastructure fails or when drivers' attention is diverted momentarily away from the crossing warnings immediately prior to activation.

This is the first study to examine drivers' visual scanning and head check behavior on approach to urban RLCs in an on-road context. Our findings extend previous observational studies (e.g., Meeker et al., 1997; Tenkink and Van der Horst, 1990; Tey et al., 2011) by moving beyond examining driver compliance at RLCs to exploring an aspect of driver behavior that may underlie why drivers fail to comply with crossing signals, namely where they focus their visual attention on approach and how this is shaped by the wider road environment. Overall, the visual scanning findings from the current study suggest that while drivers spend the majority of the time on approach to urban crossings with their eyes on the forward roadway, they do also look at various off-road areas, even when a train is approaching. Our findings therefore lend support to previous work by Caird et al. (2002) and Pickett and Grayson (1996) which suggest that driver distraction could play a role in drivers failing to detect the signals at RLCs. Further work should investigate the mechanisms underlying driver distraction at RLCs. Is the environment too visually demanding or attention grabbing leading drivers to pay less attention to RLCs? Are the RLCs not conspicuous enough in busy urban areas? Or do drivers simply not consider RLCs to be risky enough to warrant their undivided attention?

Visual scanning behavior provides important insights into driver behavior at high workload urban crossings; however, this is only part of the picture. As part of the wider rail program, the authors are interrogating drivers' verbal protocols and post-drive interview data to build a more comprehensive picture of driver behavior on approach to urban RLCs. These analyses will provide insight into where drivers direct their cognitive as well as their visual attention on approach to urban crossings, the information cues used to identify the presence of the crossing and make a crossing decision and if behavior in this context is influenced by driver experience.

Acknowledgments

This project was funded by ARC Linkage Grant (LP100200387) to Monash University in partnership with the University of Sunshine Coast and the University of Southampton, and the following partner organizations: the Victorian Rail Track Corporation, Transport Safety Victoria, Public Transport Victoria, Transport Accident Commission, Roads Corporation

(VicRoads) and V/Line Passenger Pty Ltd. We acknowledge the support of project partners and community participants. We also acknowledge the assistance of Nebojsa Tomasevic, Christine Mulvihill, Casey Mackay, and Campbell Andrea. Paul Salmon's contribution to this research was funded through the Australian National Health and Medical Research Council postdoctoral training fellowship scheme.

References

ATSB. 2002. Level crossing accidents: Fatal crashes at level crossings. *Paper Presented at the 7th International Symposium on Railroad-Highway Grade Crossing Research and Safety*, Melbourne, Victoria, Australia.

ATSB. 2012. Australian rail safety occurrence data 1 July 2002 to 30 June 2012. Report number RR-2012-010. Australian Transport Safety Bureau, Canberra, ACT.

Caird, J.K., Creaser, J.I., Edwards, C.J., and Dewar, R.E. 2002. A human factors analysis of highway-railway grade crossing accidents in Canada. TP 13938E. Transport Canada, Ottawa, Canada.

Davey, J., Wallace, A., Stenson, N., and Freeman, J. 2008. Young drivers at railway crossings: An exploration of risk perception and target behaviours for intervention. *International Journal of Injury Control and Safety Promotion*, 15(2), 57–64.

Davey, J.D., Ibrahim, N.R., and Wallace, A.M. 2007. Human factors at railway level crossings: Key issues and target road user groups. In Anca, J.M. (Ed.), *Multimodal Safety Management and Human Factors: Crossing the Borders of Medical, Aviation, Road and Rail Industries*. Aldershot: Ashgate, pp. 199–210.

European Railway Agency. 2013. Intermediate report on the development of railway safety in the European Union. European Railway Agency Safety Unit, May 15, 2013.

Horberry, T. 1998. Bridge strike reduction: The design and evaluation of visual warnings. PhD thesis, University of Derby, UK.

Jahn, G., Oehme, A., Krems, J.F., and Gelau, C. 2005. Peripheral detection as a workload measure in driving: Effects of traffic complexity and route guidance system use in a driving study. *Transportation Research Part F: Traffic Psychology and Behaviour*, 8(3), 255–275.

Lenné, M.G., Beanland, V., Salmon, P.M., Filtness, A.J., and Stanton, N. 2013. Checking for trains: An on-road study of what drivers actually do at level crossings. In Dadashi, N., Scott, A., Wilson, J.R., and Mills, A. (Eds.), *Rail Human Factors: Supporting Reliability, Safety and Cost Reduction*. London, UK: Taylor & Francis, pp. 53–59.

Lenné, M.G., Rudin-Brown, M., Navarro, J., Edquist, J., Trotter, M.G., and Tomesevic, N. 2011. Driver behaviour at rail level crossings: Responses to flashing lights, traffic signals and stop signs in simulated driving. *Applied Ergonomics* 42(4), 548–554.

Meeker, F., Fox, D., and Weber, C. 1997. A comparison of driver behaviour at railroad grade crossings with two different protection systems. *Accident Analysis and Prevention*, 29(1), 11–16.

Patten, C.J., Kircher, A., Ostlund, J., Nilsson, L., and Svenson, O. 2006. Driver experience and cognitive workload in different traffic environments. *Accident Analysis and Prevention*, 38(5), 887–894.

Pickett, W. and Grayson, G.B. 1996. Vehicle driver behaviour at level crossings (No. 98). Transport Research Laboratory, Berkshire, United Kingdom.

Salmon, P.M., Lenné, M.G., Young, K.L., and Walker, G.H. 2013. An on-road network analysis-based approach to studying driver situation awareness at rail level crossings. *Accident Analysis and Prevention*, 53, 195–205.

Tenkink, E. and Van der Horst, R. 1990. Car driver behaviour at flashing light railroad grade crossings. *Accident Analysis and Prevention*, 22(3), 229–239.

Tey, L.S., Ferreira, L., and Wallace, A. 2011. Measuring driver responses at railway level crossings. *Accident Analysis and Prevention*, 43(6), 2134–2141.

US Department of Transportation. 2014. *Railway-Highway Grade Crossing Facts and Statistics*. Federal Highway Administration. Retrieved January 8, 2014, from: http://safety.fhwa.dot.gov/xings/xing_facts.cfm

Witte, K. and Donohue, W. 2000. Preventing vehicle crashes with trains at grade crossings: The risk seeker challenge. *Accident Analysis & Prevention*, 32, 127–139.

14

An Evaluation of a National Rail Suicide Prevention Program

Ann Mills, Alice Monk, and Ian Stevens

CONTENTS

14.1 Introduction

Over the past 10 years, there has been an average of 217 suicides per year on the railway. Figure 14.1 shows that on average 39% of suicides take place at stations and 11% occur at level crossings. The category *other locations* mostly comprises suicides on the running line, but also includes a small proportion (<2%) occurring at other railway locations, for example, bridges. Around 80% of recorded suicide attempts have a fatal outcome. Of those that do not, more than half result in major injuries, many of which are severe and life affecting.

Suicides on the railway represent by far the largest proportion of railway-related fatalities, but they represent a relatively small percentage of suicides on a national level. National suicide figures are not available as recently as railway figures and are published on a calendar year basis. Figure 14.2 shows the figures at the time of the evaluation activities.

The number of national suicides has been variable around a 10-year average of 5704; the figure for 2011 was the highest recorded of the analysis, by a notable margin. The

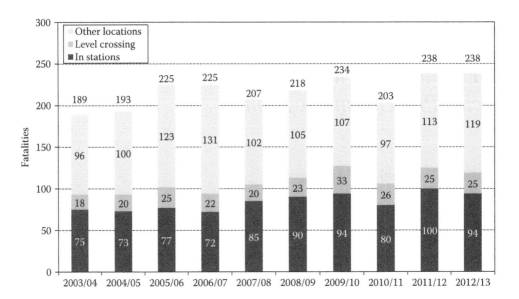

FIGURE 14.1
Suicides and suspected suicides by location.

proportion of the national total occurring on railway property was 3.7% over the presented period. Age and gender demographics of railway suicides vary somewhat from national suicides. Compared with the national profile, a greater proportion of railway suicides are male; this is particularly the case in the 15–44 years age group.

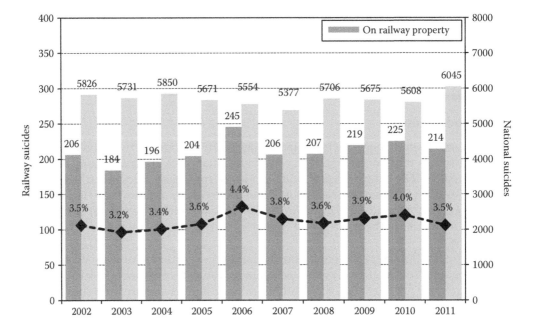

FIGURE 14.2
Railway suicide trends in the wider setting.

14.2 GB National Suicide Prevention Program

As train performance declined in terms of delays from the turn of the century, the rail industry sought to understand the causal factors. A significant number were the result of infrastructure failures. As these were addressed, nonmechanical factors became more apparent. Suicide was identified in this category. Initially the industry sought to manage trespass on the network in a bid to prevent suicides. Some early success was achieved but through time the number of suicides showed a general upward trend as did the delays they caused. The impact of this financially on the regulated railway was significant. Passenger satisfaction and journey experience was also being undermined as was the morale and health of staff who were directly involved in these events or managing them in their aftermath.

Whilst these factors were the catalyst for the industry's suicide prevention program, there was a recognition that it did not possess the skills "in house" to manage and mitigate against the complex societal and personal issues that drove them without professional support. This realization led to discussions taking place in 2009 with Samaritans, the UK's largest suicide prevention charity and experts in this arena and in 2010, a formal contractual relationship being signed with them.

So began a 5-year partnership which through the introduction of a national strategy sought to reduce railway suicides. The partnership involved the roll out of a program of prevention and post-incident support initiatives to reduce the impact of suicide. These include multiagency partnership working at national and local level, bespoke training of railway industry staff, a national public awareness campaign (Figure 14.3), a volunteer call out service providing emotional support to people in distress at railway locations, post-incident support provided by local Samaritans branches, and work to encourage responsible media reporting of suicides.

As mentioned earlier, a number of suicides occur at level crossings and signage has been placed at these sites to encourage vulnerable members of the public to call Samaritans (see Figure 14.4).

The two targets established at the outset of the program were for a reduction of rail suicide (by 20%) and for 10,000 rail staff to receive training (although the type of training or proportion of staff to be trained was not set).

Table 14.1 summarizes the program activities at the time of the evaluation and details the level at which they are being developed and delivered.

FIGURE 14.3
Posters from the public awareness campaign.

FIGURE 14.4
Signage at a level crossing.

TABLE 14.1

Summary of Program Activities

	At a National Level	At a Local Level
Partnership working	• National suicide prevention steering and working groups • Development of guidance and policies • Appointment of program support teams and leads in key organizations (Samaritans, Network Rail, TOCs) • Collation and dissemination of data centrally (by Network Rail, Samaritans, RSSB, Association of Train Operating Companies [ATOC])	• Local engagement/development of local suicide prevention plans • Station audits • Third-party engagement and outreach activities
Prevention activities	• Design and delivery of public awareness campaigns and information materials for stations and rail staff • Design and delivery of MSC and ESOB (emotional support outside branch) training (for local Samaritan branches) • Coordination of the ESOB service	• Priority location identification • Recruitment of station staff to MSC training • Public awareness (poster) campaign, Samaritans metal signs, and distribution of information for station and NR staff • Physical mitigation measures • Call out of Samaritans on identification of a vulnerable person
Postvention activities	• Development and delivery of trauma support training for management and unions • Development of driver fatality guidance • Development of guidance to prevent copycat suicides (media guidance, memorials policy)	• Recruitment to trauma support training • Post-incident visits to stations by Samaritans to support staff and public who have witnessed or been involved in fatal and nonfatal incidents

14.3 Evaluation Challenges

Undertaking an evaluation of a program of this nature is challenging for a number of reasons. As Table 14.1 shows, the program is very complex and operates at both national and local levels, includes a large number of different interventions that have multiple organizations involved in their delivery, is constantly evolving as more is learned, and partners are themselves changing their practice through their involvement.

Although a number of evaluation activities were undertaken during the first year of the program, the detailed evaluation activity reported in this study did not commence until the second year of the program, meaning that it was too late to undertake any robust before and after measures, apart from in areas for which data is available from the years prior to the introduction of the program (such as data on the number of suicides and delays caused by suicide).

A program of this size and complexity means that it can be difficult to determine and keep a record of the measures that have been implemented at individual locations and the date of their installation. Another challenge is that suicide prevention measures are usually not implemented in isolation. For example, a station may decide to display Samaritans signage on a platform. However, they may also release their platform staff to attend the Samaritans managing suicide contacts (MSC) training. For these reasons, identifying the effectiveness of individual measures is problematic.

An evaluation of this kind needs to consider "counterfactuals." The program is embedded within a wider environment inside which a range of activities are taking place. These may impact on the program, for example, activities to restrict access to the track or British Transport Police (BTP) activities to speed up response to incidents. It is also important to consider the impact of the social context in which the program exists, and the impact at macro and micro level of societal changes such as the recession.

The evaluation of complex interventions is always challenging because it can be difficult to establish clear cause and effect relationships between the interventions and their outcomes/impacts (attribution). One approach that is increasingly being used in these circumstances is a "Theory of Change" evaluation framework, which seeks to map the pathways between different elements of a program and their intended outcomes. Figure 14.5 shows an example theory of change map from this study.

The theory of change maps developed was used to both identify the different sources of data that can be used to establish "progress" toward the intended outcomes and impacts, and as a framework for bringing the different sources of data together in the final analysis. Complex evaluations using theory of change maps involve the triangulation of different sources of data involving a mix of both qualitative and quantitative data.

The evaluation of the Network Rail/Samaritans program addressed the following overarching questions:

- To what extent have program activities, and activities enabled or supported by the program, led to a reduction in loss of life from suicide on the railways?
- What evidence is there that the program has reduced disruption to services and distress, or lowered the cost resulting from suicides, to the railway industry?
- Has the program contributed to improved partnership and interagency collaboration in the prevention and mitigation of the impact of suicide, across the rail industry (nationally and locally)?

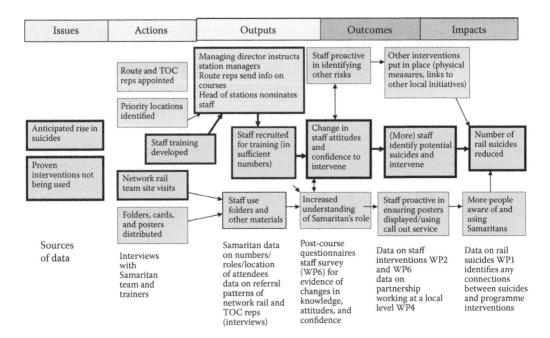

FIGURE 14.5
Example theory of change map for managing suicide contact training.

- What is the evidence that individual elements of the program have been effective in themselves, and contributed toward the overarching objectives of the program?
- To what extent has the program supported the implementation and promotion of best practice in prevention and mitigation of the impact of suicide across the rail industry?
- What has been learned that can support future strategy?

To support these overarching evaluation questions, a number of work packages involving data collection and analysis were being undertaken. Table 14.2 provides detail on the main data collection activities.

14.4 Results

The full results of this evaluation are reported in RSSB (2013a) and the following sections provide a selection of the key findings that answer the first overarching questions mentioned previously.

14.4.1 Program Coverage

The number of locations and individuals involved in the program activities has grown year on year. For example, Figure 14.6 shows that the number of staff that has attended the Samaritans MSC training per year has increased. At the date of writing this chapter,

TABLE 14.2

Summary of Evaluation Activities

Work Package Number	Work Package Area	Main Activities
WP1 and 2	• Prevention—has there been a reduction in suicide numbers? • Postvention—has there been a reduction in disruption and distress?	• Statistical analysis of suicide data and performance data • Collation of information on where program interventions have taken place and their impact • Collection of counterfactual data—information on other interventions initiated outside to reduce disruption caused by suicide, impact of societal elements
WP3 and 4	• National partnership working • Local partnership working	• Analysis of documentation from national steering group • Partner survey aimed at gathering views of the success of the partnership and any barriers to its roll out, changes to company's policies or activities as a result of the program and individual's attitudes to suicides. • Eight station site case studies—interviews with all relevant parties
WP5	Case studies of specific program subactivities	Data from other work packages on implementation and whether the subactivity has reached the intended audience, the impact it has had on those involved and whether the wider program has supported the roll out of the particular intervention
WP6	Staff attitudes and experience	Questionnaires to collect of data from front line staff on their knowledge of involvement in the program, attitudes to suicide, likelihood to intervene, and personal experiences with suicidal contacts
WP7	Overview/collation of findings	Overview of other work packages, collate overall picture, highlight areas of best practice for future strategy development and dissemination

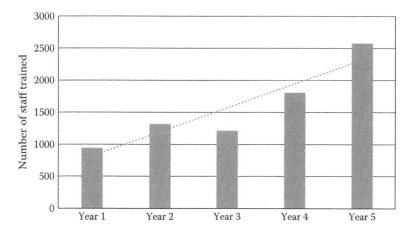

FIGURE 14.6
Number of staff attending MSC training per year.

over 8000 frontline staff had attended the MSC training. In the first year of recording interventions, there were five interventions reported. In year 5, this figure has increased to 203 reported interventions.* However, implementation has often been patchy, largely because of the size and complexity of the rail network and the number of organizations involved.

14.4.2 Preventing Rail Suicide

At the time of the evaluation, the number of rail suicides (as shown in Figure 14.1) showed no sustained reduction since the program commenced. An overall comparison of the 3 years prior and 3 years since the launch of the program in 2010 shows that there have been an average increase of seven events per year. This increase does not however represent a statistically significant increase over this period. These figures need to be seen in the context of the fact that the program has not yet been fully implemented across the country.

As Figure 14.2 shows, there is some evidence that rail suicides account for a higher proportion of national suicides than a decade ago although there has been no significant change in recent years or over the period during which the program has been implemented. National suicide figures have also been impacted by the fact that coroners are now more frequently using narrative verdicts where there is uncertainty over the person's intent which may account for variations recorded at a national level. As the Statistical Bulletin published by the Office of National Statistics (ONS) (2013) reports, a guidance note was issued to coroners in 2011 to ensure that they provide enough detail in narrative verdicts. This allowed coders at the ONS to have more information when coding which may potentially have increased events coded to intentional self-harm. For railway suicides however, in most cases there is enough evidence to support the coding of suicide; the Ovenstone criteria (RSSB, 2013a; Evans, 2013) is used to categorize cases for which there is no definitive coroner's verdict. Further work to understand how rail suicides have been coded compared to ONS-coded suicides is ongoing.

Further analysis looked at those stations considered a priority in 2012. Stations were designated as priority locations using a risk-based approach, taking into account the suicide history, delay minutes, and hazard rating based on station layout and through traffic at the station. These stations had substantially more program activities directed at them than those not designated as priority locations.

Figure 14.7 shows the number of suicides at stations designated as priority stations in 2012. The analysis of priority locations showed that 3 years before the program commenced in 2010 the number of suicides at priority locations averaged around 78 per year. The 3 years since the program there have been on average 59 at priority locations, which is closer to the long-term average at those locations.

This result may be interpreted as evidence that the designation of these stations as priority locations and the resultant program activities has led to a reduction in suicides. However, without detailed information about when and what interventions were taken at each of these stations it is difficult to ascertain concrete support to this theory. An alternative hypothesis may be that this reduction is nothing more than a reversion to the mean, in other words the drop in suicides at priority locations after they were designated could simply be the result of an unusual "spike" in suicides at these sites which then led to this designation.

---

* Include reference to Samaritan's end of period report.

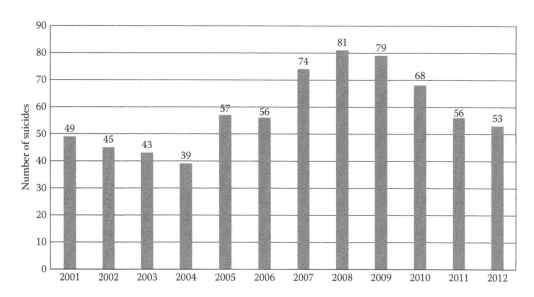

FIGURE 14.7
Railway suicides at priority locations in 2012.

14.4.3 Reducing the Impact of Rail Suicide

It is widely acknowledged that railway suicide is very disruptive and costly. In GB (Great Britain) railways there are three significant parties that bear the costs associated with suicide, these are Network Rail, BTP, and the train operating companies affected by the incident. This research estimated the average annual cost of suicides to the industry to be on average £34 million per year.

There has been no obvious downward trend in the delay and cancellation costs that arise from suicides. As none of the program activities were specifically targeted at reducing delay, this finding is directly related to the fact that there is currently no evidence of a reduction in suicides on the railway.

There is some evidence that response times to suicides are improving. This requires BTP, Network Rail, and the train operators to work together to resume the train service as quickly as possible. Response times between 2009 and 2012 show that the average time for the train service to resume following a suicide has reduced from 2 h 44 min to 1 h 59 min.

The main area in which the program may influence the disruption caused by suicide (in addition to reducing suicide) is in the activities designed to reduce staff trauma. The training aimed at managers and supervisors of staff who may be exposed to vulnerable individuals seeking to take their lives has not yet been established long enough for the changes to be fully assessed. It is interesting to note that some prevention activities, such as the MSC training and Samaritans information distributed at priority locations, were seen as being as helpful in reducing the distress caused by suicide as they were in reducing the level of suicide itself.

14.4.4 The Role of Partnership Working in Suicide Prevention

Establishment of the National Suicide Prevention Steering Group (NSPSG) and National Suicide Prevention Working Group (NSPWG) has provided the opportunity for different organizations to meet, discuss, and work collaboratively to deliver the program together.

A survey was distributed to those individuals involved directly in implementing the program at a national level (NSPSG and NSPWG), those with a supporting function (BTP) and those that facilitate implementation by virtue of their role, such as Train Operating Company (TOC) managing directors. Eighty-seven percent of respondents felt that the program had improved partnership working and 77% and 72% felt that program activities had reduced staff distress and service disruption following a suicide, respectively.

Responses to the partnership survey suggested that respondents felt that relevant organizations are working well together to prevent suicides and reduce service disruption following an incident. However, only 37% of respondents felt that identified good practice was being effectively implemented on a national basis suggesting that communication between the steering group, working group, and other partners at a local level could be improved to ensure that industry strategy and activities are effectively rolled out. Qualitative responses suggested that there was a need for a coherent communications strategy to support strategy identified at the national level and delivery of the program at the local level. On a positive note those respondents working at the local level felt that local groups were an effective means of achieving partnership working and that priority location identification was a useful way of targeting program activities.

Respondents were also asked about which organizations they felt could have the most influence on suicide reduction and the perceived efforts of different organizations in this aim. Over 75% felt that Network Rail, Samaritans, and BTP can influence the occurrence of suicides significantly, and 67% felt that the National Health Service had a key part to play. All of the representatives from TOCs recognized their role in influencing the occurrence of suicides, however over 40% felt that their potential influence was less than the previously mentioned organizations. When asked about the current efforts in suicide prevention, the majority of respondents felt that the level of effort made by Samaritans and BTP was about right, and however that the TOCs, Network Rail, trade unions, and the National Health Service needed to do more.

14.4.4.1 Staff Perceptions of the Program and Individual Suicide Prevention Measures

The data in this section is based on the results to a frontline staff survey distributed in 2013 to train drivers, driver manager's, platform staff, BTP officers, local operations managers (LOM), and mobile operations managers (MOM). There were 732 respondents to the survey from a range of different ages, ethnicity, religion, relationship status, and geographical location.

14.4.4.2 Participation in Program Activities and Priority Locations

Table 14.3 shows the number of respondents to the survey who had participated in individual program activities. Participation in program activities did vary according to route. There were no significant differences in participation in program activities according to age, religion, and ethnicity.

Respondents that worked at priority locations were significantly more likely to have participated in the following activities compared to respondents that did not work at priority locations:

- MSC training
- Local suicide preventions plan
- Station audits/risk assessments
- Engagement with organizations outside of the rail industry

TABLE 14.3

Number of Respondents That Had Participated in Program Activities

Program Activity	No	Yes	Planned for the Future
Managing suicidal contacts training	421 (69.5%)	171 (28.2%)	14 (2.3%)
Samaritans information for staff	431 (71.6%)	164 (27.2%)	7 (1.2%)
Local suicide prevention plans	511 (85.5%)	77 (12.9%)	10 (1.7%)
Station audit or risk assessment	494 (83.0%)	90 (15.1%)	11 (1.8%)
Engagement with organizations outside the rail industry	455 (76.0%)	138 (23.0%)	6 (1.0%)

They were also more likely to have the utilized Samaritans emotional support outside of the branch, posters, metal signs, and information for staff at their stations.

14.4.4.3 Perceptions of the Program and Individual Activities

The survey showed that there was a willingness of staff to be involved in the program. Table 14.4 shows the perceived effectiveness of individual program activities. There were a small minority of staff who held the belief that the program activities had led to an increase in the number of suicides. Overall, perceptions about the individual activities were reasonably positive in terms of reducing the number of suicides. Respondents were most positive about physical barriers and the managing suicidal contacts training.

14.4.4.4 MSC Training

The results of the survey showed that a high proportion (69.9%) of frontline staff would make an intervention if they saw someone they thought might take their life on a platform. It was positive that such a high percentage of staff would intervene if they suspected someone may take their life. There was no significant difference in the percentage of staff that would make an intervention between respondents that had and had not attended the Samaritans MSC training. However, respondents that had attended the training were significantly more likely to identify someone on a platform that may take their life.

Just over a quarter of respondents had not reported an intervention they had made. Some of the reasons given for not reporting an intervention were that staff did not feel it was necessary to report it, intervening was just part of their normal day job, the intervention was recorded elsewhere, or their manager was not around at the time. There

TABLE 14.4

Effectiveness of Individual Program Activities

	Would Increase the Number of Suicides	Do Not Know	Would Decrease the Number of Suicides
Physical barriers (n = 513)	38 (7.4%)	137 (26.7%)	338 (65.9%)
Managing suicidal contacts training (n = 490)	27 (5.5%)	154 (31.4%)	309 (63.1%)
Posters (n = 502)	33 (6.6%)	152 (30.3%)	317 (63.1%)
Local suicide prevention plans (n = 481)	23 (4.8%)	174 (36.2%)	284 (59.0%)
Engagement with organizations outside of the rail industry (n = 480)	26 (5.4%)	173 (36.0%)	281 (58.5%)
Samaritans information for staff (n = 488)	30 (6.1%)	183 (37.5%)	275 (56.4%)

were still barriers cited to intervening. Examples of some of the barriers to making interventions were: "I would be worried be that the individual may become aggressive," "I would be worried that I would not be successful and they may still attempt suicide," and "I would be worried that my suspicions were incorrect and they didn't really need help." Respondents who had attended the MSC training were less likely to cite barriers to intervening.

The survey also showed that respondents that had attended the MSC training were more likely to take desirable actions when making an intervention and less likely to take undesirable actions. Desirable actions included: approaching the person, introducing yourself, asking their name, and encouraging the person to talk about their troubles and listening carefully. Examples of undesirable actions are physically restraining someone and giving them advice about their problems. Post-incident, respondents that had attended the MSC training were also more likely to take desirable actions. Desirable actions included asking colleagues about how they were feeling about the incident, using Samaritan services or an employee assistance program and requesting Samaritans emotional support outside of the branch service. An undesirable action would be to allow the family or friends of the deceased to leave a memorial at the location.

Overall the survey has shown that respondents who had attended the MSC training were more likely to identify a vulnerable person on a platform and take positive actions if they make an intervention and after a suicide incident. It was positive that a high percentage of staff would make an intervention but the MSC training did not mean that staff was more likely to make an intervention. There are also still improvements that could be made in terms of reporting of interventions and addressing barriers to intervening. Improving the number of interventions and quality of actions taken when intervening will impact on the number of suicides on the railway and the impact of suicides in terms of reducing costs and delays to services.

14.4.5 Attitudes to Suicides

Respondents were asked to rate the extent to which they agreed or disagreed with the validated attitude to suicides (ATTS) question set (Renberg and Jacobsson, 2003). This scale was developed to measure attitudes toward suicide in the general population and was adapted to make it suitable for front line rail staff. The scale did not have a consistent factor structure. Therefore a factor analysis was conducted to identify the key themes in the question set and these are shown in Table 14.3 (for more details refer to RSSB, 2013b). A mean average was calculated for each factor identified in the factor analysis (see Table 14.3). Averages close to 5% strong agreement and averages close to 1 represent strong disagreement (Table 14.5).

Obligation to prevent, preventability, and acceptance of suicide factors are all positive attitudes that should result in motivation to try and prevent suicide, and demonstrate sympathy toward suicidal people. The staff is generally quite strong on these attitudes, indicating a good foundation for the program.

Analysis of the ATTS scale identified a number of influencing factors on attitudes toward suicide such as religion, gender, ethnicity, marital status, previous experience of suicide, job role, and organization. It found that people who had experienced a suicide attempt at work, or in their personal life (by a friend, family member, or acquaintance) generally had ratings indicating a higher acceptance and openness about the issue of suicide. These respondents also had a high rating on the factor related to having a strong obligation to prevent suicide.

TABLE 14.5

Overview of ATTS Factor Ratings

ATTS Factors	N	Mean	Standard Deviation	Desirability of Scores
Obligation to prevent	692	4.11	0.864	High scores are desirable because they indicate a stronger feeling of obligation to prevent suicide
Preventability	682	3.24	0.763	Higher scores are desirable because they indicate that suicide is more preventable
Openness/acceptance of suicide	672	3.03	0.841	Higher scores are desirable because they indicate a high level of acceptance of suicide. Lower scores might be indicative of hostile attitudes toward people who are suicidal
Reasons why people take their own life	682	2.91	0.555	Higher scores indicate more agreement that people take their own lives because of mental health problems, inter-personal conflict, or as a cry for help
Unpredictability of suicide	675	2.86	0.446	Lower scores are more desirable because they indicate that suicide is more predictable. People with higher scores may feel suicide cannot be prevented because it cannot be predicted
Suicide as a taboo subject to talk about	671	2.72	0.427	Lower scores are more desirable because they indicate that suicide is a subject that should be talked about
Suicide as a long-lasting issue	676	2.42	0.574	Lower scores are more desirable because they indicate that once a person has suicidal thoughts they can be overcome. People with lower scores may feel more motivated to try and prevent suicide because they do believe that the person can recover

Respondents who had not experienced a suicide at work had higher ratings on the factor indicating that they were more likely to agree that suicide is a subject that should not be talked about.

The survey results show that the majority of front line staff is generally willing to take part in or use the program activities. Respondents' willingness to engage in program activities was positively linked to having a strong "obligation to prevent suicide." However, a reluctance to engage in program activities was associated with the attitude that suicide is a long-lasting issue (hard to address) and a difficult subject to talk about. These results were significant but the correlation coefficients were weak, indicating that these were not strong relationships.

There was also evidence that people who feel more obligated to prevent suicide, and feel more strongly that it is preventable, may be more likely to take positive actions following a suicide while people who feel that suicide cannot be overcome may be less likely to take positive actions.

As the surveys only represented a "snap shot" of attitudes to suicide at a single point of time, it is not possible to directly compare attitudes before and after people have taken part in program activities. However, a comparison of attitudes indicated that people who have participated in program activities generally reported holding more positive attitudes toward suicide and its prevention than those who had not participated. Overall, the MSC training has not had a dramatic impact on staff attitudes. As already reported attitudes varied depending on many factors including religion, ethnicity, previous experience of suicide, etc. It is likely that the impact of these other factors outweighed the influence of a 1-day training course.

14.5 Conclusion

Evaluation of a program of this nature is challenging. Reliance on simple indices such as the change in number of suicides—particularly when they are so low in number—is not sufficient and a more sophisticated approach that seeks to identify the relationship between the interventions and the full range of expected outcomes/impacts is required.

Early analysis of the program's coverage has shown that there are wide variations in the level of implementation across the country. This variation in delivery of program activities is likely to have had an influence on the overall effectiveness of the program to date, but is currently being addressed through establishing a stronger leadership team within Network Rail and more full time "route representatives" to support the implementation of activities on the ground.

At this stage, there is no evidence that the overall level of rail suicide, either in absolute terms or proportional to overall national suicide levels, has been reduced since the program was set up. Nor is there currently evidence of impacts in terms of delay and disruption caused by suicide, although the time taken to respond to suicide, and resume normal services has been reduced as a result of changes in BTP practice. Providing evidence of change as a result of program activities is challenging on account of the high level of fluctuation in the number of suicides. A small number of incidents can result in considerable levels of disruption and delay when taking place at a peak travel time. It has also been difficult, because of the dispersed nature of the network and program activities, to gain accurate data on where the program has, and has not, been fully implemented.

The research clearly identified that the promotion of partnership and more "joined up" response to rail suicides at both a national and local level is a key underlying principle of the program. There is considerable qualitative evidence to suggest that the program has promoted closer collaboration between different organizations seeking to address rail suicide. However, the evaluation identified the need to communicate the program's strategy in a more consistent fashion. The industry partner survey also identified the need to engage and work more closely with the public health authorities.

Staff is generally strong on attitudes relating to the obligation to prevent suicides, preventability, and acceptance of suicides which are all positive attitudes that should result in motivation to try and prevent suicides and sympathy toward suicidal people. Those staff who participated in program activities generally reported holding more positive attitudes toward suicide and its prevention and there is also strong evidence that the MSC training has led attendees to better understand how to approach vulnerable individuals they have identified and indeed encouraged staff to intervene to prevent suicides on a number of occasions. Given the large costs associated with suicides, the prevention of a small number of suicides per year would be sufficient for the scheme to be cost beneficial.

14.6 Program Development Post Evaluation

Five years on from the inception of the program, and following the evaluation, it now appears very different:

- The original drivers which were biased toward financial and performance loss are now secondary to reducing the traumatic impact suicides have on rail staff and passengers alike
- The industry now boasts professionals within its ranks to determine the approach and direction of the program
- BTP have become more prominent in the prevention program with a dedicated unit part funded by Network Rail proactively engaging with public health authorities
- Within the industry there is a growing understanding that physical mitigation measures alone, whilst important will not stop suicides occurring on the railway and that a holistic community approach to prevention has to be adopted through collaborative relationships with local health authorities, with the industry being viewed as a key player in the nation's attempt to reduce suicides in the United Kingdom as a whole. This is reflected in their relationships with the Departments for Transport and Health and the National Suicide Prevention Alliance

A follow-up evaluation project for the program is currently underway. A key focus of this project will be to conduct in-depth case studies on new measures. This will involve collecting qualitative data to indicate the effectiveness of new measures and identify any negative impacts they may have, for example, on driver sighting at platforms. In addition to this, the impact of measures implemented at a large enough number of stations will be analyzed statistically to indicate their effectiveness. The study will also look at the latest suicide figures, how these compare to the national suicide statistics, and how suicide figures vary across the country. There will be another frontline staff survey that will include the ATTS question set. The results will be compared to those from the initial evaluation to understand how the program has evolved and whether this has impacted on the attitudes and behaviors of frontline staff.

References

Evans, A. 2013, The economics of railway safety, *Research in Transport Ergonomics*, 43(1), 137–147.

Office for National Statistics. 2013, *Statistical Bulletin—Suicides in the United Kingdom 2011*, Retrieved March 3, 2014, from http://www.ons.gov.uk/ons/dcp171778_295718.pdf

Renberg, E. and Jacobsson, L. 2003, Development of a questionnaire on attitudes towards suicide (ATTS) and its Application in a Swedish population, *Suicide and Life-Threatening Behavior*, 33(1), 52–64, The American Association of Suicidology.

RSSB. 2013a, *Annual Safety Performance Report 2012/13—A Reference Guide to Safety Trends on GB Railways*, Retrieved March 3, 2014, from http://www.rssb.co.uk

RSSB. 2013b, *Improving Suicide Prevention Methods on the Rail Network in GB 2013—Annual Report*, Retrieved March 3, 2014, from http://www.rssb.co.uk

Section III

Road Domain

Introduction

The rail and road domain is by far the largest subsector of transportation, employing about nine out of 10 transport workers and also representing a significant percentage of the GNP (gross national product) of the entire world economy. Given its size, it is, however, also responsible for the most important problems connected with the entire transport sector: from environmental problems caused by harmful emissions to strictly ergonomic issues, and to the psychosocial stress induced on whole populations, groups, or individuals users of transport systems.

The 10 chapters within Sections II and III of the book are divided into two groups, respectively, related to the road transportation (Chapters 10 through 14) and rail transportation (Chapters 15 through 19). You will notice how, starting from fields of application that are often very specific, but which have references to wider issues, it is possible to face the broader issue of the pervasiveness of new technologies and its impact on the evolution of social and cultural aspects of contemporary society.

In particular, the research and experiences described in these chapters relate largely to the behavior of drivers, cyclists, pedestrians, users, and to the relationships established between them, and with respect to the means of transport, infrastructure design and driving assistance systems that are increasingly advanced and complex.

Neville Stanton

15

Analyzing Eco-Driving with the Decision Ladder: The First Step to Fuel-Efficient Driving for All

Rich C. McIlroy and Neville A. Stanton

CONTENTS

15.1 Introduction

Many modern vehicles come with a means for providing feedback and advice to the driver about fuel efficiency, be they traditionally powered (i.e., by internal combustion engine [ICE]) or powered by more novel means (e.g., hybrid, electric, hydrogen). A fairly well-established finding is that the way in which a car is driven significantly affects the amount of fuel used—a difference that varies from around 15% in ICE vehicles (Evans, 1979; Waters and Laker, 1980) to as much as 30% in electric vehicles (Bingham et al., 2012). To support such fuel-efficient driving styles would clearly be a worthwhile activity; however, there exists great variance in the ways different vehicle manufacturers go about this, in terms of the actual type of information presented, and in the way it is presented.

This chapter therefore presents the first step toward the design of a fuel-efficiency support system that does not provide feedback about current efficiency levels, but aims to support the very behaviors that characterize fuel-efficient driving. To achieve this goal it is first necessary to understand those behaviors, hence the focus of the current research is the analysis of the decision-making processes made when driving in an economical fashion.

Following a review of both the academic literature and of publicly available web-based eco-driving resources, four specific driving activities that can each have a significant effect on fuel efficiency were identified. These activities were modeled using Rasmussen's decision ladders (Rasmussen, 1974), an analysis technique that models activities in decision-making terms. Then followed a number of interviews with experienced "eco-drivers" (i.e., subject matter experts) with the resulting information serving to amend, supplement, and validate these decision ladder models. One of the completed models is then discussed, largely in terms of the skills, rules, and knowledge (SRK; Rasmussen, 1983) theoretical framework, and in terms of its contribution to the design of a new, in-vehicle information system. A more detailed report of this research has also been published as a journal article (McIlroy and Stanton, 2015); this includes a discussion of all of the completed decision-ladder models.

15.2 SRK and Decision Ladders

15.2.1 The SRK Taxonomy

The SRK taxonomy theoretical framework (Figure 15.1) describes three different levels of cognitive control with which actors interact with their environment (Rasmussen, 1983). Skill-based behavior (SBB) involves automatic, direct interaction with the environment; rule-based behavior (RBB) involves associating familiar perceptual cues in the environment with stored rules for action and intent; knowledge-based behavior (KBB) involves

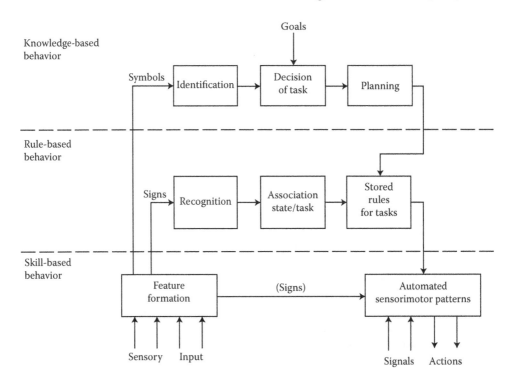

FIGURE 15.1
Graphical representation of the SRK taxonomy. (Adapted from Rasmussen, J. 1983. *IEEE Transactions on Systems, Man, and Cybernetics*, 13(3), 257–266.)

analytical problem solving based on symbolic reasoning and stored mental models (Vicente, 2002). Typically, interaction in unfamiliar or unanticipated events, and novice interaction (akin to interaction in unfamiliar events) will proceed using KBB, whereas expert interaction, and interaction in highly routine and familiar situations will more often proceed with SBB. The theory can also be applied to the process of learning, insofar as an individual, starting as a novice, will usually interact with a task at the KBB level. As experience grows, behavior will progress through RBB to SBB, whereby actions become routine and automatic. In this sense the theoretical framework bears resemblance to earlier descriptions of learning from the field of psychology, such as the conversion of declarative to procedural knowledge (Anderson, 1976, 1983).

Declarative knowledge refers to information in individual fragments that are stored separately, for example, knowledge of facts, events, and relationships, while procedural knowledge represents knowledge of how to do things, for example, complex motor skills, cognitive skills, and strategies. Behavior based on declarative knowledge requires effort-ful and time-consuming integration of knowledge fragments (Anderson, 1993). With pro-cedural knowledge, on the other hand, the retrieval of information required to guide behavior is said to be fast and automatic (Pirolli and Recker, 1994). As Anderson (1993) explains, it is the conversion of declarative knowledge to procedural knowledge, through the amalgamation (or aggregation, in Rasmussen's words) of individual pieces of informa-tion into coherent concepts, or higher-level chunks that guide action, that characterizes skill development. These distinctions clearly resonate with the SRK philosophy; where KBB requires the operator to perform complex reasoning, reflecting on and interpreting information displayed in the interface (using declarative knowledge), perceptual-motor reasoning (skill- and rule-based) needs only recognition of familiar aspects of the task or problem to guide behavior (Glaser, 1984). Such similarities between the SRK and ear-lier descriptions of human cognition are by no means accidental; Rasmussen et al. (1990) expressly state that the SRK taxonomy "is compatible with the main-line of conceptualiza-tion within cognitive science and psychology (declarative vs. procedural knowledge ...)" (Rasmussen et al., 1990, p. 106).

15.2.2 Decision Ladders

The decision ladder, first described in detail by Jens Rasmussen in 1974, aims to provide a model of human data-processing that can be used "to facilitate the matching of the for-matting and encoding of data displays to the different modes of perception and process-ing used by human process controllers" (Rasmussen, 1974, p. 26). The diagrams depict the decisions that actors are required to make at different stages of a particular decision-making process (see Figure 15.2) and contain two different types of nodes; the rectangular boxes represent information-processing activities, the circles represent the resultant state of knowledge. For example, the information-processing activity labeled as diagnose state leads to knowledge of the current system state. The left portion of the diagram is con-cerned with an analysis of the situation and diagnosis of the current state of affairs, while the right side deals with the definition, planning, and execution of an action. The top of the diagram represents the evaluation of options and the consideration of specific goals pertaining to the task at hand.

The sequential arrangement of the rectangles (information-processing activities) and circles (states of knowledge) characterizes both the process of decision making through which a novice operator would progress, and the decision-making steps required during unanticipated and novel situations (i.e., at the knowledge-based level of cognitive control).

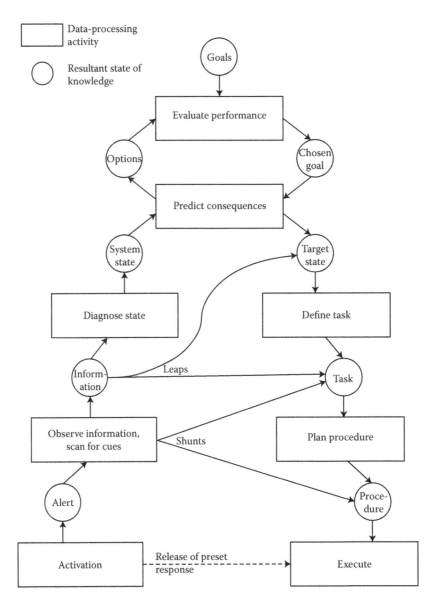

FIGURE 15.2
Decision ladder. (Adapted from McIlroy, R. C., and Stanton, N. A. 2015. *Ergonomics*, 58(6), 1–17. Doi: 10.1080/00140139.2014.997807.)

In these situations, and with novice actors, the top part of the diagram may be circulated around more than once. In these instances the decision maker may have to consider the various options available to him or her, and what affect each of these options will have on the chosen goal of the activity. Furthermore, there may be multiple, conflicting goals present in the decision-making task; each will require consideration.

Experienced actors rarely follow the linear sequence depicted in the decision ladder unless an unexpected situation or event is encountered; they are likely to take shortcuts. There are two types of shortcut defined in the literature (Jenkins et al., 2009; Vicente,

1999): shunts and leaps. Shunts connect data-processing activities to nonsequential states of knowledge while leaps connect two states of knowledge. The arrows in the center of Figure 15.1 represent these shortcuts. For example, in certain situations the process of diagnosing the system state may lead directly to the knowledge that a set procedure is required; such a shortcut is an example of a shunt. An example of a leap would be the association of knowledge of the current system state with the knowledge of a task that needs to be performed in order to, for example, get the system back to normal system operations. These shortcuts are often driven by rules and heuristics, learned through, for example, formal training and informal experience. Experienced actors may also enter the decision ladder at different nodes; they do not necessarily have to enter at activation and exit at execute. For example, an experienced actor may enter the decision ladder with an understanding of the current system state. From this they may infer, from past experience, the action required to achieve his or her given goal. Similarly, the activity may not necessarily flow from left to right, but can occur from right to left. For example, knowledge of the desired target state may lead an actor to observe for more information and cues to understand how this state may be achieved.

These shortcuts are indicative of rule-based behaviors (RBBs); they represent instances in which familiar perceptual cues in the environment are associated with stored rules for action and intent. SBB, the fast, automatic response to stimuli in the environment, is represented on the decision ladder by the arrow connecting activation with execute. Here, upon activation of the decision-making process, a preset response is released, resulting in the execution of a particular activity.

As previously described, the full decision ladder, when annotated for a given decision-making process, will represent the way in which an actor analyzes the situation, evaluates and selects goals, and plans and executes a task when using knowledge-based reasoning (i.e., follows the sequential path in its entirety), with all possible information inputs and options; this represents a prototypical model of activity (Jenkins et al., 2010). Rather than representing any one particular instance of an activity and the decisions therein (this would be a typical model of activity), the prototypical model aims to capture all possible elements that may affect the decision-making process (though not all will be used in any given situation). For example, Jenkins et al. (2010) describe the process of developing a decision ladder by means of asking a subject matter expert about a specific instance in which the activity of interest was performed. This supported development of a model of typical activity, that is, a particular example of an event that, in the case of Jenkins et al. (2010) has happened in the past. This typical model was then supplemented with all the additional and alternative information that may have been used, and the information that could be used in similar situations. This converts the typical model into the prototypical model.

According to Elix and Naikar (2008), the decision ladder approach can be used to inform the design of an interface; they do not, however, go into great detail on how this is to be achieved. Jenkins et al. (2010) go further in explaining how the generated prototypical models support an understanding of the relationships between the elements in the decision-making process. It is suggested that understanding the decisions to be made and the information sources that guide these decisions will help a designer to design an interface that more fully supports the operator in their task. Rasmussen et al. (1994) also make this point, stating that a designer must have a satisfactory understanding of the decision-making process of the potential user if they are to provide the correct information in the correct volumes in the interface. The process of developing the decision ladder supports such an understanding.

15.3 Modeling Fuel-Efficient Driving

For the analysis of the decision-making processes when driving in an economical fashion, it is necessary to first select specific situations, and in turn decision-making events, that have the most significant effect on fuel economy. This selection process serves to constrain, and give focus to the analysis. Hence a review of the available information on eco-driving was undertaken.

15.3.1 Activity Identification

Information on the driving styles that characterize a more economical use of fuel in the road vehicle is widely available, both in the academic literature, and across a plethora of more publicly available websites. Hooker (1988) offered one of the first descriptions of the specific driving styles that characterize economical driving. His research revealed that it is the style of acceleration and the timings of gear selections that have the greatest effect on fuel use in the vehicle. This is still the case in modern vehicles; Barkenbus (2010) states that eco-driving is characterized by (among other things) smooth acceleration, shifting up to the highest gear possible as early as possible (within the boundaries of safety), and anticipating the traffic flow and road layout ahead so as to avoid sudden starts and stops (i.e., to drive as smoothly as possible).

The concept of anticipation for eco-driving also features heavily in the more publicly available media, including specific eco-driving websites (e.g., ecodrive.org, 2013; Travelfootprint.org, 2013), motoring organizations (e.g., The AA, 2013), car manufacturers (e.g., Ford, 2013; Renault, 2013), local government (e.g., Devon County Council, 2013), and national and international nongovernmental organizations (e.g., Energy Saving Trust, 2013; United Nations, 2013). These resources offer advice not only on the style of driving that characterizes lower fuel consumption, but on the general maintenance of the vehicle as well. For example, removing unnecessary weight from the vehicle (e.g., not keeping the golf clubs in the car when they are not to be used), avoiding the use of air conditioning, and maintaining the recommended tyre pressures will all have a beneficial effect on fuel economy. This research is, however, only concerned with the types of driving styles and behaviors that characterize fuel-efficient use of the vehicle, that is, the driving task itself, hence these maintenance and peripheral use-related considerations were not included in our research.

This leaves us with two primary classes of driving behavior that significantly affect fuel economy. Behaviors related to use of the vehicle's gears, and behaviors related to use of the vehicle's accelerator and brakes. The second point can be further simplified to only use of the accelerator pedal; to minimize use of the vehicle's hydraulic brakes the driver must anticipate the road scene ahead in order to remove his foot from the accelerator pedal such that coasting down to a required speed can be achieved. This allows for smoother driving and, over the course of a route, reduces the amount of accelerator pedal usage (and therefore fuel usage).

Though the issue of gear choice behaviors is an important one in terms of the use of fuel in a manual transmission, ICE vehicle, we did not include this class of behavior in our analyses for two main reasons; first, the aim is to develop a system that is equally useful in both ICE vehicles and electric vehicles (which do not have gears in the same way ICE vehicles do); second, to reduce complexity and maintain focus. Hence only those behaviors associated with use of the accelerator pedal were considered.

TABLE 15.1

Eco-Driving Activities Selected for Analysis

Driving Activity	Description
Acceleration	Either from a standstill or from a lower speed to a higher speed. Though advice on fuel-efficient acceleration varies across information sources, there is a consensus that harsh, abrupt acceleration should be avoided
Deceleration (full stop more likely)	For example, when approaching a stop sign or traffic light at red. Early release of the accelerator pedal to take advantage of the vehicle's momentum to carry it to the stopping event is advised, that is, to minimize use of the brake pedal
Deceleration (full stop less likely)	For example, when approaching a bend in the road or going from a higher speed limit to a lower one. Again, early release of the accelerator pedal is advised in order to take advantage of the vehicle's momentum to carry it down to the required speed. Again, to minimize use of the brake pedal
Headway maintenance	Though this does not have a direct effect on fuel economy, the indirect effect of maintaining a sufficient distance to the lead vehicle allows for early responses to upcoming events and affords the driver a better view of the road ahead (i.e., it is less blocked by the lead vehicle) therefore again supporting early responses to upcoming road events. This is also largely about minimizing the need for brake pedal depression

Source: Adapted from McIlroy, R. C., and Stanton, N. A. 2015. *Ergonomics*, 58(6), 1–17. Doi: 10.1080/00140139. 2014.997807.

Based on the information provided in the academic and public literature, and on the aforementioned criteria, four specific activities were identified for modeling; these are presented in Table 15.1 alongside a brief description of why each is important in terms of fuel efficiency.

15.3.2 Eco-Driving Decision Ladder Validation

A decision ladder model was developed for each of the four activities listed in Table 15.1. The first iteration of the analysis was based on information gathered from web resources (e.g., specific eco-driving websites) and from the academic literature on eco-driving. A focus group was held at the University of Southampton's Transportation Research Group, the participants of which were four researchers, including the two authors of this chapter, each of whom possessed a working knowledge of human factors in road transport. Note, however, that none of the members of the focus group was an expert in eco-driving specifically. The group served both to validate the choice of activities, and to discuss the resultant models. It provided a platform for the discussion of the first iteration of the analysis. Table 15.2 provides a summary of the four participants' relevant information.

TABLE 15.2

Focus Group Participant Information

Participant	Gender	Age	Years Driving	Years Involved in Road Transport Research
1	Male	53	37	20
2	Male	27	4	2
3	Female	28	11	6
4	Female	25	8	2

Source: Adapted from McIlroy, R. C., and Stanton, N. A. 2015. *Ergonomics*, 58(6), 1–17. Doi: 10.1080/00140139.2014.997807.

Though the focus group discussions were useful for an initial attempt at model validation, the participants were not subject matter experts (i.e., they did not have specific eco-
driving experience). As such, a number of interviews with experienced "eco-drivers" were
arranged to further validate the models.

Participants were initially sought from two eco-driving websites: ecomodder.com and
hypermiler.co.uk. These websites provide a platform for those interested in both the technologies and behaviors associated with fuel-efficient driving, offering news of new technologies, advice on saving fuel when driving, and providing a space for the community
to discuss experiences and practices. A request for participation was posted to the forums
hosted on each website. From this, two individuals contacted the current authors; one was
a member of the forums on ecomodder.com, the other on hypermiler.co.uk. Two more
participants were contacted through the ECOWILL project, details of which can be found
from www.ecodrive.org. This European-wide project aims at providing information on
eco-driving to the general public, as well as undertaking formal, academic research into
various eco-driving aspects, including research involving driving instructors trained and
experienced in teaching eco-driving techniques to individuals.

In all cases, participation was entirely voluntary, without any payment (monetary or
otherwise). Due to the geographically dispersed nature of the participants (one each in the
United States, Germany, Scotland, and England), face-to-face interviews were not possible;
hence three interviews were conducted using Skype™, with the other conducted over the
telephone (as per this participant's preference). Each interview lasted approximately 1 h.
Relevant participant information is provided in Table 15.3.

To elicit information regarding each specific driving situation, a procedure similar to
that described in Jenkins et al. (2010) was followed. Each expert was introduced to the
decision ladder model and asked about his goals for each activity. The left-hand side of the
diagram was then populated with information regarding the cue or cues responsible for
bringing to attention the need for some action. Then the expert was asked to list the sources
of information he uses to build an understanding of the current state of the system, that is,
what cues in the environment will later go on to affect his decision-making process. The
top section of the diagram was populated through a discussion of the options available to
the driver and how these impact on the chosen goal, be it efficiency or otherwise. Then the
target state was discussed; this largely related to the selection of accelerator pedal position
at particularly points along the roadway. Finally, the task required to achieve this target
state and the necessary procedure were discussed.

TABLE 15.3

Interviewee Information

Participant	Gender	Age	Years Driving	Years Eco-Driving	Motivation	Primary Car Driven
1	Male	45	30	27	Financial and environmental	2003 Honda Civic Hybrid
2	Male	72	>50	30	Financial and "as a game"	Kia C'eed 1.6 diesel (year unknown)
3	Male	45	27	7	Environmental and through work	2004 Ford c-max
4	Female	42	25	9	Environmental and through work	2005 Audi A3

Source: Adapted from McIlroy, R. C., and Stanton, N. A. 2015. *Ergonomics*, 58(6), 1–17. Doi: 10.1080/00140139.
            2014.997807.

Following the discussions it became clear that "deceleration (full stop less likely)" was too broad a category, insofar as the information used to guide performance when approaching a road curvature was sufficiently different to the information used in other slowing events to warrant its own decision ladder. As such, this model was broken down into two separate models: "deceleration for road curvature" and "deceleration for other slowing event." For the purposes of brevity, only the "deceleration for road curvature" decision ladder will be discussed in detail here (Figure 15.3; see McIlroy and Stanton (2015) for descriptions of the remaining models).

15.4 Interpreting the Model

As this research is interested in the decision-making processes specific to eco-driving in particular situations, the goal of the activity being modeled was identified as "to decelerate from a higher speed to a lower speed in order to negotiate a road curvature whilst maintaining safety and minimizing overall fuel consumption for the journey."

The first annotated step on the left-hand side of Figure 15.3 is the alert; this indicates that the driver has been alerted to the curvature in the road ahead (i.e., it has been seen). The driver then scans for cues, from both within and outside of the vehicle. In terms of useful information, the driver may attend to, for example, the speedometer and tachometer, other road users, the road layout, markings, and signage both before and (if possible) after the road curvature, as well as physical movement (i.e., vestibular cues), visual momentum (i.e., the passing of the road scene outside the vehicle), the sounds of the engine, and car–road interactions (e.g., tyre noise at moderate-to-high speeds). These information sources allow the driver to establish an understanding of the road environment, the state of the driver's own vehicle (e.g., speed, acceleration, weight characteristics), the weather conditions, and the behavior of other road users.

In the top part of the diagram the driver may cycle through the potential options for action, and consider the effect that the current system state will have on these possibilities. For example, based on an understanding of the system state, the driver can estimate the effect of engine braking and different levels of hydraulic (i.e., traditional, brake–pedal initiated braking), and regenerative braking (where this is applicable) on the state of the system as a whole. For the purposes of this analysis, a primary goal is to be able to decelerate, in the most fuel-efficient manner, down to a speed that is appropriate to safely negotiate the road curvature. This is achieved by minimizing the use of the hydraulic brake pedal, or conversely, maximizing the use of the vehicle's momentum to carry it to the corner. Of course, safety will always be paramount in an on-road situation. The state of the system may therefore impact on the ability to turn the corner efficiently, for example in icy conditions or in conditions of heavy traffic flow.

The right-hand side of this upper section also has two other, potentially conflicting goals, namely to conform to social pressure and to reach the destination as quickly as possible. Each one of the subject matter experts raised both of these issues. One can imagine various situations in which speed is critical, from the emergency (e.g., a pregnant woman, going into labor, being rushed to hospital) to the relatively mundane (e.g., rushing home from work in order to get back before the plumber arrives). In terms of social pressure, this can come from both within and outside the vehicle. For those pressures coming from within the vehicle, one can imagine, for example, a situation in which a young driver succumbs to peer pressure to drive more aggressively (an established finding, particularly for

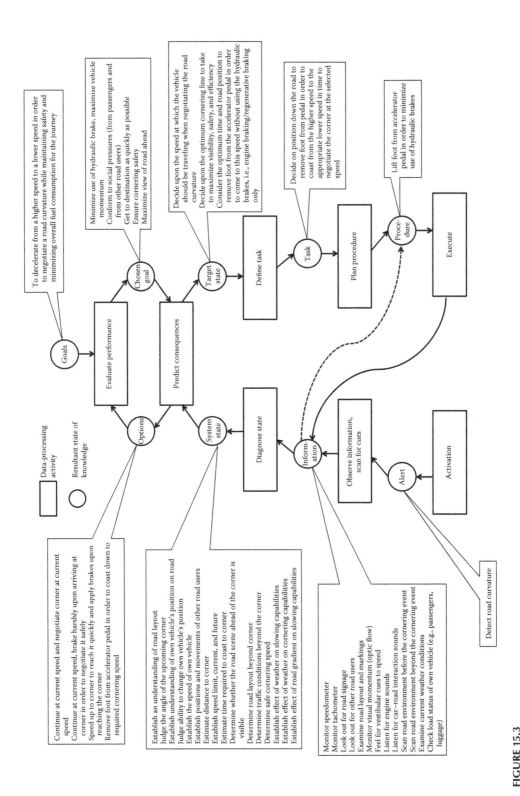

FIGURE 15.3
Decision Ladder for "deceleration for road curvature." (Adapted from McIlroy, R. C., and Stanton, N. A. 2015. *Ergonomics*, 58(6), 1–17. Doi: 10.1080/00140139.2014.997807.)

young men, e.g., Conner et al., 2003). Pressures coming from outside the vehicle relate to the behavior of other road users, for example, other drivers' use of their horns to influence the driver's behavior, or the act of driving very close to the rear of the driver's vehicle to encourage the driver to go faster (see, e.g., Åberg et al., 1997 for a discussion on the effect of the social environment on driver behavior and perceptions).

Moving down the right hand side of the diagram, the target state (assuming the goal of fuel-efficient negotiation of the corner) can be understood in terms of the use of the accelerator pedal, or more specifically, the time and road position (dependent on current speed) at which the foot should be removed from the accelerator pedal in order to coast, from the current speed, down to the required cornering speed. This involves an understanding of the current speed, the ideal speed for cornering, and the deceleration characteristics of the vehicle when using only engine braking (i.e., without the use of the hydraulic brake). This knowledge of the target state necessarily leads on to an understanding of the task, that is, when to remove the foot from the accelerator pedal, and the procedure, that is, remove the foot and minimize hydraulic brake use.

15.5 Discussion

The way in which an individual will progress from the alert stage to the execute stage will depend on a number of factors, for example, the characteristics of the driver (e.g., novice or expert) or the information available at a specific location (e.g., signage may differ, visibility may be different depending on time of day or weather). These shortcuts (i.e., the shunts and leaps) are often associated with actors of different experience; novices unfamiliar with a situation are usually expected to progress linearly through the diagram in its entirety (with notable effort), whereas experts may use a particular cue in the environment on which to base immediate action. Though it is experience that most commonly guides the shortcuts through the model, it may be possible to encourage them through the careful design of information presented to the driver in the vehicle. A primary aim of doing so would be to transform a cognitive task into a perceptual task. The question is, therefore, how do we support skill-based control in the novice eco-driver?

For some guiding principles we can turn to theory of direct manipulation interfaces (DMI) (e.g., Hutchins et al., 1986). This approach emphasizes the need to represent objects of interest and to allow the users to act directly on what they can see in the display; it both provides an "attempt to display the domain objects of interest and allow the operator to act directly on those objects" (Rasmussen and Vicente, 1989, p. 527) and allows the operator "to rely on the perceptual cues provided by the interface to control the system" (ibid, p. 525). Note that these quotes come not from DMI proponents, but from the creators of ecological interface design (EID; Rasmussen and Vicente, 1989; Vicente and Rasmussen, 1992), a design approach intimately linked with the SRK taxonomy. Both design approaches argue for the benefits of taking advantage of the human sensorimotor system, that is, to encourage behavior at the skill-based level. In terms of the task under analysis here, the fuel-efficient cornering of the road vehicle, one could imagine a system that informs the driver, through some salient stimulus, of the particular behavior required as well as the time at which that particular behavior should be performed (given the goal of fuel efficiency). When one considers that the task in question is largely related to use of the accelerator pedal (note that the "task" box on the right-hand side of Figure 15.3 talks of removing the foot from the pedal at a given road position), the possibility of providing tactile or haptic feedback through the accelerator pedal becomes one that satisfies both the tenets of EID (and, in turn, the SRK taxonomy) and those of the DMI approach.

In order to support SBB, an information system should provide information that encourages the driver to take a shortcut through the decision ladder as low down in the diagram as reasonably possible; in this case, this would likely support a "leap" from the alert that a corner ahead has been detected, to the knowledge of the task, namely to lift the foot from the accelerator pedal. This would support interaction via time–space signals, a necessary means for encouraging SBB, as the stimulus could come at a particular point on the road; this would be determined by a combination of spatial data and speed data (i.e., the faster a car is traveling, the earlier the signal should come to support a full coasting phase) and calculated using already-present information from car radar systems and satellite navigation information. Furthermore, following the suggestion that the operator should be able to act directly on the display, this time–space signal could be presented through the accelerator pedal, as a vibration (e.g., Birrell et al., 2013), or as an additional counterforce applied to the pedal (e.g., Mulder et al., 2010). This type of system, one that combines the action and control surfaces (i.e., the area onto which an action is performed is one and the same as the area from which information is garnered), would satisfy the theoretical arguments of both the SRK taxonomy and the DMI approach and should, in theory, support SBB in the driver.

15.6 Conclusions

This chapter has discussed the first step toward the design of an in-vehicle, eco-driving information system: the preceding analysis phase. The most influential in-vehicle activities or behaviors for fuel consumption, identified through a review of the academic literature and of more publicly available web-based resources, were modeled using Rasmussen's (1974) decision ladders. One of these, the "deceleration for road curvature" decision ladder, was presented and discussed in terms of the possibility for designing an in-vehicle information system that will support drivers, particularly those currently lacking in eco-driving expertise, to perform the eco-driving activities at the skill-based level of cognitive control. The model was also discussed in terms of the tenets of DMI, with the concept of combining action and control surfaces with accelerator–pedal-based haptic feedback offering a potential avenue for future research. While there are already examples of these kinds of systems in the extant literature (e.g., Birrell et al., 2013; Hajek et al., 2011; Mulder et al., 2011), the current research provides the first attempt to theoretically ground these efforts in existing descriptions of human control behavior and approaches to system design (see McIlroy and Stanton, 2015).

Acknowledgment

This research was jointly funded by the UK's Engineering and Physical Sciences Research Council (EPSRC) through the Transport and Environment program, and Jaguar Land Rover PLC.

References

Åberg, L., Larsen, L., and Beilinsson, L. 1997. Observed vehicle speed and drivers' perceived speed of others. *Applied Psychology*, 46(3), 287–302.

Anderson, J. R. 1976. *Language, Memory, and Thought*. Hillsdale, New Jersey: Erlbaum.

Anderson, J. R. 1983. *The Architecture of Cognition*. Cambridge, Massachusetts: Harvard University Press.

Anderson, J. R. 1993. *Rules of the Mind*. Hillsdale, New Jersey: Erlbaum.

Barkenbus, J. N. 2010. Eco-driving: An overlooked climate change initiative. *Energy Policy*, 38(2), 762–769.

Bingham, C., Walsh, C., and Carroll, S. 2012. Impact of driving characteristics on electric vehicle energy consumption and range. *IET Intelligent Transport Systems*, 6(1), 29–35.

Birrell, S. A., Young, M. S., and Weldon, A. M. 2013. Vibrotactile pedals: Provision of haptic feedback to support economical driving. *Ergonomics*, 56(2), 282–292.

Conner, M., Smith, N., and McMillan, B. 2003. Examining normative pressure in the theory of planned behavior: Impact of gender and passengers on intentions to break the speed limit. *Current Psychology*, 22(3), 252–263.

Devon County Council. 2013. Eco-driving film. Retrieved from http://www.devon.gov.uk/index/video/videotransport/ecodrivingvid.htm

ecodrive.org. 2013. The golden rules of eco-driving. Retrieved from http://www.ecodrive.org/en/what_is_ecodriving-/the_golden_rules_of_ecodriving/

Elix, B., and Naikar, N. 2008. Designing safe and effective future systems: A new approach for modelling decisions in future systems with cognitive work analysis. In *Proceedings of the 8th International Symposium of the Australian Aviation Psychology Association*. Sydney, Australia: Australian Aviation Psychology Association.

Energy Saving Trust. 2013. Driving. Retrieved from http://www.energysavingtrust.org.uk/Travel/Driving

Evans, L. 1979. Driver behavior effects on fuel consumption in urban driving. *Human Factors*, 21, 389–398.

Ford. 2013. Driving to lower fuel consumption and emissions. Retrieved from http://www.ford.co.uk/OwnerServices/FuelEconomyandEnvironmentalProtection/FuelEfficientEcoDriving Tips

Glaser, R. 1984. Education and thinking: The role of knowledge. *American Psychologist*, 39, 93–104.

Hajek, H., Popiv, D., Just, M., and Bengler, K. 2011. Influence of a multimodal assistance supporting anticipatory driving on the driving behavior and driver's acceptance. In M. Kurosu (Ed.), *Human Centered Design, HCII 2011, LNCS 6776* (pp. 217–226). Berlin: Springer-Verlag.

Hooker, J. N. 1988. Optimal driving for single-vehicle fuel economy. *Transportation Research Part A: Policy and Practice*, 22(3), 183–201.

Hutchins, E. L., Hollan, J. D., and Norman, D. A. 1986. Direct manipulation interfaces. In D. A. Norman and S. W. Draper (Eds.), *User Centered System Design: New Perspectives on Human-Computer Interaction* (pp. 87–124). Hillsdale, New Jersey: LEA.

Jenkins, D. P., Stanton, N. A., Salmon, P. M., and Walker, G. H. 2009. *Cognitive Work Analysis: Coping With Complexity*. Farnham, England: Ashgate Publishing Limited.

Jenkins, D. P., Stanton, N. A., Salmon, P. M., Walker, G. H., and Rafferty, L. 2010. Using the decision ladder to add a formative element to naturalistic decision-making research. *International Journal of Human-Computer Interaction*, 26(2–3), 132–146.

McIlroy, R. C., and Stanton, N. A. 2015. A decision ladder analysis of eco-driving: The first step towards fuel-efficient driving behavior. *Ergonomics*, 58(6), 1–17. Doi: 10.1080/00140139.2014.997807.

Mulder, M., Abbink, D. A., van Paassen, M. M., and Mulder, M. 2011. Design of a haptic gas pedal for active car-following support. *IEEE Transactions on Intelligent Transportation Systems*, 12(1), 268–279.

Mulder, M., Pauwelussen, J. J. A., van Paassen, M. M., and Abbink, D. A. 2010. Active deceleration support in car following. *IEEE Transactions on Systems, Man, and Cybernetics—Part A: Systems and Humans*, 40(6), 1271–1284.

Pirolli, P., and Recker, M. 1994. Learning strategies and transfer in the domain of programming. *Cognition and Instruction*, 12, 235–275.

Rasmussen, J. 1974. The human data processor as a system component. Bits and pieces of a model. Riso-M-1722. Roskilde, Denmark.

Rasmussen, J. 1983. Skills, rules, and knowledge; signals, signs, and symbols, and other distinctions in human performance models. *IEEE Transactions on Systems, Man, and Cybernetics*, 13(3), 257–266.

Rasmussen, J., Pejtersen, A., and Goodstein, L. P. 1994. *Cognitive Systems Engineering*. New York: Wiley.

Rasmussen, J., Pejtersen, A. M., and Schmidt, K. 1990. *Taxonomy for Cognitive Work Analysis*. Roskilde, Denmark: Risø National Laboratory.

Rasmussen, J., and Vicente, K. J. 1989. Coping with human errors through system design: Implications for ecological interface design. *International Journal of Man-Machine Studies*, 31(5), 517–534.

Renault. 2013. Eco Driving Tips. Retrieved from https://www.renault.co.uk/discover-renault/environment.html

The AA. 2013. Eco-driving advice. Retrieved from http://www.theaa.com/motoring_advice/fuels-and-environment/drive-smart.html

Travelfootprint.org. 2013. Reduce your Travel foot print—Eco-driving. Retrieved from http://www.ecolane.co.uk/

United Nations. 2013. United Nations Environment Programme: Ecodriving. Retrieved from http://www.unep.org/transport/Programmes/Ecodriving/

Vicente, K. J. 1999. *Cognitive Work Analysis: Towards Safe, Productive and Healthy Computer-Based Work*. Mahwah, New Jersey: Lawrence Erlbaum Associates.

Vicente, K. J. 2002. Ecological interface design: Progress and challenges. *Human Factors*, 44, 62–78.

Vicente, K. J., and Rasmussen, J. 1992. Ecological interface design: Theoretical foundations. *IEEE Transactions on Systems, Man, and Cybernetics*, 22, 589–600.

Waters, M. H. L., and Laker, I. B. 1980. Research on fuel conservation for cars. Report No. 921. Crowthorne, England.

16

If You Can't Beat Them, Join Them: Using Road Transport Methods to Communicate Ergonomics Methods' Outcomes to Aid Road Design

Miranda Cornelissen, Paul M. Salmon, Neville A. Stanton, and Roderick McClure

CONTENTS

16.1 Performance Variability in Road Transport

Performance variability is one of the most important facets of behavior for human factors and safety research. Road transport is a complex sociotechnical system (Larsson et al., 2010; Salmon et al., 2012; Salmon and Lenné, 2015). Many components such as road users, vehicles, infrastructure, and the environment interact and circumstances and demands vary. Performance variability is therefore a natural occurrence in road transport.

Driving a car under windy conditions is an example of a situation where performance variability is required in road transport. While driving, a driver will have to compensate for disturbances of the wind, for example, by steering in the opposite direction, and disturbances from other traffic, for example, by maintaining a safe gap. However, the exact pattern of the disturbances caused by the wind and subsequent traffic and driver actions cannot be predicted. Rather, drivers have to vary their performance in real time to safely negotiate the road transport system (Vicente, 1999). It is not possible to specify, *a priori*, a normative description of how to deal with windy conditions.

Despite its current importance in human factors and safety science, performance variability in a road transport context has received limited attention and is not well understood (Larsson et al., 2010). Much of the research to date has focused on performance variability as a negative phenomenon, such as human error (Hale et al., 1990; Hollnagel et al., 1999) and accident causation (England, 1981; Jenkins et al., 2011). Road safety education, vehicle, and infrastructure design are subsequently proposed to constrain variability (Larsson et al., 2010). This is contrary to the principles of performance variability which argue that variability is required for and contributes to the success of complex sociotechnical systems, but sometimes combines and leads to failure, and hence, performance variability should be managed and not constrained (Hollnagel, 2004, 2009).

16.1.1 Modeling Performance Variability

Recent research has addressed the need to better understand road user performance variability at intersections (Cornelissen et al., 2012c). The Strategies Analysis Diagram (SAD; Cornelissen et al., 2012a) was developed to better describe performance variability aiming to improve road transport designs that manage road user variability based on a detailed understanding of the interaction of road users, vehicles, and the road transport system (Cornelissen et al., 2013, 2015).

The SAD builds upon Cognitive Work Analysis (CWA; Vicente, 1999) and is a modification of the third phase of the framework, Strategies Analysis, to provide a more structured method for modeling performance variability. CWA is a popular systems theory-based modeling method used for designing and evaluating complex sociotechnical systems. It comprises five phases and through these phases describes systems based on its constraints and the behavior possible within those constraints (Vicente, 1999). First, Work Domain Analysis (WDA) describes system constraints from the purpose of the system down to the physical objects of which the system is made up to achieve this purpose. Second, the Control Task Analysis (ConTA) describes situational constraints acting upon the system and decision-making requirements necessary for the system to fulfill its functions. Third, Strategies Analysis describes different ways in which activities can potentially be carried out within the system's constraints. Fourth, Social Organization and Cooperation Analysis describes communication and coordination demands resulting from organizational constraints acting upon the system. Fifth, Worker Competencies Analysis describes the level of cognitive control required by various actors, human or technical, within the system.

16.1.2 Modeling Performance Variability in Road Design Process

A safe systems approach has been acknowledged as the underlying philosophy of contemporary road safety strategies (Koornstra et al., 1992; Johansson, 2009, Corben et al., 2010b). Despite this, systemic applications considering more than just individual road user groups (e.g., drivers) or single countermeasures (e.g., traffic light sequencing) are sparse.

Global road safety campaigns such as the Swedish Vision Zero (Johansson, 2009) and the Dutch Sustainable Safety program (Koornstra et al., 1992) speak the language of system safety, but are based on traditional reductionist approaches (Emmerik van, 2001; Salmon et al., 2012). Indeed, Hughes et al. (2015) conducted a review of systems theory and road safety models and concluded that current road safety approaches have some way to go before they fully align with systems theory. The kinetic energy model underpins many global road safety strategies, for example. This model condenses the road system into an equation of mass of an object and its speed at any instant in time (Corben et al., 2010b).

Evaluations of road transport designs are often restricted to quantitative simulation of operational performance (cf. Cunto and Saccomanno, 2008) and risk or safety analysis (cf. Gross et al., 2013). Human behavior is often only considered in simulator studies (Rudin-Brown et al., 2012) or on road studies (cf. Gstalter and Fastenmeier, 2010) focusing on single road user groups or tested when a design has already been finalized or built (cf. Waard et al., 1995).

Applying system-based methods such as CWA and SAD during the road design process provides the opportunity to proactively address design issues using a desktop method before an intersection is built in the real world. It also provides the opportunity to deliver road design concepts that align with the systems approach and support all road users by understanding and adequately managing performance variability based on the interaction of the different road user groups, vehicles, the infrastructure, and the environment. Being able to verify a wide range of behaviors possible by different road users early on in the design life cycle provides a safety and cost benefit as concepts can be comprehensively evaluated and modified before they are implemented in the real world.

16.1.3 Lost in Translation?

While system-based methods have a key role to play in road transport design and evaluation, the outcomes of analyses are often abstract representations that may hamper the communication between ergonomists and road transport designers. Outputs include, for example, hierarchies of goals (Hierarchical Task Analysis; Annett, 2004) or complex networks and diagrams (e.g., CWA and Event Analysis of Systemic Teamwork—EAST; Stanton et al., 2005).

A review of human factors methods that had some element of time or space in their analysis output was conducted to evaluate whether they could represent ergonomics-based analysis conclusions in a real world context (Cornelissen et al., 2012b). The methods were applied to analyze a right-hand turn and outputs were mapped onto an intersection schematic. Initial results demonstrated that link and network analysis-based methodologies were best able to represent the complexity and interaction of road user tasks; however, these methods were not intended to be used for this function. Indeed, whether they are best suited as a visualization metaphor remains ambiguous as it is unclear how to combine such methods with analyses output of methods from different theoretical origins (Cornelissen et al., 2012b).

Without a visualization method translating abstract analyses to real-world concepts that resonate with transport designers, it will remain difficult to conceptualize outcomes and communicate their relevance to design teams. This issue represents a key barrier to the integration of human factors in road system design life cycles.

16.1.4 Looking across the Waters

Perhaps the translation solution is not in ergonomics methods. One alternative approach may be to utilize road transport design methods to find a solution that will resonate with transport designers and serve as a communication vessel for systems-based ergonomic method analyses. The field of safe travel (FST; Gibson and Crooks, 1938) is a road transport theory that has links to both ergonomics and road safety. It is a theory of driving relating the driver to the road transport environment and has underpinned the application of ergonomics methods such as CWA in road safety. Whilst it has not received significant attention from road safety professionals, its road safety origins may remove some of the barriers that other models face when being used in practice.

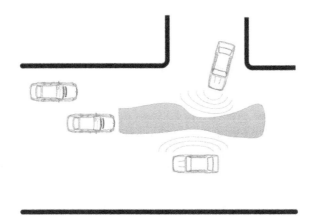

FIGURE 16.1
Field of safe travel. (Adapted from Gibson, J. J. and Crooks, L. E. 1938. *The American Journal of Psychology*, 51(3), 453–471.)

The FST theory describes driving as a type of locomotion through a terrain or field of space with the goal to move an individual from one space to another while modifying locomotion to avoid obstacles. The FST stretches out in front of a road user, exists within the boundaries of the road, and consists of the possible paths a road user may take unobstructed (see Figure 16.1).

Gibson was a forefather of ecological psychology of which CWA is a product. The FST has been cited to explain the relevance of constraint-based approaches to describe complex systems such as road transport. Applications of CWA and the FST have mainly been to design in-vehicle warning devices (Jenkins et al., 2007) and have primarily focused on the first phase of CWA, WDA. The theory was used to visualize obstacles, system constraints, or the FST to help drivers appropriately modify and avoid obstacles (Stoner et al., 2003; Jenkins et al., 2007; Young and Birrell, 2010).

While applications to date have predominantly focused on interface design, it is these authors' opinion that the FST has the potential to clearly and effectively translate the outputs of ergonomic analyses in the road transport design context.

In particular, we see significant potential in using FST to communicate the outputs of SAD analyses. Application of the SAD, describing possible paths within the FST and interaction between constraints and the paths possible, is essential to describe performance variability in road transport. It is anticipated that CWA helps describe the FST of the road user based on the interaction of road users with each other, vehicles, and the road transport environment. The FST then acts as a road transport theory-based visual mediator between CWA output and road transport evaluations and design.

16.2 CWA and FST to Communicate Design Outcomes

This chapter will explore the use of the FST in conjunction with CWA to communicate the findings of ergonomic analyses to road transport designers, based on a desktop evaluation of a new intersection design. The evaluation is conducted on a concept design to exemplify how such methods can be used to examine design concepts early on in the design life cycle. This is a critical part of human factors integration in any system (Stanton et al., 2013).

16.2.1 The Intersection Concept Design

The concept design examined, named the cut-through intersection, was developed based on road transport safe-system principles as part of a major road safety project aiming to improve intersection safety through infrastructure design (Corben et al., 2010a). The intersection design is presented in Figure 16.2. It has traffic islands fitted in the middle of the intersection which create a cut-through lane in the middle. This lane serves to protect right-hand turning traffic from oncoming traffic and change the angle at which traffic meet to reduce the severity of crashes. An additional traffic island separates right-hand turning traffic wanting to use this cut-through lane from straight through traffic upon approach. The intersection has a similar footprint to the traditional intersections it will be replacing and due to its circular shape, no left-hand turning slip lane (which is common in Australian major intersections) is present. Left turning traffic merges with and diverges from straight through traffic in the intersection.

Other aspects of traditional intersections remain largely unchanged. The intersection is signalized, carries multiple lanes of traffic, and speed limits on approach are between 60 and 80 km/h.

The current analysis will focus on a right-hand turn, as this is the maneuvre the proposed intersection design aims to make safer. In Australia, road users travel on the left-hand side

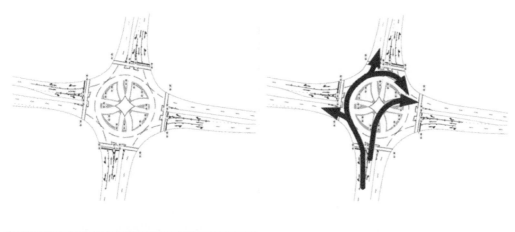

FIGURE 16.2
Concept intersection design and an example of the traditional design it intends to replace. (Reprinted from *Accident Analysis & Prevention*, Vol. 74, Cornelissen, M. et al., Assessing the "system" in safe systems-based road designs: Using cognitive work analysis to evaluate intersection designs, 324–338, Copyright 2015, with permission from Elsevier.)

of the road. The designers intend that for right-hand turns, road users traveling on the main road (i.e., drivers, motorcyclists, and cyclists) approach the intersection as close to the median as possible, position themselves to the right of the traffic island on approach and use the cut-through lane to finalize their right-hand turn once they receive a green light. Pedestrians and cyclists that turn right using off-road facilities approach the intersection on the left-hand side and use the pedestrian crossings to travel to the far right opposite corner.

16.2.2 Data and Analysis

CWA and SAD analyses were undertaken to identify the constraints influencing behavior at both cut-through and existing intersection and the range of strategies road users could employ to make a right-hand turn. The CWA and SAD analysis of the cut-through design was based on CWA of traditional intersections. Data for the traditional intersection analysis was derived from an on-road study of different road users' (drivers, motorcyclists, cyclists, and pedestrians) behavior at intersections (cf. Cornelissen et al., 2013). The analysis was complemented by information on those elements that were different as obtained from design documents and discussions with designers. For example, system constraint changes include the removal of the slip lane, addition of traffic islands, and changes in traffic light design. The analysis for subsequent phases was then completed to determine the consequences of changes to system constraints for situational constraints, decision-making requirements, and courses of action possible within the constraints. As CWA is a formative method, one analyst subsequently explored additional nodes or relationships to ensure that a comprehensive coverage of possible system behavior was included. The CWA output was examined and the conclusions were then mapped onto the FST by one of the analysts. Visualizations that fit within the theoretical framework of the FST were explored and verified by the other analysts.

16.3 CWA and SAD: The Ergonomic Analysis of the Intersection Design

The abstraction hierarchy (AH), see Figure 16.3 for an extract, was used for the WDA, to assess and visualize changes in system constraints from the traditional to the cut-through intersection. System constraints are the constraints defined by the system's purpose, functions that need to be carried out to achieve the purpose and physical objects that allow the functions to be carried out.

The AHs showed that the higher levels (functional purpose, values and priorities, and purpose-related functions) were the same for both intersections. The main differences between the two lay in the physical objects of which each is comprised. For example, the slip lane to turn left is no longer available in the cut-through concept and the lane markings across the intersection are now replaced by traffic islands.

While the system constraints appear to be largely similar, the effect it has on road user performance variability becomes apparent when the other phases of CWA are applied. While the constraints may be similar, their physical distribution through the intersection environment is different. The effect that this has on road user behavior can be assessed by examining the situational constraints, using the contextual activity template (CAT) during the second phase of CWA. The CAT is used to describe when certain functions can be executed (e.g., on approach, in the intersection, or when exiting the intersection).

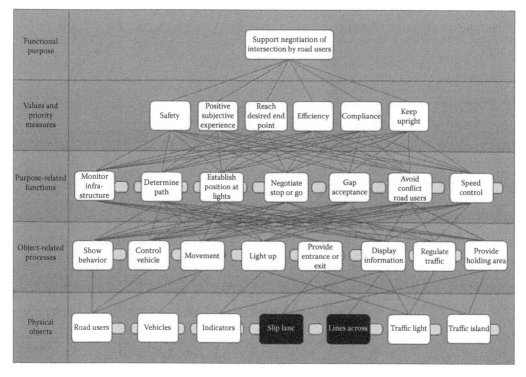

FIGURE 16.3
AH displaying changes in system constraints between cut-through intersection and traditional intersection. (Reprinted from *Accident Analysis & Prevention*, Vol. 74, Cornelissen, M. et al., Assessing the "system" in safe systems-based road designs: Using cognitive work analysis to evaluate intersection designs, 324–338, Copyright 2015, with permission from Elsevier.)

The CAT shows that the distribution of the physical objects through the cut-through intersection, for example, traffic islands, results in changes to the place and time when decisions have to be made. For example (see Figure 16.4), when turning right the decisions to "determine a path" across the intersection, "determine a lane" and the action "take a lane" now have to occur on approach before road users encounter the additional traffic island separating them from straight through traffic. Once in the intersection the traffic islands located within the intersection prevent changes to these decisions. The traffic islands furthermore lead to multiple instances of "gap acceptance" decisions. They do, however, occur under better visual circumstances so that road users can gather information more accurately to make their gap acceptance decision.

The effect that modifications to the intersection have on road user decision making is analyzed using decision ladders. Decision ladders help analyze the different options road users have to execute a certain function, the information elements used to decide between different options and subsequently how functions can be executed (Rasmussen, 1974).

The options road users have to turn right have changed due to the cut-through lane and the circular shape of the intersection. While designers intend for road users on the road to take the cut-through lane and for those off the road to use the pedestrian crossings, alternative options are made possible due to the design changes. In the cut-through design, for example, road users can travel the long way around such as when negotiating a traditional roundabout.

(a)

(b)

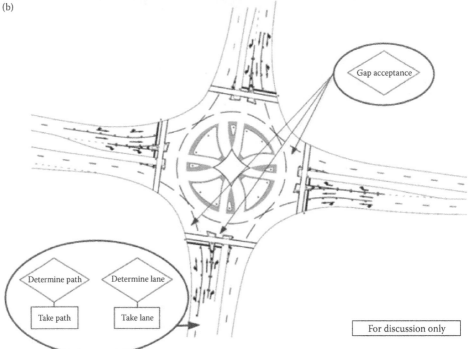

FIGURE 16.4
Situational constraints on road user behavior. (a) Situational constraints on road user behavior in traditional intersection; (b) situational constraints on road user behavior in cut-through intersection. (Reprinted from the *Accident Analysis & Prevention*, Vol. 74, Cornelissen, M. et al., Assessing the "system" in safe systems-based road designs: Using cognitive work analysis to evaluate intersection designs, 324–338, Copyright 2015, with permission from Elsevier.)

Road users that turn right using the cut-through lane have to make the decisions to "determine a path and lane" earlier on approach due to the traffic islands that separate straight through and right turning traffic on approach. The information elements required for those decisions, however, for example, traffic lane arrows painted on the tarmac, are positioned as in a traditional intersection and not made available earlier on approach when required. The SAD was used to evaluate further consequences of these changes.

To evaluate the variability in road user behavior possible within the new cut-through intersection system, the SAD is used; see Figure 16.5 for an extract. The SAD is used to describe different ways in which functions can possibly be executed within a system's constraints.

The latitude for behavior is different in the cut-through intersection when compared with the traditional intersection. For example, traffic islands in the cut-through intersection make it easier for drivers, motorcycle riders, and cyclists to "avoid conflict with other

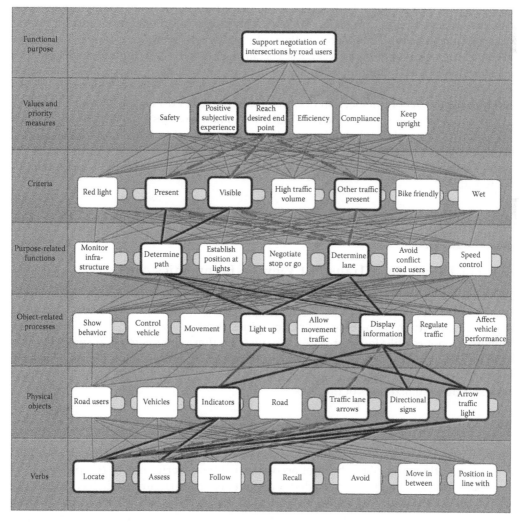

FIGURE 16.5
SAD extract with scenario highlighted.

road users" while they are in the intersection as they are protected by the traffic islands as physical barriers. The same traffic islands also influence the range of behaviors possible by pedestrians and cyclists. Traffic islands afford protection and movement of road users. For pedestrians and cyclists, traffic islands are generally designed for this purpose at pedestrian crossings on the outer ends of the intersection. With traffic islands now placed in the middle of the intersection, however, the same affordances allow pedestrians and cyclists to cross the intersection diagonally. Such actions are likely when traffic volumes are low or when traffic is stopped and road users feel that this strategy satisfies both efficiency and safety values. Understanding emergent behavior and the safety, compliance, and efficiency effects is an important element of performance variability-based evaluations.

The SAD was also used to evaluate the complementary nature and redundancy of road user behaviors. Above, it was concluded that traffic lane arrows were provided to road users in the cut-through intersection after they reached the point where the decision "to determine and take a lane" had to be made. The SAD analysis revealed that road users have alternative strategies that they can use to make the decision. For example, road users may have to rely on other information, for example, other road user indicators, or predict what the lane layout will be based on other information elements, for example, traffic light layout. This creates higher workload, less certainty of making the right decision, attempts to last minute changes and subsequently leads to these road users displaying less predictable behavior to other road users. Understanding such trade-offs is essential in understanding the impact of intersection design on different road users.

From the SAD analysis it can be concluded that small changes in system can have large consequences for road user behavior and its variability possible. Some of the behavior that emerged was not intended to be present nor taken into account by the designers at the concept stage but would have had safety consequences if the design were to be built in the real world.

16.4 The FST: A Road Transport View on Ergonomics Analysis

The analyses above have important ramifications for intersection design; however, as discussed the implications are very much expressed through systems theory and CWA language and output. Both represent philosophies and frameworks that road safety practitioners are not readily familiar with. The question that remains to be answered is how these insights can be communicated to road transport designers using the FST. First, we will consider links between the CWA and the FST elements after which we will introduce some specific visualizations of findings.

16.4.1 FST Elements Described by CWA

The FST comprises paths that a road user can take unimpeded. This is often represented as a space in time and place, but not accompanied by descriptions of the paths within that space. The SAD analysis provided a comprehensive insight into behavior possible within the cut-through intersection. As a result, the paths possible within the FST can be described in terms of functions that have to be executed to achieve the purpose (e.g., take lane), options road users have to execute those functions (e.g., enter cut-through lane) and courses of actions to decide between and execute the chosen option (e.g., assess

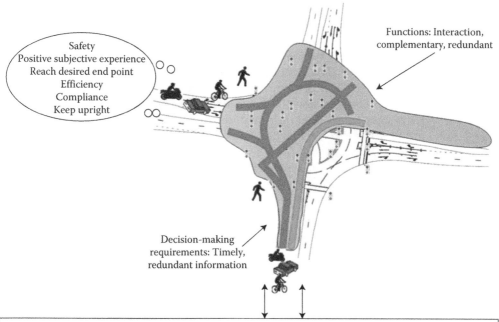

Safety
Positive subjective experience
Reach desired end point
Efficiency
Compliance
Keep upright

Functions: Interaction, complementary, redundant

Decision-making requirements: Timely, redundant information

Moving obstacles: Road users and vehicles, road user attributes (e.g., signals, PPE, friendly or unfriendly), vehicle attributes (e.g., lights, indicators, mirrors, doors, wheels)
Road layout: Road, lane, traffic island, slip roads, footpath, bike path, bus lane, zebra crossing
Markings and signage: Lane markings, traffic lane arrows, lines across, stop line, other road markings, directional signs, regulatory signs, other signs, cats eyes
Traffic light infrastructure: Traffic light, arrow light, traffic light sensor, traffic light button, activation light
Enforcement: Road safety camera, rules and regulations
Conditions: Weather and road surface conditions, road works break downs, traffic conditions (volume, speed, directions, room to move, interfere path)
Conditions operator: Conventionality, size of intersection, experience, familiarity, time pressure
Condition of infrastructure: Available, visible, present, activated
Type of road user: Motorized, two wheeler, on or off road, vulnerable

FIGURE 16.6
FST defined further by ergonomics analysis.

traffic lane arrows). This allows for more comprehensive understanding of the space within the FST, as well as how road users select and take paths within the FST in an intersection.

The original FST theory describes the FST as being influenced by objects that have to be avoided or influence the visual task of driving. Using the CWA and SAD analysis, the influences on the FST include physical objects and system constraints, situational constraints, decision-making requirements, functions and criteria such as type of road user, availability of physical objects, changes in circumstances or values. All these constraints influence the paths possible and the paths that are selected within the FST (Figure 16.6).

16.4.2 Situational Representation of FST

The FST is a spatial field but is not fixed in time and moves with the road user through space (Gibson and Crooks, 1938). Situational constraints have been shown to influence road user behavior as described above in the CWA. How the situational constraints influence road user performance variability becomes clearer when using situational representations of the FST (see Figure 16.7). While approaching the intersection, the FST the possible paths a road user can take unobstructed is rather large. It stretches out into multiple directions and the performance variability between different road users and under different circumstances is thus greatly varied. The closer the road user gets to the intersection, the more constrained the FST becomes as certain decisions have been made which subsequently limit the variability possible (e.g., the cut-through lane has been taken). Exiting the intersection, the FST stretches out straight in front of the road user and variability is temporarily less. This kind of situational representation helps describe the intersection journey in terms of performance variability and provides a way to examine the workload of different road users at different points in the intersection, the interaction between information elements and decisions road users are making, and whether this is provided to best support the road user.

16.4.3 Emergent Behavior Visualized in FST

The paths possible within the FST may contain paths that were not intended to occur by the designers. The CWA and SAD analysis of the cut-through intersection revealed emergent behavior that was not considered by the designers. The FST can be used to communicate this visually (see Figure 16.8). The designers intended that drivers, motorcycle riders, and cyclists use the cut-through lane to turn right. The CWA and SAD analysis, however, demonstrated that road users have a range of other options, including using the intersection as a roundabout. Cyclists and pedestrians can further use the traffic islands in the middle of the intersections to cross diagonally rather than use the pedestrian crossings on the outside as intended. The paths that road users can take and the shape of their FST are much larger than initially considered. If designers do not consider this behavior at the concept design stage then this form of emergent behavior could potentially create conflicts between different road user groups. If not picked up it can lead to unsafe situations once the intersection has been built and modifications or road user campaigns will have to be implemented ad hoc.

16.5 Discussion and Conclusions

Translating ergonomics research in practice is difficult and one of the reasons for this is the difficulty in communicating complex analysis findings with practitioners. As a response to this, the aim of this chapter was to explore the value of ergonomics systems analysis methods in conjunction with road transport theory to communicate performance variability evaluations of intersection designs. The intention was to provide a suitable framework for effectively communicating systems analyses of road design concepts in a way that design improvements can be made.

CWA and SAD have been successfully applied to describe, evaluate, and propose how to manage performance variability in a road transport context (Cornelissen et al., 2013, 2015). One of its potential barriers to success has been the communication of the findings. This book chapter outlined how CWA and SAD help describe the different elements of the FST theory and how subsequently the FST can be employed to visualize findings and tell a visual story to transport designers. CWA provided the analytical rigor to analyze the

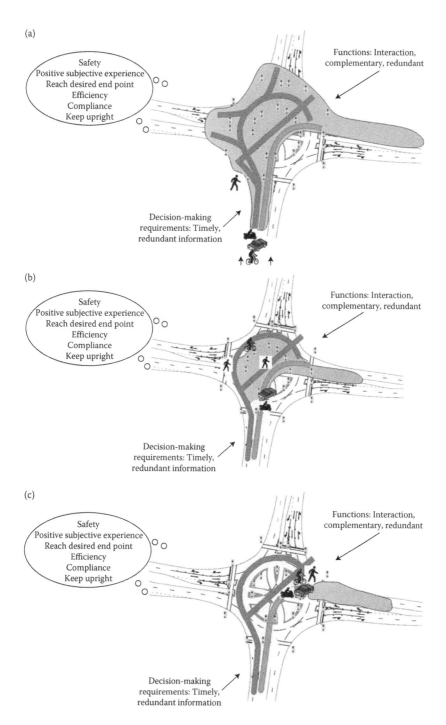

FIGURE 16.7

Situational constraints throughout different stages when turning right in cut-through intersection. (a) Situational constraints on and paths possible within field of safe travel turning right when approaching the intersection; (b) situational constraints on and paths possible within field of safe travel turning right when in the intersection; (c) situational constraints on and paths possible within field of safe travel turning right when exiting the intersection.

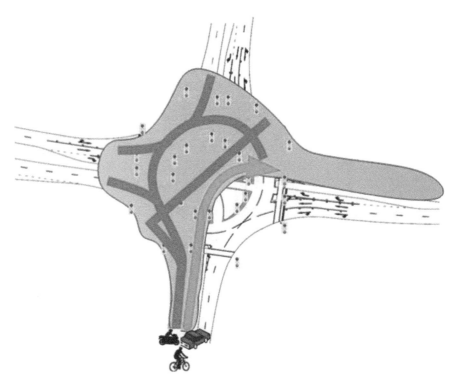

FIGURE 16.8
Emergent behavior in cut-through intersection.

interaction between the intersection design, road users, vehicles, and the environment. The FST provided a visual platform to communicate the findings while also benefiting further performance variability-based analysis using its dynamic representations for example.

In addition, using the FST in this context allowed to further describe the FST theory in a performance variability context which can further aid our understanding of performance variability in road transport. The FST originally viewed driving and walking as locomotion either with or without a tool. The SAD analysis, however, demonstrated that the task of conducting a right-hand turn at intersections is different for different road user groups under different circumstances. For example, different road users have different options for function execution (e.g., off-road or on-road facilities) and situation constraints on execution of purpose-related functions differ (e.g., cyclist determine their lane position continuously throughout the entire intersection). Different road users have different information requirements (e.g., vulnerable road users assess road user behavior to protect their safety and chose their path through the intersection accordingly) and employ different and differing number of courses of action (e.g., motorcycle riders and cyclists can use additional courses of action to determine if they should filter to the front or position behind other vehicles).

Descriptions of paths possible within the FST and understanding the selection and execution of the different paths following the CWA and SAD analysis provide greater explanatory power to the FST in a performance variability context. Small changes in constraints can have large and different effects on possible road user behavior and its variability, and thus on the FST. For example, evaluation of the cut-through intersection concept design showed that redistribution of traffic islands throughout the intersection changed the different options for cyclists and decision-making processes to determine a lane and path for all road users on the road. Understanding the influence of different intersection and road user aspects and their interaction on the FST is important. This enables the use of the FST in a performance variability context and for it to inform road design adequately early on and throughout the entire design life cycle.

Considering the impact of road transport designs on the FST for different road users can aid performance variability-based evaluations. It allows, for example, to evaluate the range of variability that road users can, and often have to, take into account and the consequence of this for workload and interaction with the intersection and other road users (e.g., cyclists having a range of options to navigate an intersection each with different safety and efficiency trade-offs resulting in a high workload and low levels of predictability on approach). This can lead to design decisions regarding distribution of information and decisions in such a way that workload on road users is best distributed and information is provided at appropriate times in the intersection journey (e.g., providing dedicated turning options for cyclists that meet safety and efficiency requirements). Distribution of decisions is effective if the right information is provided at the right time, not all decisions are simply transferred to a different location which just transfers the problem to a different location in the intersection journey, and other elements such as speed and other road user interactions are altered in line with the new distribution of decision making.

This research provides a first step in better integrating ergonomics methods during concept design phases of the road transport design life cycle. Following this exploration, a full-scale application of CWA and the FST are encouraged. Also, further development and testing of the compatibility of the FST visualizations with road transport design processes should be conducted, including during other stages of the design life cycle. Future applications can also aid further formal development of the FST theory as well as performance variability theory in a road transport context.

Acknowledgments

Dr. Miranda Cornelissen's contribution to this book chapter was primarily conducted during her PhD candidature which is funded by a Monash graduate scholarship and a Monash international postgraduate research scholarship.

Professor Paul Salmon's contribution to the article was funded through his ARC Future Fellowship (FT140100681).

Dr. Miranda Cornelissen acknowledges that this work was published while working at Roads and Maritime Services but that the views expressed in this study are the individual's and do not necessarily represent the views of the agency.

References

Annett, J. 2004. Hierarchical task analysis. In D. Diaper and N. A. Stanton (eds.), *The Handbook of Task Analysis for Human-Computer Interaction* (pp. 67–82). Mahwah, New Jersey: Lawrence Erlbaum Associates.

Corben, B., Candappa, N., Van Nes, N., Logan, D., and Peiris, S. 2010a. *Intersection Safety Study: Meeting Victoria's Intersection Challenge. Task 5 Generation of Intersection Designs within the Safe System Context*. Melbourne, Australia: Monash University Accident Research Centre.

Corben, B., Logan, D., Fanciulli, L., Farley, R., and Cameron, I. 2010b. Strengthening road safety strategy development "Towards Zero" 2008–2020—Western Australia's experience scientific research on road safety management SWOV workshop 16 and 17 November 2009. *Safety Science*, 48(9), 1085–1097. Doi: http://dx.doi.org/10.1016/j.ssci.2009.10.005.

Cornelissen, M., Salmon, P. M., Jenkins, D. P., and Lenné, M. G. 2012a. A structured approach to the strategies analysis phase of cognitive work analysis. *Theoretical Issues in Ergonomics Science*, 14(6), 1–19. Doi: 10.1080/1463922x.2012.668973.

Cornelissen, M., Salmon, P. M., McClure, R., and Stanton, N. A. 2013. Using cognitive work analysis and the strategies analysis diagram to understand variability in road user behavior at intersections. *Ergonomics*, 56(5), 1–17. Doi: 10.1080/00140139.2013.768707.

Cornelissen, M., Salmon, P. M., Stanton, N. A., and McClure, R. 2012b. Human factors now showing in 3D. In N. A. Stanton (Ed.), *Advances in Human Aspects of Road and Rail Transportation* (pp. 653–662). Boca Raton, Florida: CRC Press.

Cornelissen, M., Salmon, P. M., Stanton, N. A., and McClure, R. 2015. Assessing the "system" in safe systems-based road designs: Using cognitive work analysis to evaluate intersection designs. *Accident Analysis & Prevention*, 74, 324–338. Doi: http://dx.doi.org/10.1016/j.aap.2013.10.002.

Cornelissen, M., Salmon, P. M., and Young, K. L. 2012c. Same but different? Understanding road user behavior at intersections using cognitive work analysis. *Theoretical Issues in Ergonomics Science*, 14(6), 1–24. Doi: 10.1080/1463922x.2012.678282.

Cunto, F. and Saccomanno, F. F. 2008. Calibration and validation of simulated vehicle safety performance at signalized intersections. *Accident Analysis & Prevention*, 40(3), 1171–1179. Doi: 10.1016/j.aap.2008.01.003.

Emmerik van, A. 2001. A systems approach to road safety. In *Proceedings of the 24th Australasian Transport Research Forum*. Hobart, Tasmania: Tasmania Department of Infrastructure. https://trid.trb.org/view.aspx?id=712233; http://atrf.info/papers/2001/2001_VanEmmerik.pdf.

England, L. I. Z. 1981. The role of accident investigation in road safety. *Ergonomics*, 24(6), 409–422. Doi: 10.1080/00140138108924864.

Gibson, J. J. and Crooks, L.E. 1938. A theoretical field-analysis of automobile-driving. *The American Journal of Psychology*, 51(3), 453–471.

Gross, F., Lyon, C., Persaud, B., and Srinivasan, R. 2013. Safety effectiveness of converting signalized intersections to roundabouts. *Accident Analysis & Prevention* 50(0). Doi: 10.1016/j.aap.2012.04.012.

Gstalter, H. and Fastenmeier, W. 2010. Reliability of drivers in urban intersections. *Accident Analysis & Prevention*, 42(1), 225–234. Doi: 10.1016/j.aap.2009.07.021.

Hale, A. R., Stoop, J., and Hommels, J. 1990. Human error models as predictors of accident scenarios for designers in road transport systems. *Ergonomics*, 33(10–11), 1377–1387. Doi: 10.1080/00140139008925339.

Hollnagel, E. 2004. *Barriers and Accident Prevention*. Aldershot: Ashgate Publishing.

Hollnagel, E. 2009. *The ETTO Principle: Efficiency-Thoroughness Trade-Off*. Surrey: Ashgate.

Hollnagel, E., Kaarstad, M., and Lee, H. 1999. Error mode prediction. *Ergonomics*, 42(11), 1457–1471. Doi: 10.1080/001401399184811.

Hughes, B. P., Newstead, S., Anund, A., Shu, C. C., and Falkmer, T. 2015. A review of models relevant to road safety. *Accident Analysis & Prevention*, 74, 250–270.

Jenkins, D. P., Salmon, P. M., Stanton, N. A., Walker, G. H., and Rafferty, L. A. 2011. What could they have been thinking? How sociotechnical system design influences cognition: A case study of the Stockwell shooting. *Ergonomics*, 54(2), 103–119.

Jenkins, D. P., Stanton, N. A., Walker, G. H., and Young, M. S. 2007. A new approach to designing lateral collision warning systems. *International Journal of Vehicle Design*, 45(3), 379–396.

Johansson, R. 2009. Vision Zero—Implementing a policy for traffic safety. *Safety Science*, 47(6), 826–831. Doi: http://dx.doi.org/10.1016/j.ssci.2008.10.023.

Koornstra, M. J., Mathijssen, M. P. M., Mulder, J. A. G., Roszbach, R., and Wegman, F. C. M. 1992. *Naar een duurzaam veilig wegverkeer [Towards Sustainable Safe Road Traffic]*. Leidschendam: SWOV.

Larsson, P., Dekker, S. W. A., and Tingvall, C. 2010. The need for a systems theory approach to road safety. *Safety Science*, 48(9), 1167–1174. Doi: 10.1016/j.ssci.2009.10.006.

Rasmussen, J. 1974. *The Human Data Processor as a System Component: Bits and Pieces of a Model*. Roskilde, Denmark: Danish Atomic Energy Commission.

Rudin-Brown, C. M., Lenné, M. G., Edquist, J., and Navarro, J. 2012. Effectiveness of traffic light vs. boom barrier controls at road–rail level crossings: A simulator study. *Accident Analysis & Prevention*, 45(0), 187–194. Doi: http://dx.doi.org/10.1016/j.aap.2011.06.019.

Salmon, P. M. and Lenné, M.G. 2015. Miles away or just around the corner? Systems thinking in road safety research and practice. *Accident Analysis & Prevention*, 74, 243–249. Doi:10.1016/j.aap.2014.08.001.

Salmon, P. M., McClure, R., and Stanton, N. A. 2012. Road transport in drift? Applying contemporary systems thinking to road safety. *Safety Science*, 50(9), 1829–1838. Doi: 10.1016/j.ssci.2012.04.011.

Stanton, N. A., Salmon, P. M., Rafferty, L., Walker, G. H., Baber, C., and Jenkins, D. P. 2013. *Human Factors Methods: A Practical Guide for Engineering and Design* (Second edition). Aldershot, United Kingdom: Ashgate Publishing Limited.

Stanton, N. A., Salmon, P. M., Walker, G. H., Baber, C., and Jenkins, D. P. 2005. *Human Factors Methods: A Practical Guide for Engineering*. Aldershot: Ashgate Publishing Limited.

Stoner, H. A., Wiese, E. E., and Lee, J. D. 2003. Applying ecological interface design to the driving domain: The results of an abstraction hierarchy analysis. *Paper Presented at the Human Factors and Ergonomics Society 47th Annual Meeting*, Santa Monica, California.

Vicente, K. J. 1999. *Cognitive Work Analysis: Toward Safe, Productive and Healthy Computer-Based Work*. Mahwah, New Jersey: Lawrence Erlbaum Associates, Inc.

Waard, D. D., Jessurun, M., Steyvers, F. J. J. M., Reggatt, P. T. F., and Brookhuis, K. A. 1995. Effect of road layout and road environment on driving performance, drivers' physiology and road appreciation. *Ergonomics*, 38(7), 1395–1407. Doi: 10.1080/00140139508925197.

Young, M. S. and Birrell, S. A. 2010. Ecological IVIS design: Using EID to develop a novel in-vehicle information system. *Theoretical Issues in Ergonomics Science*, 13(2), 225–239. Doi: 10.1080/1463922x.2010.505270.

17

Hands and Feet Free Driving: Ready or Not?

Victoria A. Banks and Neville A. Stanton

CONTENTS

17.1 Introduction

In 2013, the World Health Organization declared that approximately 1.24 million people per annum die as a result of road traffic accidents with half of these considered to be vulnerable road users: motorcyclists (23%), pedestrians (22%), and cyclists (5%). If the benefits of vehicle automation can outweigh potential costs, then it may prove to be beneficial in both economic, societal, and environmental terms (Stanton and Marsden, 1996; Young et al., 2011; Khan et al., 2012). Highly automated vehicles therefore have great potential in improving the safety of our roads. Even so, a recent study by Banks and Stanton (2015a) found that while the implementation of an autonomous emergency brake feature significantly reduced the number of accidents in a driving simulator study, it was detrimental to driver decision making with the weakening and disintegration of links between information processing nodes (e.g., Parasuraman et al., 2000). These studies indicate that despite improving road safety, there may be unintended behavioral changes on behalf of the driver as the level of automation increases in the driving task and more control delegation is handed to the vehicle.

One reason for this may be due to the changing nature of the driver's role within an already complex system of control-feedback loops as they move from being an active operator to a passive monitor of system operation (Byrne and Parasuraman, 1996). For example, an "almost driverless" system would see 100% of the physical driving task being completed by automated subsystems (Banks et al., 2013) but this does not mean that the role of the driver becomes redundant. Instead, they remain an important agent of control given that they will still continue to receive feedback from both the vehicle and wider environment that enables them to anticipate changes in the environment using feedforward information which may be far more superior to their automated counterparts due to learning and experience (see Figure 17.1). However, high levels of automation in the

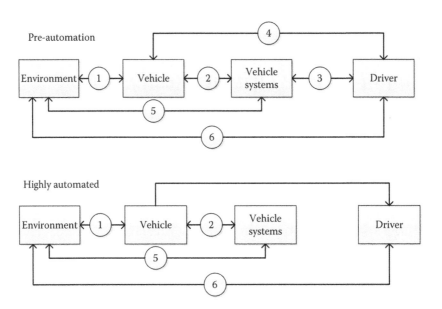

FIGURE 17.1
Control-feedback loops in driving.

driving task leave drivers vulnerable to disengagement (Cuevas et al., 2007; Bekier et al., 2012) from the primary task (driving) and more likely to engage in secondary tasks (i.e., in-vehicle entertainment) (Carsten et al., 2012; Jamson et al., 2013). With the likelihood of "eyes-off-road" time increasing as both the level of automation and the duration at which it is used increase, any failure on part of the automation may delay appropriate driver response (e.g., Young and Stanton, 2007). Failure of an automated longitudinal control system has been associated with inappropriate braking responses in both driving simulator (e.g., Young and Stanton, 2007) and test-track studies (e.g., Rudin-Brown and Parker, 2004). Furthermore, failure to an automated system of lateral control has previously been related to issues of driver complacency (e.g., Desmond et al., 1998). It is therefore becoming increasingly important to appreciate the driver's ability to undertake their *new* supervisory role as the average motorist becomes less actively involved in traditional vehicle handling especially when considering that a vast amount of research over the past 30 years has shown that drivers do not always respond in the way that engineers anticipate in the design of automated assistance (e.g., Hoedemaeker and Brookhuis, 1998; Rudin-Brown and Parker, 2004; Young and Stanton, 2007). This means that while fully automated cars are technologically feasible and have been for some time (Brookhuis and de Waard, 2006), during the intermediate phases of automation, the driver must remain active and in-the-loop (Hoeger et al., 2008) and work cooperatively with their automated counterparts (Soualmi et al., 2014).

Despite the allocation of function between the driver and automated subsystems being key in facilitating and developing driver-automation cooperation (Hoc, 2000), the industry has continued to be plagued by criticism for inadequately acknowledging the role of the driver and how it may change once these systems have been deployed (Banks et al., 2013). It would seem that we do not fully understand or appreciate the complexities of driver-automation cooperation in modern day cars (Weyer et al., 2015). This poses many challenges for systems designers to ensure that the interaction between humans and automated systems is designed appropriately (Strand et al., 2014) to ensure that the negative effects typically associated with being out-of-the-loop are minimized (Endsley and Kiris, 1995; Wickens and Hollands, 2000).

17.2 Adopting a Systems View Approach to Driving Automation

Systems engineering can be seen as an interdisciplinary approach to the field of engineering that integrates both technical and human-centered approaches to look more closely at work processes, optimization, and risk management. This holistic approach is concerned with how the functioning and performance of a joint cognitive system, such as driving (Salmon et al., 2008), can be best described and further understood. This viewpoint stems from the belief that every "agent" within a system plays a critical role in the successful completion of a task and more importantly, "agents" can be both human and nonhuman (Stanton et al., 2006; Salmon et al., 2009). Although early research into automation seemed to focus most heavily upon autonomy, current research now focuses upon satisfying the requirements of joint activity, including human–machine teamwork (Klein et al., 2004). An interdisciplinary approach such as this is extremely complex because the "behavior" or "interaction" that occurs between system components is not always well defined or understood. The aim of systems engineering therefore is to better define and characterize "system agents" and the interactions that occur between them. One way of doing this is through the application of distributed cognition (Hutchins, 1995a); an approach that aims to provide a clearer understanding of task partitioning between the driver and automated subsystems and recognizes that the cognitive processes normally completed by the driver can be shared across this system (Hollnagel, 2001).

From this perspective, situation awareness (SA; Endsley, 1995) is formulated through a myriad of individual components and cannot be predicted based solely upon one of these individual components or the mere combination of individual SA from different agents (Salas et al., 1995). This idea is particularly relevant to vehicle automation because the driver uses assistive aids to help build a "picture" of what is happening in the world (Walker et al., 2010). It is possible therefore to apply Endsley's (1995) three-stage model of SA to a system (e.g., Stanton et al., 2006) although there is a need therefore to move away from traditional notions of SA that currently dominate ergonomics to one that focuses upon entire systems (Gorman et al., 2006; Salmon et al., 2008; Sorensen et al., 2011; Walker et al., 2010). This is because there are very few complex tasks that can be performed on a completely individualistic basis (Perry, 2003; Walker et al., 2010). Distributed situation awareness (DSA; Stanton et al., 2006) offers a compatible approach that assumes SA is a system level phenomenon rather than individual-orientated (Salmon et al., 2008; Stanton et al., 2006). DSA outlines that SA can be held by human and nonhuman agents, that different agents view their environment differently and that at an individual level SA overlap will be dependent upon the goals of each agent. DSA also recognizes that communication can be nonverbal and that SA loosely holds systems together whereby one agent has the ability to compensate for degraded SA in another (Stanton et al., 2006).

Up until now, it is an approach that has been successfully applied to a number of domains including ship navigation (Hutchins, 1995b), airline cockpits (Hutchins and Klausen, 1996; Sorensen et al., 2011), engineering practice (Rogers, 1993), and air traffic control (Halverson, 1995). The application of distributed cognition to driving is a new and relatively unexplored medium yet there appears to be great benefit in doing so in terms of automation development and system safety.

17.2.1 Proposed Framework for the Application of Distributed Cognition to Driving Automation Research

The authors propose that in order to achieve a comprehensive understanding of distributed cognition in driving, the following steps can be taken to explore the design and allocation of system function for both preexisting automated technologies as well as in the development of future automated systems (see Figure 17.2). This approach combines traditional task analysis with qualitative research methods in a systems design framework.

17.2.1.1 Phase 1

17.2.1.1.1 Design Idea, Concept, or Prototype

The purpose of this stage is to identify the automated feature of interest and provide general information relating to its purpose and the subsystem components essential for its build and functionality.

> **EXAMPLE**
>
> The combination of Stop and Go Adaptive Cruise Control, an extension of Adaptive Cruise Control, and some form of Lane Keep Assist would see much of the driving task being completed autonomously enabling the driver in essence, to become "hands and feet free". Although a combined system of Automated Longitudinal and Lateral Control is not an entirely new concept (Young and Stanton, 2002), there has been a significant increase in manufacturers introducing their own versions that fit such a specification over recent years (e.g., General Motor's Super Cruise, Fleming, 2012; Mercedes Distronic Plus with Steering Assist, Daimler, 2013). Where Stop and Go Adaptive Cruise Control is capable of bringing the vehicle to a complete stop through the addition of radar that can operate at slower speeds and over shorter distances (Stanton et al., 2011), future Lane Keep Assist technologies may be able to automatically maintain lane position.

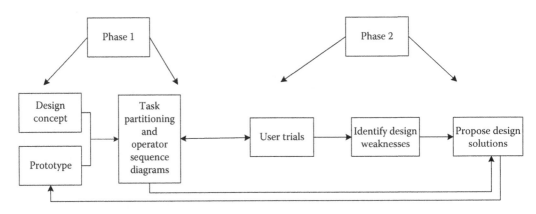

FIGURE 17.2
Systems design framework.

17.2.1.1.2 Task Partitioning and Operator Sequence Diagrams

This form of task analysis provides an insight into "who" owns "what" information, "who" performs "which" function, and so on. Once automated subsystems are activated by the driver, they assume that partial or full control has been assigned to the system. Task partitioning enables us to see how the workload begins to shift from the driver to the automated system (Banks et al., 2013). From a design point of view, and echoing the viewpoint of Endsley and Kaber (1997), it must be ensured that the driver knows exactly "who" is expected to do "what" during the driving task. Thus, in order to design a human-centered product, the designer should augment the task which the product is designed for. In this way, it is possible to establish "who" can do "what" at differing levels of automation. Much like Hutchins described two roles within a pilot's cockpit (pilot flying and pilot not flying), the driving system contains two primary actors capable of controlling the vehicle (driver and automation). A simple mapping exercise similar to the one outlined by Banks et al. (2014a) can achieve the desired output and show how the driver and automated subsystems can work in parallel or independently of one another to provide insight into how physical and cognitive work may be distributed between the driver and automation.

The first step of this process is to outline the functionality of the overall system.

EXAMPLE

Once the driver issues the command for longitudinal control to be automated, the longitudinal controller would begin to hold, represent, and modify information from the changing environment in order to reach the shared goal of the system network (in this case to maintain a desired speed and gap that is preset by the driver). Similarly, the shared goal of the system network that sees lateral control automated is to safely stay within the confines of a lane and avoid deviation. The lateral controller would begin scanning the road environment for lane markings, much like how the driver evaluates the pathway ahead. If no markings are found, the lateral controller would reach its system limits and alert the driver to regain control via human–machine interface (HMI) and possible auditory signals depending on individual manufacturers design parameters. If, however, lane markings are successfully identified, the vehicle could be controlled automatically by the lateral controller.

The second step of this process is to construct operator sequence diagrams (OSD). These offer one way of visualizing distributed cognition within the driving system by showing how work is distributed or "communicated" amongst different system agents in a much more overt manner than the description above (Brookhuis et al., 2008; Cuevas et al., 2007).

EXAMPLE

Figure 17.3 visually represents how workload is divided between system agents of a combined subsystem approach (i.e., automated longitudinal and lateral control systems). Based upon this representation, it would appear that automated subsystem components become central to the functionality of the driving system as the driver delegates increasing levels of control to them. Much of the additional information that is added into the driving task as a result of automation implementation remains firmly embedded within the automated subsystem architecture with only the most relevant information being shared with the driver via the HMI relating to system status.

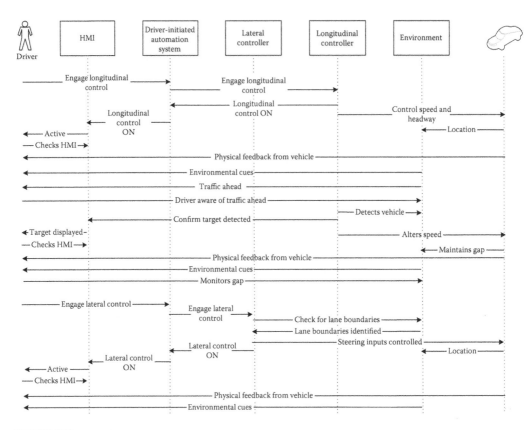

FIGURE 17.3
Basic schematic representation of distributed cognition in an automated system of longitudinal and lateral control.

In order for the system to function effectively, all system agents will need to have an awareness of how each other (a) works and (b) have some intelligence regarding specific functional limitations. In essence, automated subsystems will monitor the behavior of each intelligent counterpart and adapt accordingly. All system agents will use and process information from the environment and from other system agents in order to respond accordingly (DSA; Stanton et al., 2006).

Although task analysis is a popular and widely used method to assist in the design and development of automated technologies (Putkonen and Hyrkkänen, 2007), it remains a challenge to capture both the cognitive and behavioral elements of a task and there is still some debate over whether it adequately represents cognition (Patrick, 1992). For example, the processes outlined within the OSD representation are likely to be continuous (i.e., continual monitoring and risk assessments will be performed to take into account the ever-changing environment) and also highly adaptable and not constrained to the processes outlined in representational methods. Constraining complex behavior in this way is of course limiting yet phase 1 of the systems design framework offers a reasonable approximation of distributed cognition in specific driving tasks. This means that they provide a good foundation for future investigation at the earliest stages of system development.

17.2.1.2 Phase 2

17.2.1.2.1 User Trials

In order to validate and extend the representation of OSDs, it is essential that these assumptions are experimentally tested through the conductance of user trials. Generating further insight into driver–vehicle–world interactions is thought to be possible through use of verbal protocol analysis (VPA; Ericsson and Simon, 1993) and has been as an extension methodology by Banks et al. (2014b), Banks and Stanton (2015a,b) in both driving simulation and on-road testing. In an extensive review of VPA methodology, Ericsson and Simon (1993) suggest that verbalizations give an insight into the contents of a subject's working memory. This means that verbal reports can generate a rich information source that would otherwise be inaccessible by any other form of data collection (Russo et al., 1989). It offers one way of recording human thought processing when completing a task in real-time meaning that information relating to human–machine interaction is present (Hughes and Parkes, 2003).

> **EXAMPLE**
>
> VPA was used by Banks and Stanton (2015a) to investigate the processes underlying driver decision making at different levels of automation. It revealed that human thought processing was directly affected by both the level and type of automation introduced into the driving task. Analysis revealed that while automation did not alter the decision-making pathway (e.g., the processes between initial hazard detection and response remained the same), it did appear to significantly weaken or sever the links between information processing nodes.

17.2.1.2.2 Identify Design Weaknesses

Based upon the findings of experimental user trials, it may be possible to identify system weaknesses that were not previously identified in the processes outlined in phase 1. For example, more specific issues relating to human–vehicle interaction in driving may be identified through driver protocols and subjective measures of workload (e.g., Hart and Staveland, 1988) and trust (e.g., Jian et al., 2000).

> **EXAMPLE**
>
> Driver verbalizations and subjective reports of mental workload and stress revealed evidence of driver–vehicle coordination problems within an early prototype system of automated longitudinal and lateral control with an automatic overtake function (Banks and Stanton, 2015b). This was an unexpected finding for system developers as they had overlooked the potential for miscommunication to occur between the driver and automated system.

17.2.1.2.3 Propose Design Solutions

Recommendations for alternative strategies should be raised following the appraisal of results if required. Any change to systems design should adopt task augmentation, modeling, and user trial strategies to ensure the success of later prototypes.

> **EXAMPLE**
>
> A recent study by Banks and Stanton (2015c) issued design recommendations for the HMI of a combined system of automated longitudinal and lateral control following a

comprehensive thematic analysis of driver protocols that identified issues with ambiguity and inadequate warning mechanisms in the proposed HMI prototype.

17.3 Discussion

While phase 1 of the systems design framework holds great potential in describing system–level interaction in a relatively short space of time and can help define numerous hypotheses for future research direction, phase 2 seeks to empirically validate the assumptions made through the conductance of user trials. Although specific functionality issues cannot be easily addressed in these representations, as long as functionality issues are considered, OSDs may prove to be a useful HMI design and allocation of system functions in the development of future automated systems. What these methodologies do however demonstrate is that ironically, driver task loading does not appear to reduce with increased automation as commonly presumed. Quite possibly workload will actually increase as the driver is required to monitor and anticipate both the road environment, the behavior of other road users, and the automated aspects of vehicle control, synthesizing the wider literature on malleable attention (Young and Stanton, 2002).

References

Banks, V. A., Stanton, N. A. 2015a. Contrasting models of driver behaviour in emergencies using retrospective verbalisations and network analysis. *Ergonomics*, 58(8), 1337–1346.

Banks, V. A., Stanton, N. A. 2015b. Discovering driver-vehicle coordination problems in future automated control systems: Evidence from verbal commentaries. In *Proceedings of the 6th International Conference on Applied Human Factors and Ergonomics AHFE 2015*, Las Vegas, pp. 2497–2504, July 26–30, 2015.

Banks, V. A., Stanton, N. A. 2015c. Keeping the driver in control: Automating automobiles of the future. *Applied Ergonomics*, 53(B), 389–395.

Banks, V. A., Stanton, N. A., Harvey, C. 2013. What the crash dummies don't tell you: The interaction between driver and automation in emergency situations, In *Proceedings of the 16th International IEEE Conference on Intelligent Transportation Systems (ITSC 2013)*, The Hague, pp. 2280–2285, October 6–9.

Banks, V. A., Stanton, N. A., Harvey, C. 2014a. Sub-systems on the road to vehicle automation: Hands and feet free but not "mind" free driving. *Safety Science*, 62, 505–514.

Banks, V. A., Stanton, N. A., Harvey, C. 2014b. What the drivers do and do not tell you: Using verbal protocol analysis to investigate driver behaviour in emergency situations. *Ergonomics*, 57(3), 332–342.

Bekier, M., Molesworth, B. R. C., Williamson, A. 2012. Tipping point: The narrow path between automation acceptance and rejection in air traffic management. *Safety Science*, 50(2), 259–265.

Brookhuis, K., de Waard, D. 2006. The consequences of automation for driver behaviour and acceptance. *Proceedings of the International Ergonomics Association (IEA)*. Maastricht, The Netherlands: Elsevier.

Brookhuis, K. A., van Driel, C. J., Hof, T., van Arem, B., Hoedemaeker, M. 2008. Driving with a congestion assistant; mental workload and acceptance. *Applied Ergonomics*, 40(6), 1019–1025.

Byrne, E. A., Parasuraman, R. 1996. Psychophysiology and adaptive automation. *Biological Psychology*, 42(3), 249–268.

Carsten, O., Lai, F. C. H., Barnard, Y., Jamson, A. H., Merat, N. 2012. Control task substitution in semi automated driving: Does it matter what aspects are automated? *Human Factors: The Journal of the Human Factors and Ergonomics Society*, 54(5), 747–761.

Cuevas, H. M., Fiore, S. M., Caldwell, B.S., Strater, L. 2007. Augmenting team cognition in human–automation teams performing in complex operational environments. *Aviation, Space, and Environmental Medicine*, 78, B63–B70.

Daimler, 2013. DISTRONIC PLUS: Warns and assists the driver [Company website]. Retrieved from http://www.daimler.com/dccom/0-5-1210218-1-1210321-1-0-0-1210228-0-0-135-0-0-0-0-0-0-0-0.html.

Desmond, P. A., Hancock, P. A., Monette, J. L. 1998. Fatigue and automation-induced impairments in simulated driving performance. *Transportation Research Record*, 1628, 8–14.

Endsley, M. 1995. Measurement of situation awareness in dynamic systems. *Human Factors*, 37, 65–84.

Endsley, M. R., Kaber, D. B. 1997. Out-of-the-loop performance problems and the use of intermediate levels of automation for improved control system functioning and safety. *Process Safety Progress*, 16(3), 126–131.

Endsley, M. R., Kiris, E. O. 1995. The out-of-the-loop performance problem and level of control in automation. *Human Factors: The Journal of the Human Factors and Ergonomics Society*, 37(2), 381–394.

Ericsson, K. A., Simon, H. A. 1993. *Protocol Analysis: Verbal Reports as Data*. Cambridge, Massachusetts: MIT Press.

Fleming, B. 2012. New automotive electronics technologies. *IEEE Vehicular Technology Magazine*, 7, 4–12.

Gorman, J.C., Cooke, N.J., Winner, J.L., 2006. Measuring team situation awareness in decentralized command and control environments. *Ergonomics*, 49(12–13), 1312–1325.

Halverson, C.A., 1995. Inside the cognitive workplace: New technology and air traffic control. PhD thesis, Department of Cognitive Science, University of California, San Diego.

Hart, S. G., Staveland, L. E. 1988. Development of NASA-TLX (Task Load Index): Results of empirical and theoretical research. *Advances in Psychology*, 52, 139–183.

Hoc, J. M. 2000. From human-machine interaction to human-machine cooperation. *Ergonomics*, 43(7), 833–843.

Hoedemaeker, M., Brookhuis, K. A. 1998. Behavioural adaptation to driving with an adaptive cruise control (ACC). *Transportation Research Part F: Traffic Psychology and Behaviour*, 1(2), 95–106.

Hoeger, R., Amditis, A., Kunert, M., Hoess, A., Flemish, F., Krueger, H. P., Pagle, K. 2008. Highly automated vehicles for intelligent transport: Have-it approach. *Presented at the 15th World Congress on Intelligent Transport Systems and ITS America's 2008 Annual Meeting*, New York City, USA, November 16–20.

Hollnagel, E., 2001. Extended cognition and the future of ergonomics. *Theoretical Issues in Ergonomics Science*, 2(3), 309–315.

Hughes, J., Parkes, S. (2003). Trends in the use of verbal protocol analysis in software engineering research. *Behaviour & Information Technology*, 22(2), 127–140.

Hutchins, E., 1995a. How a cockpit remembers its speed. *Cognitive Science*, 19(3), 265–288.

Hutchins, E., 1995b. *Cognition in the Wild*. Cambridge, Massachusetts: MIT Press.

Hutchins, E., Klausen, T., 1996. Distributed cognition in an airline cockpit. In D. Middleton and Y. Engeström (Eds.), *Communication and Cognition at Work*, Cambridge: Cambridge University Press, pp. 15–34.

Jamson, A. H., Merat, N., Carsten, O. M. J., Lai, F. C. H. 2013. Behavioural changes in drivers experiencing highly-automated vehicle control in varying traffic conditions. *Transportation Research Part C*, 30, 116–125.

Jian, J., Bisantz, A. M., Drury, C. G. 2000. Foundations for an empirically determined scale of trust in automated systems. *International Journal of Cognitive Ergonomics*, 4(1), 53–71.

Khan, A. M., Bacchus, A., Erwin, S. 2012. Policy challenges of increasing automation in driving. *IATSS Research*, 35(2), 79–89. International Association of Traffic and Safety Sciences.

Klein, G., Woods, D., Bradshaw, J., Hoffman, R., Feltovich, P. 2004. Ten challenges for making automation a "Team Player" in joint human-agent activity. *IEEE Intelligent Systems*, 19(6), 91–95.

Parasuraman, R., Sheridan, T. B., Wickens, C. D. 2000. A model for types and levels of human interaction with automation. *IEEE Transactions on Systems, Man, and Cybernetics. Part A, Systems and Humans: A Publication of the IEEE Systems, Man, and Cybernetics Society*, 30(3), 286–97.

Patrick, J. 1992. *Training: Research and Practice*. London: Academic Press.

Perry, M. 2003. Distributed cognition, In J.M. Carroll (Ed.), *HCI Models, Theories and Frameworks*, San Francisco, California: Morgan-Kaufmann, pp. 93–224.

Putkonen, A., Hyrkkänen, U. 2007. Ergonomists and usability engineers encounter test method dilemmas with virtual work environments. In D. Harris (Ed.), *Engineering Psychology and Cognitive Ergonomics*. Berlin: Springer-Verlag, pp. 147–156.

Rogers, Y. 1993. Coordinating computer-mediated work. *Computer-Supported Cooperative Work*, 1, 295–315.

Rudin-Brown, C. M., Parker, H. A. 2004. Behavioural adaptation to adaptive cruise control (ACC): Implications for preventive strategies. *Transportation Research Part F: Traffic Psychology and Behaviour*, 7(2), 59–76.

Russo, J. E., Johnson, E. J., Stephens, D. L. 1989. The validity of verbal protocols. *Memory & Cognition*, 17, 759–769.

Salas, E., Prince, C., Baker, D. P., Shrestha, L. 1995. Situation awareness in team performance: Implications for measurement and training. *Human Factors: The Journal of the Human Factors and Ergonomics Society*, 37(1), 123–136.

Salmon, P.M., Stanton, N. A., Walker, G. H., Baber, C., Jenkins, D. P., McMaster, R., Young, M. S., 2008. What really is going on? Review of situation awareness models for individuals and teams. *Theoretical Issues in Ergonomics Science*, 9(4), 297–323.

Salmon, P.M., Stanton, N.A., Walker, G.H., Jenkins, D.P. 2009. *Distributed Situation Awareness: Advances in Theory, Measurement and Application to Teamwork*. Aldershot: Ashgate.

Sorensen, L. J., Stanton, N. A., Banks, A. P. 2011. Back to SA school: Contrasting three approaches to situation awareness in the cockpit. *Theoretical Issues in Ergonomics Science*, 12(6), 451–471.

Soualmi, B., Sentouh, C., Popieul, J. C., Debernard, S. 2014. Automation-driver cooperative driving in presence of undetected obstacles. *Control Engineering Practice*, 24, 106–119.

Stanton, N. A., Dunoyer, A., Leatherland, A. 2011. Detection of new in-path targets by drivers using stop and go adaptive cruise control. *Applied Ergonomics*, 42(4), 592–601.

Stanton, N. A., Marsden, P. P. 1996. From fly-by-wire to drive-by-wire: Safety implications of automation in vehicles. *Safety Science*, 24(1), 35–49.

Stanton, N. A., Stewart, R., Harris, D., Houghton, R. J., Baber, C., McMaster, R., Salmon, P. M. et al. 2006. Distributed situation awareness in dynamic systems: Theoretical development and application of an ergonomics methodology. *Ergonomics*, 49(12–13), 1288–1311.

Strand, N., Nilsson, J., Karlsson, I. C., Nilsson, L. 2014. Semi-automated highly automated driving in critical situations caused by automation failures. *Transportation Research Part F*, 27(B), 218–228.

Walker, G. H., Stanton, N. A., Baber, C., Wells, L., Gibson, H., Salmon, P., Jenkins, D. 2010. From ethnography to the EAST method: A tractable approach for representing distributed cognition in air traffic control. *Ergonomics*, 53(2), 184–197.

Weyer, J., Fink, D., Adelt, F. 2015. Human-machine cooperation in smart cars. An empirical investigation of the loss-of-control thesis. *Safety Science*, 72, 199–208.

Wickens, C. D., Hollands, J. G. 2000. *Engineering Psychology and Human Performance* (3rd Ed.). Upper Saddle River, New Jersey: Prentice-Hall.

World Health Organization. 2013. Global status report on road safety: Supporting a decade of action. Available at: http://www.who.int/violence_injury_prevention/road_safety_status/2013/en/ [Accessed November 26, 2014].

Young, M.S., Birrell, S. A., Stanton, N. A. 2011. Safe driving in a green world: A review of driver performance benchmarks and technologies to support "smart" driving. *Applied Ergonomics*, 42(4), 529–532.

Young, M. S, Stanton, N. A. 2002. Malleable attentional resources theory: A new explanation for the effects of mental underload on performance. *Human Factors*, 44(3), 365–375.

Young, M. S., Stanton, N. A. 2007. What's skill got to do with it? Vehicle automation and driver mental workload. *Ergonomics*, 50(8), 1324–1339.

18

Telematics, Urban Freight Logistics, and Low Carbon Road Networks

Guy Walker and Alastair Manson

CONTENTS

18.1 Introduction

18.1.1 Telematics

Eighty percent of the UK population now lives in an urban area. This drives increasing urban freight movements which, in turn, interact with dense street networks and a strong planning incentive to maximize their capacity and reduce vehicle emissions (Hesse and Rodrigue, 2004). Telematics, in the form of route guidance, is a key enabler for this. There is good evidence for the positive benefits telematics can have (e.g., Asvin, 2008; Dutton, 2011; Giannopoulos, 1996, etc.) but as this technology continues along the s-curve toward full market saturation there are some fundamental questions that still need to be explored. Are some urban road network topologies more energy efficient when paired with telematics technology than others? If so, to what extent might it influence a telematics strategy? Do all drivers have to have complete knowledge of traffic conditions? How realistic is this assumption anyway? Is it safe to assume that having invested in telematics drivers will adhere to route guidance information in all cases? Research (e.g., Bonsall, 1992; Bonsall and Palmer, 1999; Chorus et al., 2006; Karl and Bechervaise, 2003; Lyons et al. 2008, etc.) shows that between 30% and 50% of drivers do not: what happens then? Clearly, the success of telematics technology is heavily contingent on factors like these and this study provides an initial exploration.

18.1.2 Street Patterns and Network Types

Conventional transport network analysis methods use planar representations to reduce a complex transport network into a set of fundamental elements: nodes that represent junctions and links that represent roads (Lowe, 1975). A two-dimensional set of systematically organized points and lines like this are referred to as a planar graph. These are the basis upon which various forms of spatial analysis normally proceed (e.g., Bowen, 2012; O'Kelly, 1998). As for urban centers themselves, these tend to evolve as a product of the area's rate of growth, period of formation, location, topography, climate, culture, and so on (Thomson, 1977). Planar graphs of street patterns reveal this individuality, as do the large number of descriptive terms applied to them (Table 18.1).

A long-standing goal in telematics research has been to define certain road network "typologies" (e.g., Reggiani et al., 1995), a task made more difficult by the inconsistent use of terms/concepts such as those shown in Table 18.1. Despite this, however, it is possible to discern a much smaller recurring subset of network patterns. Brindle (1996) argues for as few as two "fundamental" urban street layouts, the grid and the tributary. Table 18.1, however, concurs with the work of Marshall (2005) in which four archetypal street patterns can be identified: the linear, tributary, radial, and grid patterns (as shown in Figure 18.1).

18.1.3 Network Metrics

Network diagrams provide a visual representation that, in simple cases, makes it very easy to discern one archetype from another. In more complex real-world examples, the visual complexity makes this task difficult to perform reliably and objectively. In these cases it is possible to turn to a number of formal metrics drawn from graph theory. These enable the connectivity of street patterns to be calculated. Three metrics, the Beta Index, the Gamma Index and network depth, are used for this purpose.

The Beta Index is a simple equation used to determine the relationship between the total number of links and the total number of nodes in a network and is calculated using the following equation:

$$\beta = e/v$$

where e is number of links, v number of nodes and $0.5 < \beta < (v - 1)/2$.

The Beta Index provides a measure of linkage intensity or "the number of linkages per node" (Lowe, 1975). Beta values generally lie between 0.5 and 3, with networks of values >1 consisting of some of the nodes within the network having more than one route between them and the network being considered well connected. Beta values <1 indicate that the network is not as well connected, there being only one route between nodes.

TABLE 18.1

Descriptive Terms Applied to Urban Street Patterns

Author	Descriptive Terms
Unwin (1920)	Irregular
	Regular
	Rectilinear
	Circular
	Diagonal
	Radiating lines
Moholy-Nagy (1969)	Geomorphic
	Concentric
	Orthogonal connective
	Orthogonal modular
	Clustered
Lynch (1981)	Star (radial)
	Satellite cities
	Linear city
	Rectangular grid cities
	Baroque axial network
	The lacework
	Other grid (parallel, triangular, hexagonal)
	The "inward" city (medieval, Islamic)
	The nested city
	Current imaginings (megaform, bubble floating, underground, etc.)
Satoh (1998)	Warped grid
	Radial
	Horseback
	Whirlpool
	Unique structures
Frey (1999)	The core city
	The star city
	The satellite city
	The galaxy of settlements
	The linear city
	The polycentric net

Source: Marshall, S., *Streets and Patterns*, Spon Press, Oxon, 2005.

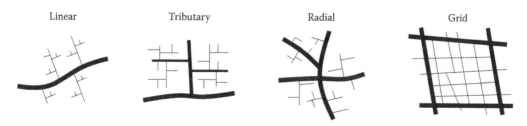

FIGURE 18.1
Urban street pattern archetypes.

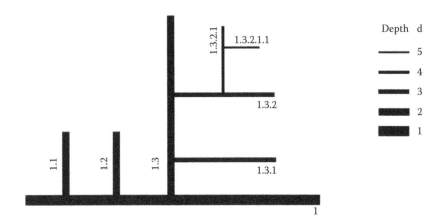

FIGURE 18.2
An example of what is meant by the "depth" of a network, with the major route being represented as the thickest line and the minor routes being represented with the thinnest line.

The "Gamma Index" helps to identify "the ratio between the actual and the maximum possible number of links" in the network (Lowe, 1975). This essentially determines whether or not every node is connected by a link, and is derived from the following equation:

$$\gamma = e/(3(v-2))$$

where e is number of links, v number of nodes and $0 < \gamma < 1$.

For example, if $\gamma = 0.5$, this means that only 50% of the maximum possible number of links in a network are provided and that all of the nodes are not fully connected.

The "depth" of a network is a relatively simple concept. In essence, it accounts for the relative distance between the most minor route and the most major route. It establishes the idea of a hierarchy and the different interconnections that exist between levels. For example, minor roads providing access to houses are "deep" whilst major routes are "shallow." An example of this is shown in Figure 18.2 where the thinner the link, the deeper the road. This particular network, taken from Marshall (2005), demonstrates that the network has a depth of five which is a deep level of penetration and results in potentially more links being provided between origins and destinations.

Network depth is important if the goal is to impartially compare different street patterns. If one network was to have a depth of four and another a depth of two, the so-called deeper network would have more connectivity. This would allow the road user to travel between points in a shorter period of time as there would be more route choice options for a road user to exploit. Different street patterns (with different Beta and Gamma coefficients) can be legitimately compared if the network depth is kept constant.

18.1.4 The Problem

Abstracting street patterns to the level of planar networks puts them on the same level as communications and other network types, for which there is a rich literature in the fields of graph theory (Harary, 1994), sociometry (Leavitt, 1951; Monge and Contractor, 2003), and complexity (Watts and Strogatz, 1998). We know from this literature that network type is a strong contingency factor in how they perform (Leavitt, 1951; Pugh et al., 1968; Watts and

Strogatz, 1998, etc.). This is in opposition to the tacit theories underlying telematics, which tend to assume that more technology, more route guidance, and the greater the driver's knowledge of the wider traffic conditions, the better the network will perform (Nijkamp et al., 1997). We know from the wider literature on communication networks that this is not necessarily the case. Of course, roads are not communications networks in the way they have been previously studied, so in order to explore the hypothesis that street pattern is an important contingency factor in the benefits to be accrued from telematics technology, we apply instead an agent modeling approach called traffic microsimulation. Microsimulation enables us to create a set of virtual street layouts based on real towns and cities, populate them with virtual traffic having differing levels of telematics, and observe the outcomes in terms of journey length, duration, cost, and, most importantly, carbon emissions. This is an exploratory study aimed at discerning the direction of the observed effects, and using this to propose different telematics strategies and guidance.

18.2 Method

18.2.1 Design

The study uses traffic microsimulation to test the interaction between different levels of vehicle telematics and the outcomes achieved within different street patterns. There are four dependent variables: journey duration, journey length, journey cost, and carbon output. These dependent variables are contingent upon two independent variables: driver knowledge of the traffic conditions in the network provided by telematics, and street pattern type. Driver knowledge had three levels. Hundred percent telematics represented urban delivery vehicles in which every driver had complete knowledge of the traffic conditions on the network and was required to act upon the guidance given (a "best-case" telematics implementation). Fifty percent telematics represented other non-freight vehicles in the network equipped with telematics/route guidance/sat nav, etc. but which, in accordance with the literature, only complied with/took notice of 50% of the information provided. Zero percent telematics represented vehicles with no telematics, with the drivers having only immediate local knowledge of traffic conditions based on what they can see. Street pattern had four levels: linear, radial, grid, and tributary. The street patterns were based on real urban locations, with network depth held constant in order to control for nonsystematic biases in connectivity. Network demand was based on real-life traffic count data from the relevant sites.

18.2.2 Real-Life Urban Networks

As determined in the previous section, there are four common forms of urban transport network layout: linear, radial, grid, and tributary. It was necessary to develop microsimulation models which closely reflect the layouts of these urban networks and it was considered important to relate the layouts to real-life towns and cities. This gives a more realistic approach and a more credible set of results. The Beta and Gamma coefficients were used to calculate the connectivity of the network archetypes shown in Figure 18.1 in order to select real-life road networks exhibiting the same properties. This approach allows networks to be categorized by a visual and a statistical approach. The depth of each model was set to three in order to control for the effects of network magnitude.

The linear network required locating a small town in which there is one main road with the town located along its length. A settlement identified as meeting these criteria is Aviemore, a town located in the Scottish Highlands. The radial network is one that has several roads intersecting or converging at a center, analogous to the spokes of a bicycle wheel. These features can often be found in a larger town which has been formed at a cross-roads. These criteria are met by Dalkeith, a town located just to the south of Edinburgh. The grid network, as its name suggests, is a network with straight roads intersecting other straight roads at right angles to form a collection of squares or blocks. The characteristics of this network type are found in the center of Glasgow. Finally, the tributary network is analogous to tributary rivers, with the smaller rivers feeding the bigger rivers. In road networks, it is the small roads that connect to the larger roads with "network depth 1" only connecting to "depth 2" and "depth 2" only connecting to "depth 3," and so on. The result of this is that only the shallower roads are busy and the deeper roads are not used as shortcuts. An area meeting the description of a tributary network is Livingston, a so-called "new town" in central Scotland. Planar graphs of these real-life street networks are shown in Figure 18.3.

Beta and Gamma values are calculated for each of the real-life towns to check the extent to which they conform to the linear, radial, grid, and tributary archetypes, as shown in

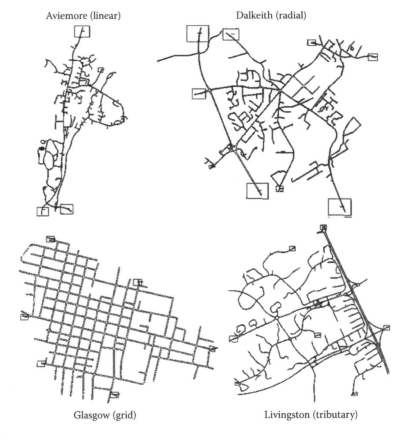

FIGURE 18.3
Traffic microsimulation models of Aviemore (tributary), Dalkeith (radial), Glasgow (grid), and Livingston (tributary).

TABLE 18.2

Values Obtained for Beta and Gamma for the Standard Layouts Shown in Figure 18.4

	Network	Nodes (v)	Links (e)	Beta	Gamma (%)
Archetype	**Linear**	19	35	1.84	69
Real Life	Aviemore	140	257	1.84	62
Archetype	**Radial**	32	65	2.03	72
Real Life	Dalkeith	140	268	1.91	65
Archetype	**Grid**	33	73	2.21	78
Real Life	Glasgow	140	278	1.99	67
Archetype	**Tributary**	9	17	1.89	81
Real Life	Livingston	140	265	1.89	64

Table 18.2. The real-life networks are considerably larger than the archetypes yet it can be noted how the network metrics align, showing how the underlying network structures are equivalent in type if not size. The Beta values for the linear and Aviemore networks, for example, are exactly equal as are the Beta values for the tributary and Livingston networks, where differences do exist (and they are relatively modest), the rank order of the network types is still preserved.

18.2.3 Development of Network Models

Planar graph representations of Aviemore, Dalkeith, Glasgow, and Livingston were extracted from ArcGIS (a mapping and spatial analysis software) and imported into S-Paramics (the traffic microsimulation software) in order to create a basic model of each. An attempt was made to ensure the modeled network was as true to life as possible but some simplifications were required in order to isolate the effects of street patterns and

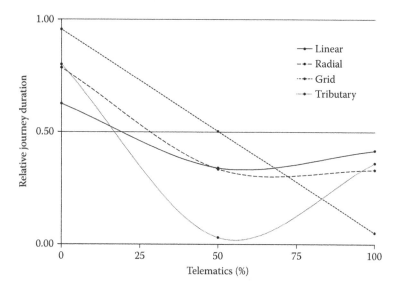

FIGURE 18.4

Interactions between street pattern, driver knowledge of network conditions provided by vehicle telematics, and relative journey length. A 100% telematics represents urban freight vehicles.

route guidance from specific and localized variations in the traffic situation. As such, junctions had simple priorities applied with the bigger roads having priority over the smaller roads. Traffic lights were avoided all together. There were also no buses or bus routes applied as not all networks had buses arriving at similar times and some networks had bus lanes whereas others did not. A number of steps were taken to calibrate the models to the real-world context despite these simplifications being applied. A number of parameters in the model were adjusted to ensure this. First, the signpost distance refers to the distance ahead of a hazard at which the modeled road users become aware of it, and was standardized to 80 m to allow the traffic to use both lanes of the entrances and exits of roundabouts (anomalous behavior would arise if not). Visibility was also varied. This relates to how far back from the stop-line vehicles begin assessing their gap distance. This ensured that vehicles approaching roundabouts continued onto it in situations where nothing was coming and stopped in cases where something was approaching. The distance which meant vehicles followed this rule was found to be 20 m. These rules allowed more life-like driver behavior to be represented, and for a good approximation of the traffic situation actually experienced in these locations to emerge. These steps were applied consistently across all the networks in order to isolate the effects of network and vehicle type.

18.2.4 Vehicle Types

The principle vehicle-based manipulation was the amount of knowledge drivers had of wider network conditions based on the amount of route guidance provided. Three levels were implemented:

- Vehicle type 1 (urban delivery vehicle) had 100% knowledge of traffic conditions on the network via telematics and followed the route guidance information they were given 100% of the time.
- Vehicle type 2 (private car) had 100% knowledge of traffic conditions on the network via telematics but only chose to follow it 50% of the time (ignoring the other 50%).
- Vehicle type 3 (private car) had no telematics or route guidance and therefore the drivers had 0% knowledge of traffic conditions on the network beyond what they could see ahead of them.

The three levels of telematics/route guidance were implemented in the microsimulation model by manipulating feedback. Feedback is information supplied to the road user, via telematics, about current network conditions. Specifically, it lets them know of journey times on all routes so they can decide an optimum route to choose. Feedback is calculated using two aspects, the feedback interval and the feedback factor. Both were adjusted in such a way as to give rise to realistic driver behavior. The feedback interval refers to how often the information is updated and an interval of 2 min was used. This is a common value in similar models and avoids the modeled drivers making unrealistically rapid and unstable route choice decisions. The feedback factor is concerned with what percentage of delay information is taken from the previous feedback interval and was taken to be 0.5% or 50% as standard. A perturbation level of 5% was applied to all three vehicle types. This helps to account for variability in travel costs, or a driver's perception of these travel costs. As perturbation increases, the road users' concentration tends to focus more on reducing journey cost. However, by applying a small percentage it means that road users will continue to focus on reducing journey length and journey duration as the key variants but will also look to reduce their cost simultaneously.

The same mix of traffic was applied to all the networks in order to capture the interaction between "normal traffic" and urban freight delivery vehicles. Fifty percent of the road users in all the network types were "vehicle type 3" (0% telematics) with the remaining 50% being split into two: 25% of road users were "vehicle type 2" (50% telematics) and 25% of road users were urban delivery vehicles, or "vehicle type 1" (100% telematics). The normal traffic (i.e., vehicle types 2 and 3) were modeled as medium-sized saloon cars, whereas the urban delivery vehicles (i.e., vehicle type 1) were modeled as car-based vans. This is (a) because larger vehicle types would give unfavorable differential effects on networks that are less suitable for larger vehicles and (b) the literature identifies vans as the "dominant mode" in these contexts (e.g., Cherrett et al., 2012).

18.2.5 Network Demand

The demand profile controls the number of vehicles in the network, the origin and destination of the vehicles, the percentages of vehicle type and the release rate of vehicles into the network. The number of vehicles released from the origin to the destination was varied from model to model based on actual traffic count data. The goal was to bring each network to its peak PM traffic flow and hold it there for the duration of the study. To do this, peak PM values from the nearest traffic counter site (Scottish Government, 2012) were used and multiplied by 24. This total value was then split evenly between each of the origins and destinations. The release rate of the vehicles over the 24-h period was constant so that all the vehicles were not released at once and a steady flow was maintained. Each model was then subject to 30 "batch runs" between the network model hours of 1600 and 2000, which was the length of the peak PM flow period. The models had, of course, been established in this peak flow state for many hours previously hence any transient effects of the model being initially loaded with traffic were avoided. These 30 runs allowed a significant amount of data to be obtained on the outcome variables of journey duration, length, cost, carbon output, and their contingency on network type.

18.3 Results and Discussion

The output from running the S-Paramics microsimulations is a set of raw data for each individual vehicle as it progressed through the network. The software also calculates the position of each individual vehicle every half second and records its coordinates against time. This data was passed through an external programme known as "AIRE" which calculated the resulting pollution output of each individual vehicle for a specific year, which was chosen to be 2012. An average value for carbon output was obtained for each vehicle type, along with data on journey duration, length, and cost. A summary of the uncorrected average and standard deviation values obtained from the simulations is provided in Table 18.3. Table 18.4 shows the same results corrected for network size/distance, collapsing some of the key variables into average speed, cost per km, carbon output in g/km, and total emissions.

The values shown in Table 18.3 are absolute, in that no correction is made for the differing sizes of the networks. The values in Table 18.4 are corrected, and provide a number of key insights. Ignoring vehicle telemetry for a moment, it can be noted that the linear and radial networks are very similar, both being able to support average speeds in the region of 42 km/h, costs per km of approximately 27 pence, with vehicles on the network

TABLE 18.3

Summary of the Uncorrected Journey Length, Duration, Cost, and Carbon Output Data Obtained from the Modeled Networks

	Vehicle Type	Journey Length (m)		Journey Duration (s)		Journey Cost (Pence)		Carbon Output (g)	
	Telematics	Mean	SD	Mean	SD	Mean	SD	Mean	SD
Linear	0%	2152.7	30.7	183.3	2.4	224.6	1.6	130.0	1.8
	50%	2155.3	51.6	182.7	4.0	225.2	2.5	128.5	2.9
	100%	2165.8	44.2	182.8	3.4	226.0	2.6	128.1	2.7
	Mean	2157.9		182.9		225.3		128.9	
Radial	0%	2286.4	10.5	194.6	0.8	198.8	0.8	137.7	0.8
	50%	2281.2	16.9	193.7	1.3	198.2	1.1	137.2	1.3
	100%	2285.8	17.5	193.7	1.3	198.9	1.1	137.2	1.1
	Mean	2284.5		194.0		198.6		137.4	
Grid	0%	1304.6	6.3	147.5	0.8	124.2	0.6	129.0	0.9
	50%	1316.3	9.5	142.5	1.3	124.2	0.9	122.1	1.3
	100%	1326.9	12.2	137.5	1.2	124.0	1.0	115.0	1.0
	Mean	1315.9		142.5		124.1		122.0	
Tributary	0%	3662.4	16.2	309.8	1.2	321.5	1.6	214.5	1.0
	50%	3654.8	28.7	309.0	2.4	320.6	2.9	213.8	1.6
	100%	3663.6	26.0	309.4	2.2	321.4	2.9	214.4	1.9
	Mean	3660.3		309.4		321.2		214.2	

TABLE 18.4

Summary of Corrected Speed, Cost, and Carbon Output Data

	Vehicles		Speed	Cost	Carbon		
	Total N	Telematics	km/h	Pence/km	g/km	Total kg	
Linear	3000	0%	42.28	28.05	60.39	390.00	
		50%	42.47	27.66	59.62	385.50	
		100%	42.65	27.31	59.15	384.32	
		Mean	42.47	27.67	59.72	1159.82	Total
Radial	4752	0%	42.30	26.34	60.23	654.40	
		50%	42.40	26.34	60.14	651.93	
		100%	42.48	26.26	60.02	651.95	
		Mean	42.39	26.32	60.13	1958.28	Total
Grid	6000	0%	31.84	75.79	98.88	773.99	
		50%	33.25	70.47	92.76	732.60	
		100%	34.74	65.32	86.67	690.02	
		Mean	33.28	70.53	92.77	2196.61	Total
Tributary	5995	0%	42.56	15.99	58.57	1285.97	
		50%	42.58	16.00	58.50	1281.77	
		100%	42.63	15.97	58.52	1285.29	
		Mean	42.59	15.99	58.53	3853.03	Total

each emitting approximately 60 g/km of carbon. The tributary is the same aside from cost, which is the lowest of all the road networks at 15.99 pence/km. The slowest (33.28 km/h), most expensive (70.53 pence/km), and carbon intensive (92.77 g/km) network is the grid but it is within this network that the urban delivery vehicles (with 100% telemetry) performed the best. In this situation telemetry is raising average speeds by 2.9 km/h, reducing costs by a not insignificant 10.47 pence/km and, most importantly, reducing carbon emissions by 12.21 g/km. The carbon value is of course determined by the physical size of the networks and the journey lengths therein, so in these examples the larger tributary network (i.e., Livingston) has the highest total emissions (3.8 tonnes of carbon per modeled peak PM), however, the network with the shortest average journey lengths (i.e., Glasgow/grid) has the second worst carbon outputs (2.2 tonnes).

There are smaller differential effects present in the other network types that are also important. Although smaller at the level of individual vehicles, when multiplied by the number of vehicles in the networks (several thousand) and the number of times peak PM hour conditions occur (every weekday evening), these differences begin to magnify significantly. For example, in the tributary network, a per-vehicle difference in carbon emissions of only 0.7 g as a result of telematics still multiplies to an additional daily peak PM carbon output of approximately 4.2 kg, or approximately 1 tonne per year. Multiplied again by the number of settlements with tributary street patterns, these initially marginal differences start to accumulate rapidly. With this in mind, the following sections shift the focus from absolute values to relative values in order to discern the direction of these various effects, and what they might mean for an overall telematics strategy.

18.3.1 Telematics versus Journey Duration

In order to provide a visually tractable representation of how the data is behaving, and show the relationships for each network on the same graph, it is necessary to convert absolute values to relative values. These were determined by finding the difference between the actual value and the minimum value and dividing this by the difference. This, therefore, shows the direction of the relationships between road network type and how different levels of vehicle telematics within it perform.

Figure 18.5 shows the relationship between different levels of telematics and the relative journey duration experienced by each type of road user for each of the four models. The graph shows that urban delivery vehicles (with 100% telematics) operating within linear and tributary road networks are worse off than "normal traffic" with 50% knowledge/acceptance of route guidance information, albeit not as bad as the population of vehicles with no telematics. Indeed, the results show that the same outcomes on journey duration are achieved at both 100% telematics penetration and approximately 20%, the difference being that the latter is considerably more costly than the former. The trend line obtained for the "radial" network only slightly differs from the "linear" and "tributary" networks in respect of the duration increasing between 50% and 100%, and leveling off thereafter. This shows there is no further benefit of telematics in this context. The relationship for the linear, radial, and tributary networks seems to show that there is an optimum level of knowledge about the traffic conditions on the network. For the optimization of journey durations, a level of knowledge consistent with 100% telematics is not required. This principle does not hold for the "grid" network. Here the trend line shows increasing benefits as more telematics is provided, reaching a maximum benefit at 100% telematics penetration. The broader principle to be extracted here is that 50% knowledge of the traffic conditions on the network are optimum for networks characterized by one or two critical

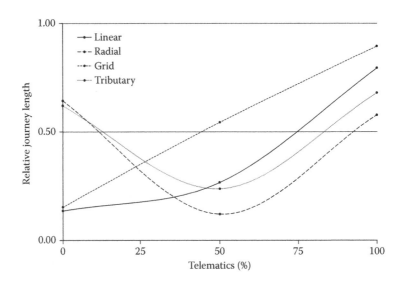

FIGURE 18.5
Interactions between street pattern, driver knowledge of network conditions provided by vehicle telematics, and relative journey duration. A 100% telematics represents urban freight vehicles.

routes between the majority of the origins and destinations. Hundred percent telematics penetration reduces journey durations in networks where there are a large range of routes available to the road user. In this study, urban freight vehicles with 100% telematics extract maximum journey time benefits in grid networks.

18.3.2 Telematics versus Journey Length

Figure 18.2 shows how the journey length taken by road users relates to the amount of telematics each vehicle has, but is contingent on street pattern type. In the linear network, the journey length increases at a slow rate between 0% and 50% telematics penetration before it increases at a much greater rate between 50% and 100%. This relationship is very different from the radial and tributary networks, which both have a long journey length for road users with 0% but falling markedly between 0% and 50% before rising again very steeply between 50% and 100%. As with the results for journey duration, there is an optimum level of telematics for radial and tributary networks (around 50%), with the same network performance achieved at 0% telematics penetration as achieved at 100%. Urban delivery vehicles with 100% telematics would see little benefit in these settings. The grid network differs again. The diagonal line traced through the chart shows that the greater the level of telematics, the greater the journey length as road users exploit the more numerous opportunities to divert. Taken together, Figures 18.2 and 18.5 suggest that shorter journey durations are achieved with longer journey lengths. The broader principle to be extracted here is that urban freight vehicles in networks with more than one route between the origin and destination travel a greater distance but in a shorter time.

18.3.3 Telematics versus Journey Cost

Figure 18.6 shows that the linear network has a more or less direct relationship between telematics-mediated knowledge of network conditions and journey cost (i.e., a straight

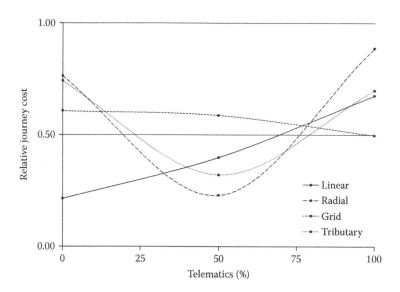

FIGURE 18.6
Interactions between street pattern, driver knowledge of network conditions provided by vehicle telematics, and relative journey cost. A 100% telematics represents urban freight vehicles.

line). It shows that as the acceptance of telematics information increases from 0% to 100%, the journey cost also increases. Interestingly, then, in this network type 0% telematics penetration is the optimum value for cost to be optimized. This is not the case for radial and tributary networks. Both of these undergo a reduction in cost between 0% and 50% with an increase then occurring between 50% and 100%. Once again, the optimum level of telematics penetration in a radial or tributary network is 50%, with further increases not only having a negative effect on journey cost but 100% telematics penetration (as per the urban freight vehicles) yields the same outcome as 0%. The grid network again performs differently to the other three network types. The relationship is linear (i.e., a straight line) between 0% and 50% telematics penetration, before tailing off slightly as 100% telematics penetration is reached. What this means is that 100% telematics penetration is required in grid networks for meaningful journey cost savings to emerge, 0% for linear networks and 50% for radial and tributary networks.

18.3.4 Telematics versus Carbon Emissions

The crux of the analysis is to see what effect these contingent values of journey length, duration, and cost ultimately have on carbon emissions. Figure 18.7 presents the results of this analysis. It can be seen that the inear, radial and tributary networks all follow a similar trend, with carbon emissions decreasing rapidly between 0% and 50% telematics but with varying levels of diminishing further benefits as the telematics rate approaches 100%. The linear and radial networks level off beyond 50%, suggesting little (if any) further benefits of increasing telematics penetration. The results suggest that the carbon emissions from the tributary network worsen with increases beyond 50% telematics penetration. The grid network is once again quite distinct. It contains a directly proportional relationship between telematics penetration and carbon emissions, with the maximum value occurring at 0% and the minimum value at 100%. If reducing carbon emissions are the goal of urban

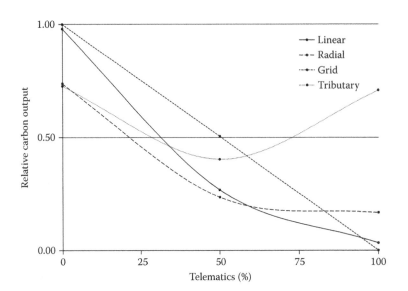

FIGURE 18.7
Interactions between street pattern, driver knowledge of network conditions provided by vehicle telematics, and relative carbon output. A 100% telematics represents urban freight vehicles.

freight vehicles, then telematics has the biggest role to play in grid networks, and apart from tributary networks, there are some benefits to be extracted before 100% telematics is achieved.

18.3.5 Optimization Values

Based on the data and relationships obtained from the previous sections, particularly Figures 18.6 through 18.8, it is possible to create Table 18.5 to show the potential network performance trade-offs involved in minimizing carbon emissions using telematics. A mean level of driver knowledge of network conditions (or acceptance thereof) is given, this representing a simple value by which the best compromise of journey duration, length, cost, and carbon variables is achieved.

Table 18.5 shows that, in theory, urban street networks can be designed to be sustainable with the smallest pollution outputs, designed to reduce the journey duration of road users traveling through them, or designed to reduce traveler costs. In practice, however, street networks have evolved and cannot be changed on the scale necessary to optimize

TABLE 18.5

Optimum Levels of Telematics Penetration for Journey Duration, Length, Cost, and Carbon Emissions for Linear, Tributary, Radial, and Grid Street Patterns

Network	Duration (%)	Length (%)	Traveler Costs (%)	CO_2 (%)	Mean (%)
Linear	50	0	0	100	38
Radial	50	50	50	100	63
Grid	100	0	100	100	75
Tributary	50	50	50	50	50

these factors. This is where telematics comes in. Telematics interacts with network types to modify their inherent performance. The interaction is not a simple one. It is contingent on the level of telematics information provided to, and accepted by, drivers combined with the topology of the network itself. Where some networks require varying values of telematics penetration to optimize various aspects of the network, some run at the optimum level for the majority of characteristics under one level of telematics. This can be seen in Table 18.5 with the tributary and radial networks, which both run at their most efficient levels for all four characteristics with a 50% telematics rate. The table also shows that for a grid network, 100% telematics penetration is the optimum value. Linear networks, however, do not reach their optimum level of efficiency for any one level of telematics. Table 18.5, therefore, becomes a useful tool when attempting to design a telematics strategy in order to modify the inherent characteristics of urban road networks. Stated simply, some levels of driver knowledge of network conditions (provided via telematics) are more optimal than others, and it depends on the street pattern urban delivery vehicles are operating within.

18.3.6 Fundamental Relationships

The results shown and discussed above convey the idea that for each output characteristic (i.e., journey length, cost, duration, and carbon emissions) different vehicles perform differently depending on the network knowledge provided/accepted by drivers via telematics. What if a particular urban freight context does not conform to the archetypes presented? In this case it is possible to increase the generalizability of the results with recourse back to the connectivity coefficients discussed earlier. These can be applied to any transport network, of any size or type, in order to reveal its underlying level of connectivity. A fundamental relationship emerges: as the telematics penetration rate increases from 0% to 100% in a road network with Beta (β) ≥ 1.9, journey costs, journey duration, and carbon output tend to improve (or at least do not worsen). In networks with $\beta < 1.9$, there is no added benefit of providing anything more than 50% telematics penetration. Figure 18.8

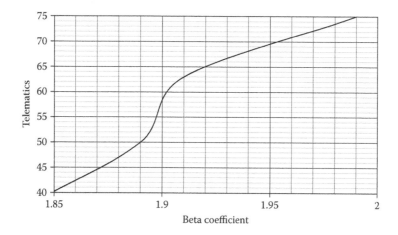

FIGURE 18.8
Relationship between the Beta coefficient (β) and the level of telematics required to optimize journey length, duration, cost, and carbon outputs.

illustrates the relationship between β and telematics within the range of data observed. Figure 18.8 represents a simple diagnostic for answering the question: "what are the benefits of increasing drivers' knowledge of the wider network conditions, via telematics, in a particular operating environment?"

18.4 Conclusions

Contrary to general belief, this study shows that there is a point at which more telematics does not lead to more efficient urban logistics. Indeed, in many cases the same outcomes can be achieved with 0% telematics as it can with 100%. Simply introducing an abundance of telematics into an urban freight situation, with planners and company policy enforcing 100% compliance with route guidance, however feasible that may be, will not result in the outcomes expected. This study shows that the topography of an urban street layout is an important contingency factor in how and when to deploy this technology. Both street pattern and route guidance interact to create congestion in the network, define who gets caught in it, and to what collective effects on time, cost, distance, and carbon emissions.

This is an exploratory study but a number of potentially important implications arise from it. The first relates to previous research that shows between 30% and 50% of drivers do not comply with telematics-based route guidance no matter how much of it is provided. As the results show, middle values like these represent an optimum on many outcome variables and street network types. This raises an interesting point. Does this established 30%–50% driver acceptance of route guidance arise through repeated experience of traffic conditions on a network, the implication being that driver behaviors (and the network itself) are self-organizing? If this is the case, then is the imposition of telematics based on a false premise? Does "everyone" in the network need to know "everything?" The results of this study would seem to suggest that, for some network types, they do not. If this is the case then the same outcomes could be achieved for considerably reduced cost.

The second implication relates to the different telematics strategies that urban logistics providers could adopt, and when? Is a costly strategy of providing complete knowledge of the network to drivers, and enforcing compliance with route guidance, optimum in all situations? No. Likewise, is a laissez fair approach to planning and route guidance, based purely on ad hoc local knowledge brought to situations by individual drivers, optimum? Again, no. Optimization is contingent upon the topology of the network being traveled upon. There are clearly some situations where it would benefit outcome variables such as carbon emissions to impose compliance with route guidance, and other situations where it would not be appropriate. The results of this study are helpful in understanding what these relationships might be and what an "adaptive telematics" strategy might look like. It would be a form of route guidance that would be cognitively compatible with drivers. One in which the timing and sequence of route guidance information would be oriented around different outcome variables at different times, but in all cases offering tangible journey based "rewards" for the driver. These rewards would encourage telematics to be used in ways that exceed the current 30%–50% acceptance rate where it is beneficial to do so, and as such, to accumulate some significant marginal gains.

Acknowledgment

This research benefitted from a collaboration between Heriot-Watt University and SIAS Transport Planners, 37 Manor Place, Edinburgh, [UK]EH3 7EB, the developers of the traffic microsimulation software S-Paramics.

References

Asvin, G. 2008. *Fleet Telematics: Real-Time Management and Planning of Commercial Vehicle Operations.* Boston, Massachusetts: Springer.

Bonsall, P. 1992. The influence of route guidance on route choice in urban networks. *Transportation,* 19, 1–23.

Bonsall, P. W. and Palmer, I. 1999. Behavioural response to roadside variable message signs: Factors affecting compliance. In: Emmerick, R and Nijkampt, P. (eds). *Behavioural and Network Impacts of Driver Information Systems.* Farnham: Ashgate.

Bowen, J. T. 2012. A spatial analysis of FedEx and UPS: Hubs, spokes, and network structure. *Journal of Transport Geography,* 24, 419–431.

Brindle, R. E. 1996. *Urban Road Classification and Local Street Function.* South Vermont, Australia: Australian Road Research Board.

Cherrett, T., Allen, J., McLeod, F., Maynard, S., Hickford, A., and Browne, M. 2012. Understanding urban freight activity—Key issues for freight planning. *Journal of Transport Geography,* 24, 22–32.

Chorus et al. 2006. Use and effects of advanced traveller information services (ATIS): A review of the literature. *Transport Reviews,* 26(2), 127–149.

Dutton, G. 2011. Fleet management's magic box. *World Trade,* 24(2), 38–44.

Frey, H. 1999. *Designing the City: Towards a More Sustainable Urban Form.* London: Routledge.

Giannopoulos, G. A. 1996. Implications of European transport telematics on advanced logistics and distribution. *Transport Logistics,* 1(1), 31–49.

Harary, F. 1994. *Graph Theory.* Reading, Massachusetts: Addison-Wesley.

Hesse, M. and Rodrigue, J. P. 2004. The transport geography of logistics and freight distribution. *Journal of Transport Geography,* 12(3), 171–184.

Karl, C. A. and Bechervaise, N. E. 2003. The learning driver: Issues for provision of traveller information services. In *10th World Congress and Exhibition on Intelligent Transport Systems and Services,* Madrid.

Leavitt, H. J. 1951. Some effects of certain communication patterns on group performance. *Journal of Abnormal and Social Psychology,* 46, 38–50.

Lowe, J. C. 1975. *The Geography of Movement.* Boston: Houghton Mifflin.

Lynch, K. 1981. A Theory of Good City Form. Cambridge: MIT Press.

Lyons, G., Avineri, E., and Farag, S. 2008. Assessing the demand for travel information: Do we really want to know? In *Proceedings of the European Transport Conference,* Noordwijkerhout, Netherlands, October. http://abstracts.aetransport.org/paper/index/id/2964/confid/14

Marshall, S. 2005. *Streets and Patterns.* Oxon: Spon Press.

Moholy-Nagy, S. 1969. Matrix of Man: Illustrated History of Urban Environment. New York: Praeger.

Monge, P. R. and Contractor, N. S. 2003. *Theories of Communication Networks.* New York: Oxford University Press.

Nijkamp, P., Pepping, G., and Banister, D. 1997. *Telematics and Transport Behaviour.* Berlin: Springer-Verlag.

O'Kelly, M. E. 1998. A geographer's analysis of hub-and-spoke networks. *Journal of Transport Geography*, 6(3), 171–186.

Pugh, D. S., Hickson, D. J., Hinings, C. R., and Turner, C. 1968. Dimensions of organisation structure. *Administrative Science Quarterly*, 13(1), 65–105.

Reggiani, A., Lampugnani, G., Nijkamp, P., and Pepping, G. 1995. Towards a typology of European inter-urban transport corridors for advanced transport telematics applications. *Journal of Transport Geography*, 3(1), 53–67.

Satoh, S. 1998. Urban design and change in Japanese castle towns. *Built Environment*, 24(4), 217–234.

Scottish Government. 2012. *Road–Traffic Count*. Accessed 2012. http://www.transport.gov.scot/road/traffic-count

Thomson, J. M. 1977. *Great Cities and Their Traffic*. London: Victor Gollancz Ltd.

Unwin, R. 1920. *Town Planning in Practice: An Introduction to the Art of Designing Cities and Suburbs*. London: Adelphi Terrace.

Watts, D. J. and Strogatz, S. H. 1998. Collective dynamics of "small-world" networks. *Nature*, 393(4), 440–442.

19

End Users' Acceptance and Use of Adaptive Cruise Control Systems

Niklas Strand, I. C. MariAnne Karlsson, and Lena Nilsson

CONTENTS

19.1 Introduction

Advanced driver assistance systems (ADAS) in vehicles are associated with safety and comfort benefits for the driver (Brookhuis et al. 2001). Adaptive cruise control (ACC) is an ADAS that controls speed and distance to vehicles ahead based on speed and time-gap settings (see, e.g., Naranjo et al. 2003 or Stanton and Young 2005). Several studies concerning ACC systems have been carried out and the majority of these studies have however not focused on actual users, instead car drivers in general have been appointed as participants. Such a recruiting strategy has advantages when addressing certain research questions, while for other research questions, knowledge and experience of a specific system are essential. In comparison with studies not including actual users, studies including them are scarce. There are nevertheless a few exceptions including studies by Strand et al. (2011), Larsson (2012), Bianchi Piccinini et al. (2012), and Sanchez et al. (2012) which all focused on actual end users and their assessment of and behavioral adaptation to the systems.

Behavioral adaptation is a psychological issue that explains behaviors that occur in response to change in a road–vehicle–user system (OECD 1990). For a full account of behavioral adaptation, see the review by Saad et al. (2004). Strand et al. (2011) explored end-user experiences of ACC by means of focus group interviews and the study by Bianchi Piccinini et al. (2012) used a similar approach. Both studies showed that, overall, end users are satisfied with their ACC systems and that there are positive effects in terms of convenience as well as safety. Nevertheless, some negative effects were also identified. Strand et al. (2011) revealed that there are end users carrying mental models, which to some extent can be considered rudimentary. Mental models can be viewed as our understanding of objects and processes that guide our interaction with them (Bernstein et al. 2003) and a proper mental model of system functionality is important to cope with for instance, mode errors (Sarter and Woods 1995). Another finding was that the drivers had experienced situations, such as roundabouts, in which ACC functionality could be improved. Negative aspects related to ACC use concluded in the study by Bianchi Piccinini et al. (2012) were that end users were not fully aware of potential critical situations when driving with ACC and that improper usage of ACC occurred. They also found that tasks distracting driving were more frequent when driving with ACC, compared to manual driving. The study by Larsson (2012) reported on a survey to 130 ACC users. In this case the results indicated that use experience is crucial for the understanding of the functional limitations of ACC.

The report by Sanchez et al. (2012) addressed user acceptance of ACC and presented results based on a survey to 227 participants in the EU project euroFOT (www.eurofot-ip. eu). The majority of them were drivers of passenger cars equipped with ACC, but the study also included those who drove heavy trucks equipped with the same system. The results suggested that acceptance (defined as perceived usefulness and driver satisfaction) was very high and stable over time. ACC was rated most useful in normal traffic on motorways. Furthermore, ACC increased perceived comfort (stated by 80%) and safety (stated by 94%).

The main purpose of this study was to investigate end-users' experiences of ACC in order to assess the system with regard to drivers' use and acceptance of ACC; usability assessment; and perceived influence on driving behavior. An additional aim was to see if these aspects were affected by where ACC has been mainly experienced.

19.2 Method

19.2.1 Data Collection

Data was collected by means of an online questionnaire, set up with SPSS mrInterview software, version 5.5, patch level 3 (SPSS Inc. 2008). The questionnaire included altogether slightly more than 70 questions covering the topics: experience, usage pattern, acceptance, and perceived usability. For acceptance, a new acceptance scale was used, namely Strömberg Karlsson Acceptance Scale (Sagamihara Keio apathy scale) (Strömberg and Karlsson n.d.). The scale included 20 items covering four areas:

- *Trust and control* attempts to capture how secure the user feels with the system. It consists of three items capturing aspects of the perceived technical reliability of the system, whether the user thinks that the information that the system gives or the action it takes can be trusted, and whether the user feels in control of the system.
- *Perceived benefit* aims to find whether the user perceives the system as something useful and as something that provides benefits (in terms of convenience, joy, efficiency, etc.) to them in the task they are trying to perform.
- *Perceived effort* tries to identify the effort the users feel they have to put into gaining the benefits. It draws on classical usability in terms of ease of use, logic, coherence, etc.

- *Compliance* tests more abstract level aspects of acceptance. The dispensability item aims to capture whether the users feel that the problem that the product is trying to solve actually is a problem that needs solving, and the appropriateness item aims to capture whether the product is a suitable way of solving that problem.

All but one question was closed-ended. The major part of questions was of a Likert type (Likert 1932) with five response categories, including a neutral category. There were also a few questions with rating scales, from low to high (1–5). In Strömberg Karlsson Acceptance Scale the response scales were instead of semantic differential type. The semantic differential scales used bipolar categories, with one positive and one negative pole. See for instance McQueen and Knussen (2006) for a brief overview of rating scales, Likert scales, and semantic differential scales.

19.2.2 Analysis

Statistical analyses were conducted with IBM SPSS statistics software, version 21 (IBM Corp. 2012). First a descriptive data analysis was conducted. This was then followed by a Mann–Whitney test in which driving context (mainly within urban areas and mainly outside urban areas) was used as a grouping variable. Only significant tests are presented in Section 19.3.

19.2.3 Participants

A list of 414 end users' of ACC was handed over by Volvo Cars Corporation. They were approached by e-mail and invited to answer an online survey. Altogether 90 of them completed the survey. However, three of them were excluded from analysis since they answered that their cars were not equipped with ACC.

The respondents' ages varied between 31 and 76 years ($M = 53.18$, $SD = 10.69$). Nearly a two-thirds majority had completed higher education (higher than upper secondary school). All of them had a valid driver's license and had held their cars for between 13 and 58 years ($M = 34.75$, $SD = 10.69$). The majority drove more than 25,000 km/year. (See Table 19.1 for an overview of participants' characteristics.)

About half of the participants were mainly engaged in nonprofessional driving, and one-third drove equal amounts professionally and nonprofessionally. The remaining drove primarily in a professional role. Furthermore, roughly equal amounts of the drivers drove mainly within urban areas and outside of urban areas respectively. See Table 19.2 for overview of responses regarding context of driving.

The participants were all owners of cars from the S, V, or XC ranges manufactured by Volvo Cars. At the time of the survey, 2013 that is, all cars were fairly recent year models with the oldest being of year model 2009 and the latest 2014. The majority (65.5%) drove a car with automatic transmission. A majority rated themselves as experienced or very experienced ACC users. Only a few considered themselves to be beginners.

19.3 Results

One of the questions posed to the respondents concerned their general attitude toward ACC. Overall the respondents had a very positive attitude toward ACC. As many as 94.3%

TABLE 19.1

Participants' Characteristics

Characteristic	(*n*)	(%)
Gender		
Men	78	89.7
Woman	9	10.3
Highest education level completed		
Compulsory school	3	3.4
Upper secondary school	29	33.3
Higher education: university college; university	49	56.3
Other education after upper secondary school	6	6.9
Annual mileage (km)		
5,001–10,000	1	1.1
10,001–15,000	9	10.3
15,001–20,000	17	19.5
20,001–25,000	15	17.2
>25,000	45	51.7

was "very positive," 4.6% was "somewhat positive," and 1.1% neither negative nor positive. None rated their attitude toward ACC as negative.

Earlier studies (Strand et al. 2011) have indicated that drivers experience that other drivers react to cars driven with the support of ACC. The respondents were therefore asked to answer the question: "Do you worry about how your driving with ACC is perceived by your fellow commuters?" According to the responses to the survey, the major part of the respondents had no such worries. Instead it seems as if they are signifying a positive influence on other commuters (Table 19.3). Nevertheless, 14.9% notes that other road users have a negative perception of the car following distance. Approximately 22% provided answers suggesting that they had not received any reactions to their driving or that they had not considered the matter.

A particular topic addressed was the drivers' usage of ACC (Table 19.4). When asked how often they activate ACC under different circumstances, the respondents answered that the system is frequently used under low as well as high traffic intensity. ACC is also frequently activated when there is a queue, when it is raining, and when driving in the dark. However, there are, as indicated in earlier studies (Strand et al. 2011), also situations where drivers do not activate ACC as often. These include snowfall and slippery road conditions.

TABLE 19.2

Context of Driving

Context	(*n*)	(%)
Type of driving		
Professional	18	20.7
Nonprofessional	42	48.3
Equal amount of professional and nonprofessional	27	31
Area of driving		
Within urban area	40	46
Outside of urban area	47	54

TABLE 19.3

How Do You Think Your Fellow Commuters Have Perceived Your Driving with ACC Regarding the Following Aspects?

	Answer									
	Very Negative		Somewhat Negative		Neither Nor		Somewhat Positive		Very Positive		N/A
Aspect	(*n*)	(%)	(*n*)	(%)	(*n*)	(%)	(*n*)	(%)	(*n*)	(%)	N
1. Your car following distance	10	14.9	–	–	23	34.3	11	16.4	23	34.3	20
2. How you adapt your car following distance	–	–	7	10.6	23	34.8	9	13.6	27	40.9	21
3. How you adapt your speed	–	–	3	4.6	20	30.8	12	18.5	30	46.2	22
4. Your traffic rhythm	–	–	5	7.8	15	23.4	17	26.6	27	42.2	23

Note: Percentages given are the valid percent.

The responses also show that the higher the allowed speed limit, the more frequently ACC is used (Table 19.5). The drivers use the function the most on roads with speed limits of 100/110/120 km/h and the least on living streets.

The respondents' answers to how frequently they use ACC depending on the speed regulation reflect on the answers they provide on how satisfied they are with the function during the same speed regulations (Table 19.6): the higher the speed, the more satisfied the user; and the lower the speed, the less responses are provided.

The respondents' general attitude toward ACC is positive, 94% are very positive and no one negative. The respondents are particularly satisfied with ACC during longer trips and when driving on roads without possibilities to overtake. The respondents' rating of satisfaction when being overtaken and when overtaking was higher for those who drove mainly outside urban areas (mode = 5;5) than it was for those who drove mainly within urban areas (mode = 5;4): $U = 701.5$, $p = 0.021$, and $U = 555.5$, $p = 0.002$. There are nevertheless specific situations, such as driving on roundabouts and curves where satisfaction drops. For instance 35% of the respondents answer that they are somewhat or very dissatisfied with the function when driving on a roundabout. There are also a considerable amount of responses missing which could indicate that the drivers do not use the function in this condition. See Table 19.7 for overview of how ACC was perceived during different circumstances.

TABLE 19.4

How Often Do You Choose to Activate the ACC in the Following Circumstances?

Circumstance	M	Mdn	Mo	SD
Low traffic intensity	4.49	5	5	0.680
High traffic intensity	3.99	4	4	0.946
Night, dark	3.89	4	4	0.933
Rainfall	3.67	4	4	0.923
Queues	3.53	4	4	1.302
Fog	3.11	3	4	1.342
Snowfall	2.80	3	3	1.109
Slippery road conditions	2.69	3	3	1.194

Note: 1 = never, 5 = always.

TABLE 19.5

How Often Do You Choose to Activate ACC When the
Following Speed Regulations Apply?

Speed Regulation	M	Mdn	Mo	SD
Living street	1.25	1	1	0.766
30/40/50 km/h	2.44	2	2	1.291
60/70 km/h	3.38	3	3	1.123
80/90 km/h	4.09	4	4	0.871
100/110/120 km/h	4.56	5	5	0.659

Note: 1 = never, 5 = always.

TABLE 19.6

How Do You Perceive ACC When the Following Speed Regulations Apply?

	Answer										
	Very Dissatisfied		Somewhat Dissatisfied		Neither Nor		Somewhat Satisfied		Very Satisfied		N/A
Factor	(*n*)	(%)	(*n*)	(%)	(*n*)	(%)	(*n*)	(%)	(*n*)	(%)	(*n*)
Living street	2	7.4	3	11.1	8	29.6	3	11.1	11	40.7	60
30/40/50 km/h	–	–	9	13.2	11	16.2	12	17.6	36	52.9	19
60/70 km/h	–	–	1	1.2	7	8.6	17	21	56	69.1	6
80/90 km/h	–	–	–	–	1	1.2	16	18.8	68	80	2
100/110/120 km/h	–	–	–	–	1	1.2	9	10.5	76	88.4	1

Note: Percentages given are the valid percent.

TABLE 19.7

How Do You Perceive ACC during the Following Circumstances?

	Answer										
	Very Dissatisfied		Somewhat Dissatisfied		Neither Nor		Somewhat Satisfied		Very Satisfied		N/A
Circumstance	(*n*)	(%)	(*n*)	(%)	(*n*)	(%)	(*n*)	(%)	(*n*)	(%)	(*n*)
1. When you are being overtaken	–	–	2	2.3	11	12.8	14	16.3	59	68.6	1
2. When you overtake	1	1.2	6	7.1	12	14.3	28	33.3	37	44	3
3. Access roads	–	–	9	13	17	24.6	19	27.5	24	34.8	18
4. Curves	3	3.5	19	22.1	11	12.8	28	32.6	25	29.1	1
5. Longer trips	–	–	–	–	2	2.3	3	3.4	82	94.3	–
6. Roads without overtake possibilities	–	–	–	–	3	3.5	10	11.6	73	84.9	1
7. Traffic roundabouts	4	8.3	13	27.1	18	37.5	11	22.9	2	4.2	39

Note: Percentages given are the valid percent.

The participants were also asked to assess if and in what way their access to ACC had influenced their driving and their experience of driving (Table 19.8). According to the responses, the most common effect concerns comfort. A majority of the drivers stated that access to ACC has increased their comfort. It is also apparent that ACC has perceived safety effects. Approximately half of the respondents indicated a reduced inclination to

TABLE 19.8

How Would You Assess the Influence of Using the ACC on the Following Factors?

	Answer										
	Drastically Reduced		Somewhat Reduced		Neither Nor		Somewhat Increased		Drastically Increased		N/A
Circumstance	(*n*)	(%)	(*n*)	(%)	(*n*)	(%)	(*n*)	(%)	(*n*)	(%)	(*n*)
1. Safety	–	–	1	1.1	4	4.6	49	56.3	33	37.9	–
2. Stress level	15	17.2	40	46	15	17.2	10	11.5	7	8	–
3. Distances of travel	–	–	–	–	64	81	12	15.2	3	3.8	8
4. Fuel consumption	–	–	31	44.3	32	45.7	6	8.6	1	1.4	17
5. Compliance with speed regulation	–	–	2	2.3	20	23	53	60.9	12	13.8	–
6. Inclination to overtake	1	1.1	44	50.6	40	46	2	2.3	–	–	–
7. Annual mileage	–	–	–	–	79	95.2	4	4.8	–	–	4
8. Attention to other traffic	–	–	9	10.6	43	50.6	30	35.3	3	3.5	2
9. Comfort	–	–	–	–	2	2.3	42	48.3	43	49.4	–
10. Ability to judge following distance	–	–	–	–	40	46	36	41.4	11	12.6	–
11. How rested you are at the end of journey	–	–	–	–	24	27.9	46	53.5	16	18.6	1

Note: Percentages given are the valid percent.

overtake other vehicles, a response which is in line with earlier findings (Strand et al. 2011), and almost three out of four answered that ACC has meant that their compliance with speed regulations has increased.

Somewhat incongruent with the response regarding comfort, approximately 20% of the respondents answered that their stress level had increased rather than decreased as a consequence of their access to ACC. Less or no effects were indicated for distances traveled, annual mileage, or fuel consumption. Negative effects were only reported by 10% of the drivers and concerned their attention to other traffic. Almost 40% thought, on the other hand, that the same function had resulted in an increase of their attention. A rated increase in *"attention toward other traffic"* was more common for those who drove mainly outside urban areas (mode = 4) than it was for those who drove mainly within urban areas (mode = 3): $U = 671.0$, $p = 0.028$.

A number of studies have shown that trust and control is a key component for acceptance. The answer to survey question "does it happen that the ACC is behaving in a way you did not expect?" shows that the drivers find ACC to be a fairly predictable system (1 = always, 5 = never: $N = 87$, $M = 4.07$, $Mdn = 4$, mode = 4, $SD = 0.938$). Those who drove mainly outside urban areas (mode = 5) consider the system more predictable than those who drove mainly within urban areas (mode = 4): $U = 679.0$, $p = 0.018$. More than half or 51.7% of the drivers feel very safe when handing over the control to ACC, 46% feels safe, and only 4% feels neither safe nor unsafe. No one feels unsafe.

Finally the respondents were asked to answer a number of questions which together made up the acceptance scale by Strömberg and Karlsson (n.d.). The scale included 20 items addressing the topics: trust and control (Table 19.9); perceived benefit (Table 19.10); perceived effort (Table 19.11); and compliance (Table 19.12). Overall, the ratings indicate that acceptance of ACC was very high: according to the rating ACC is reliable, usable, and driving becomes easier and more convenient. Driving does not however necessarily become

TABLE 19.9

Responses to Items 1–3 (Trust and Control) on the Acceptance Scale

Item	Scale	N/A	M	Mdn	Mo	SD
1. ACC is…	1 = operationally reliable, 7 = prone to fuss	3	1.69	1	1	1.006
2. ACC is…	1 = reliable, 7 = arbitrary	3	1.52	1	1	0.719
3. ACC…	1 = leaves all control for me, 7 = takes over control from me	3	3.62	4	4	1.536

TABLE 19.10

Responses to Items 4–10 (Perceived Benefits) on the Acceptance Scale

Item	Scale	N/A	M	Mdn	Mo	SD
4. For driving the ACC is…	1 = usable, 7 = unusable	3	1.14	1	1	0.352
5. If I use ACC driving becomes…	1 = easier, 7 = more challenging	3	1.57	1	1	0.765
6. If I use ACC driving becomes more…	1 = convenient, 7 = inconvenient	3	1.35	1	1	0.526
7. If I use ACC driving becomes more…	1 = safe, 7 = dangerous	3	1.61	1	1	0.728
8. If I use ACC driving becomes more…	1 = fun, 7 = boring	3	2.49	2	1	1.367
9. If I use ACC during driving I perform driving more…	1 = effective, 7 = ineffective	3	2.05	2	2	0.877
10. If I use ACC the driving I perform gets…	1 = less environmental impact, 7 = greater environmental impact	3	2.62	3	2	1.211

TABLE 19.11

Responses to Items 11–18 (Perceived Effort) on the Acceptance Scale

Item	Scale	N/A	M	Mdn	Mo	SD
11. To use ACC is…	1 = easy, 7 = difficult	4	1.27	1	1	0.607
12. To orientate in ACC is…	1 = easy, 7 = difficult	4	1.34	1	1	0.720
13. ACC is built…	1 = consistently, 7 = inconsistently	4	1.43	1	1	0.666
14. To understand the information provided by ACC is…	1 = easy, 7 = difficult	4	1.33	1	1	0.543
15. To understand how I should act based on information provided by ACC is…	1 = easy, 7 = difficult	4	1.37	1	1	0.657
16. To understand how I should do to get the ACC to do what I want it to do is…	1 = easy, 7 = difficult	4	1.34	1	1	0.547
17. To learn ACC is…	1 = easy, 7 = difficult	4	1.31	1	1	0.562
18. To remember how ACC is used from time to time is…	1 = easy, 7 = difficult	4	1.22	1	1	0.470

TABLE 19.12

Responses to Items 19–20 (Compliance) on the Acceptance Scale

Item	Scale	N/A	M	Mdn	Mo	SD
19. ACC is…	1 = necessary, 7 = unnecessary	4	2.48	2	2	1.075
20. ACC is…	1 = expedient, 7 = inexpedient	4	1.28	1	1	0.502

TABLE 19.13

Recurring Themes Concerning ACC Improvements

Description	Illustrative Quotes
1. Functionality in specific traffic infrastructure (roundabouts and steep curves): aggressive accelerations; target loss; traffic infrastructure	"...today the ACC tend to accelerate to the max when the car in front turns"
	"...the car should understand that it is still behind the vehicle in front, without increasing any speed"
	"...sometimes it loses the car in front"
	"The only negative is when driving in curves..."
	"The feeling or experience when driving in a roundabout. The speed increase feels unpleasant"
2. Functionality in specific tasks relevant for driving (overtaking); harsh decelerations; incorrect target	"The only negative is /.../ and when overtaking it can target the wrong vehicle"
	"The discomfort when it sometimes activates braking in a curve when overtaking for example a truck"
3. User interface and specific settings: usability; adaptability	"...I would like a warning when the radar sees the vehicle ahead but before the speed drops. This is relevant for taking the decision to overtake, or not, before my speed drops"
	"...sometimes I would like to be able to increase the distance to the car in front even more"
	"...I often have to look at the buttons"

Note: Some quotes are relevant for more than one theme.

more fun or more effective. The drivers assess ACC not to completely "take control away" and ACC is not considered "absolutely necessary" by all.

When answering the questionnaire the participants were also given the opportunity to provide answers as to how they would like to improve ACC. Even though the respondents were very positive toward and considered ACC to work well in different situations, there were (as already indicated) some limitations and there were also a few suggestions for modifications. Three recurring themes could be distinguished from the provided answers (Table 19.13). These themes were connected to (1) the functionality of ACC in relation to specific traffic infrastructure, (2) a specific task relevant for driving, and (3) the user interface.

19.4 Discussion

In large, the results are in line with previous studies on the subject (e.g., Strand et al. 2011; Bianchi Piccinini et al. 2012; Larsson 2012; Sanchez et al. 2012). It shows for example that end users, in general, are satisfied with their ACC. More in particular, it confirmed many of the findings by Strand et al. (2011) regarding safety and comfort. The results also show that use frequency increases as do how positive the ratings of the performance of ACC when speed limits increase and road types are more adapted toward higher speeds, a result which is in accordance with Sanchez et al. (2012). However, some earlier findings could not be supported by results of the present study. For example, the results do not show that worries about other road users are a frequent matter.

The study by Bianchi Piccinini et al. (2012) presented results indicating that end users were not fully aware of the limitations of the ACC function. This study indicates that ACC is a perceived as a very predictable system. These results could be interpreted as though end users really are not aware of ACC limitations as was evident in the Bianchi Piccinini et al. (2012) study, or that drivers are aware of the limitations and therefore view ACC as a predictable system. In this particular study the respondents were experienced ACC users and their use patterns indicate that they use it under some conditions and not under other. Hence, it is feasible that the end users in this study are aware of the limitations and have adapted to it.

The study by Bianchi Piccinini et al. (2012) suggested that drivers engage in tasks that distract them from driving when the ACC is activated. In the study by Sanchez et al. (2012), 13% of the participants stated that they use ACC in order to free more time to perform other tasks, such as changing the radio channel or eating. This study present results on a related question, namely if the drivers were less attentive to other traffic (as could be a consequence of a distractive task). Half of the answers were that ACC does not affect how attentive they are to other traffic. Of the remaining answers, the majority provided answers suggesting that they are even more attentive to other traffic. On the other hand, about 10% provided answers suggesting that ACC had a negative effect on how attentive they are to other traffic (worth noting is that there were no answers stating that it had drastically reduced). The difference in responses between overall experienced ACC users is interesting but not altogether easy to explain. It is possible that the drivers do not want to admit to becoming distracted from driving; it is also possible that they are not aware that they are distracted. How drivers make use of the handing over control to an assistance system is a topic that needs further investigation.

Overall the drivers were satisfied with ACC, a result which comply with the findings of Sanchez et al. (2012), and their acceptance of the system is high. Nevertheless the drivers' responses to the different items in the acceptance scale and their comments indicate a potential for improvement. One concerns control. It is possible that a goal should be to design the function so that the drivers feel more in charge, or control, even when the ACC is performing its duties. Such reasoning is in line with Norman (2007) who stated "make people think they are in control" (193). However, it can be debated whether this is a desired strategy when designing ADAS. Instead, the strategy could be to design systems which are more transparent in order to contribute to the drivers developing a correct and meaningful mental model of the system, developing enough trust in order to hand over the control, but at the same time be aware of the limitations of the system so that they will be able to master takeover when so required.

Another item concerns effectiveness of driving where the ratings suggest that drivers' experience becomes more effective in their driving with access to ACC. However, the earlier study by Strand et al. (2011) highlighted a potential to improve ACC in situations where drivers overtake and the answers in the present study also display some improvement potential regarding such situations.

The drivers were very satisfied with ACC and even though it was considered to contribute to comfort as well as to safety, the function was not considered absolutely necessary. The respondents in the survey were experienced drivers who had driven a considerable number of kilometers without the support of ACC or other ADAS. Safety and comfort benefits of ACC are by no means insignificant, but they are most probably viewed as a bonus to a higher order need, rather than crucial for it. An investigation some 10 or 20 years from now may well tell another story. Future generations of drivers, given the present automation trend, may never have first-hand experience of driving without support systems. Rather than questioning the drivers' trust in handing over control to the system, the challenge may be to design vehicles in which the drivers are comfortable with taking over control.

The differences between the present study and earlier ones could be partly attributed to the differences in methods used, but perhaps also to sampling differences. The participants who took part of this study were very positive toward their ACC as well as rated themselves to be very experienced ACC users. At the same time the response rate was 21.7%, which means that a substantial number of the population did not provide any answers. The mean age of the respondents was fairly high and very few women answered the survey. It is possible that this profile reflects the end-user population, but it could also be the case that those who did not provide any answers represent another user group with other experiences of their ACC systems compared to those who provided answers. Due to this, the responses may fail to reflect the views of users who have found ACC less useful and therefore choose not to activate it. If so, the results are a display of the experiences of a particular user group, and not the whole ACC end-user population.

At the same time a major part of the results are consistent with earlier studies why it is feasible to assume that a considerable amount of drivers appreciate the ACC function and adapt to its limitations. There are benefits in terms of increased perceived comfort and safety. Reported changes in driving behavior in terms of less overtaking and increased compliance with speed regulations support this assumption. At the same time some drivers report feeling more stressed when driving with ACC than without and there are contradictory results regarding drivers' attention to traffic. These factors need further investigation.

Acknowledgments

This study has been carried out at SAFER—Vehicle and Traffic Safety Centre at Chalmers, Sweden. The authors would like to acknowledge valuable input on the development of the questionnaire from SHADES project members, coworkers at VTI and CHALMERS. The authors would also like to acknowledge Mats Petersson (Volvo Cars) for contributing to the selection of participants.

References

Bernstein, D. A., Penner, L. A., Clarke-Stewart, A., and Roy, E. J. 2003. *Psychology*, 6th ed. Boston, Massachusetts: Houghton Mifflin.

Bianchi Piccinini, G. F., Simões, A., and Rodrigues, C. M. 2012. Effects on driving task and road safety impact induced by the usage of adaptive cruise control (ACC): A focus groups study. *International Journal of Human Factors and Ergonomics*, 1(3), 234–253.

Brookhuis, K. A., de Waard, D., and Janssen, W. H. 2001. Behavioural impacts of advanced driver assistance systems—An overview. *European Journal of Transport and Infrastructure Research*, 1(3), 245–253.

IBM Corp. 2012. *IBM SPSS Statistics (Version 21.0) [Computer Software]*. Armonk, New York: IBM Corp.

Larsson, A. F. L. 2012. Driver usage and understanding of adaptive cruise control. *Applied Ergonomics*, 43(3), 501–506.

Likert, R. 1932. A technique for the measurement of attitudes. *Archives of Psychology*, 22(140), 1–55.

McQueen, R. A., and Knussen, C. 2006. *Introduction to Research Methods and Statistics in Psychology.* Harlow, Essex: Pearson Education.

Naranjo, J. E., González, C., Reviejo, J., Garcia, R., and de Pedro, T. 2003. Adaptive fuzzy control for inter-vehicle gap keeping. *IEEE Transactions on Intelligent Transportation Systems*, 4(3), 132–142.

Norman, D. A. 2007. *The Design of Future Things.* New York: Basic Books.

OECD. 1990. *Behavioral Adaptations to Changes in the Road Transport System.* Chicago: University of Chicago Press.

Saad, F., Hjälmdahl, M., Cañas, J. et al. 2004. Literature review of behavioural effects (Deliverable No. D1_2_1). Retrieved from EU-project AIDE—Adaptive Integrated Driver-vehicle InterfacE website: http://www.aide-eu.org/pdf/sp1_deliv_new/aide_d1_2_1.pdf

Sanchez, D., Garcia, E., Saez, M et al. 2012. Final results: User acceptance and user-related aspects (Deliverable No. D6.3). Retrieved from EU-project euro FOT website: http://www.eurofot-ip.eu/download/library/deliverables/eurofotsp620121119v11dld63_user_acceptance_and_userrelated_aspects.pdf

Sarter, N., and Woods, D. 1995. How in the world did we ever get into that mode? Mode error and awareness in supervisory control. *Human Factors*, 37, 5–19.

SPSS Inc. 2008. *SPSS mrInterview (Version 5.5) [Computer Software]*, Chicago: SPSS Inc.

Stanton, N. A., and Young, M. S. 2005. Driver behavior with adaptive cruise control. *Ergonomics*, 48(10), 1294–1313.

Strand, N., Nilsson, J., Karlsson, I.C.M., and Nilsson, L. 2011. Exploring end-user experiences: Self-perceived notions on use of adaptive cruise control systems. *IET Intelligent Transport Systems*, 5(2), 134–140.

Strömberg, H., and Karlsson, I.C.M. (n.d.). Unpublished manuscript.

Section IV

Aviation Domain

Introduction

A significant amount of human factors work, perhaps disproportionate in terms of size, has been applied to the air transport sector. Similar to the road domain, the air transport sector has undergone rapid expansion in terms of economy, scope, and technological innovation since its inception a little over 100 years ago. Human factors issues have been a constant priority throughout this time, with the focus migrating from manual instrumentation and ergonomic issues to issues surrounding the integration of complex automation in both flight deck and air traffic control systems. The focus of this work has always been to improve the safety of the air transportation system.

Reflecting the general areas of research over the years, the 10 chapters of this section of the book are divided into three main groups: aviation safety (Chapters 20 through 22), air traffic control human factors (Chapters 23 through 26), and aircraft human factors (Chapters 27 through 29). Hopefully, these chapters provide some insight into the types of issues currently being pursued by researchers and engineers.

Steven Landry

20

Using Neisser's Perceptual Cycle Model to Investigate Aeronautical Decision Making

Katherine L. Plant and Neville A. Stanton

CONTENTS

20.1 Introduction

20.1.1 Aeronautical Decision Making

Aeronautical decision making is a form of naturalistic decision making (NDM: Klein et al., 1989) in which decision makers have domain expertise and make decisions in contexts that are usually characterized by limited time, goal conflicts, and dynamic conditions. The most popular model in the NDM domain is Klein's (1998) recognition primed decision (RPD) model. In summary, this captures how experts make decisions based on recognition of past experiences that are similar to the current situation. These experiences are used to generate one workable option before considering other options, a process known as satisficing (Klein, 1998). In complex cases, evaluation of the option reveals flaws that require modification or the option is rejected in favor of the next most typical reaction. Klein (1998) highlighted dynamic conditions, that is, the changing situation, as one of the key features of NDM. As new information is received or old information becomes invalid, the situation and goals can be radically transformed. This cyclical nature of decision making is referenced in the RPD model in terms of mental simulation but this is only internal to the decision maker. The cyclical nature of a changing external environment is not fully captured in the RPD model. Similarly, the implementation of the model does not connect the internal process of the decision maker to the external environment in which decisions are made. The explanation provided by the RPD model is primarily one of the decision-making processes occurring in the head of the decision maker (Plant and Stanton, 2014a). However, decision making of any kind, especially in the dynamic conditions that characterize the NDM environment, is a product of the interaction of the processes going on in the head of the decision maker and the conditions in the external environment. As Dekker (2006) argued, in order to truly understand decision making it is essential to account for why the actions and assessments undertaken by an operator made sense to them at the time. What makes sense for a decision maker will be based on internal information in the head and external information in the environment. As such, it is proposed that Neisser's (1976) Perceptual cycle model (PCM) is a more suitable framework to model decision-making processes because it accounts for the cyclical interaction that occurs between an operator and their environment in a way that is not captured by the RPD model.

20.1.2 Perceptual Cycle Model

As illustrated in Figure 20.1, Neisser presented the view that human thought is closely coupled with a person's interaction in the world, both informing each other in a reciprocal, cyclical relationship. World knowledge (schemata) leads to the anticipation of certain types of information (top-down processing); this then directs behavior (action) to seek out certain types of information and provides a way of interpreting that information (bottom-up processing). The environmental experience (world) results in the modification and updating of cognitive schemata and this in turn influences further interaction with the environment. The role of past experience is emphasized in the PCM, as Neisser proposed that schemata are the medium in which the past affects the future, that is, information previously acquired will determine what will be sampled next. The model has been most widely applied as an explanatory framework to understand factors such as situation awareness or decision making from the perspective of individuals operating as part of larger systems. For example, in the aviation domain, Plant and Stanton (2012a) demonstrated how the PCM could explain the actions of the pilot's involved in the Kegworth plane crash. Similarly, this approach has been used in the rail domain (see Stanton and Walker, 2011; Salmon et al., 2013). Whilst the PCM offers an explanatory framework, the account of the decision-making process is at a relatively high level. Neisser (1976) described the three elements of the PCM as being: internal schemata, actions undertaken, and world information, but did not further subcategorize these elements. This chapter presents a more detailed classification scheme in order to gain a fuller understanding of the perceptual cycle that pilots engage during critical incident decision making.

20.2 Method

20.2.1 Data Collection

Twenty helicopter pilots were interviewed using the Critical Decision Method (CDM) (Klein et al., 1989). This is one of the most commonly used cognitive task analysis (CTA)

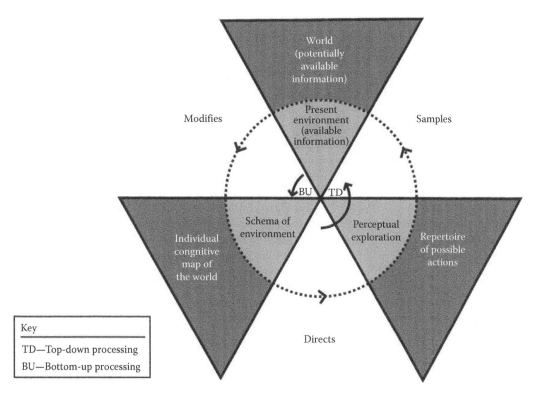

FIGURE 20.1
The Perceptual Cycle Model. (Adapted from Neisser, U. 1976. *Cognition and Reality*. W.H. Freemand and Co., San Francisco.)

methods and achieves knowledge elicitation through the use of cognitive probes as a tool for reflecting on strategies and reasons for decisions during nonroutine situations. The CDM procedure involves participants describing a critical incident they were involved with, defining a timeline of events and answering the deepening probes. The deepening probes cover factors including goals, experience, decisions, options, and information. A more detailed description of the procedure can be found in the associated literature (see Klein and Armstrong, 2005; Plant and Stanton, 2013; Stanton et al., 2013).

The 20 pilots that participated in the study were recruited through an advertisement placed on the British Helicopter Association website and via word-of-mouth. The sample consisted of 19 males and 1 female. Twenty five percent of the sample was aged between 31 and 40 years, 40% were aged between 41 and 50 years, and 35% were aged between 51 and 60 years. The pilots were all relatively experienced; flying hours ranged from 1150 to 13,000 (mean = 5942, SD = 3304, median = 5000). The pilots were employed in a variety of occupations including search and rescue, military, personal passenger transport, North Sea transport, and as test pilots. This study was granted ethical permission by the University of Southampton Research Ethics Committee.

Each pilot was interviewed at their place of work and was asked to think of a critical incident they had been involved with, which was defined as being a nonroutine or unexpected event that was highly challenging and involved a high workload in which they were the primary decision maker. Each participant provided a high-level overview of the incident and structured a timeline of events. After the incident description/timeline construction phase, the cognitive probes were asked in relation to the decision making during the incident. The interviews were audio recorded and later transcribed.

20.2.2 Data Analysis

20.2.2.1 Data Treatment

The 20 CDM interviews produced data about critical incidents that were amalgamated across the 20 interviews. The data from each interview were structured into six generic phases of incident that have been previously identified in similar data (Plant and Stanton, 2012b). The six phases were (including percentage of data in each phase): (1) pre-incident (9%), (2) onset of problem (23%), (3) immediate actions (24%), (4) decision making (24%), (5) subsequent actions (15%), and (6) incident containment (5%). In accordance with the guidelines on qualitative data analysis, the text was chunked into meaningful segments of approximately one sentence or less in length. This resulted in 904 text segments across the 20 interviews. The data analysis techniques undertaken here are based on the principles of thematic analysis which is a method for identifying, analyzing, and reporting patterns (themes) within data (Braun and Clarke, 2006). This approach offers a flexible and useful research tool that can provide a detailed account of complex data. The data were subjected to both deductive and inductive thematic analysis.

20.2.2.2 Deductive Thematic Analysis

Deductive thematic analysis was initially used to classify the data. In this process themes or patterns in the data are generated from existing theory (Boyatzis, 1998). In accordance with the broader research question of exploring decision-making processes from the perspective of the perceptual cycle, the PCM was used as the theoretical underpinning for the deductive thematic analysis. As such, three themes were defined as: schema (statements relating to the use of prior knowledge and knowing things because of experience or expectations), actions (statements of doing an action or discussing potential actions that could be taken), and world information (statements relating to potential or actually available information in the world including physical things, conditions, or states of being). Interested readers are directed to Plant and Stanton (2013), which describes the development of the coding scheme in more detail. The focus of this chapter is not to look at these themes in any detail because this has been done previously (see Plant and Stanton, 2012a, 2013, 2014b) but rather to explore themes within these high-level categories via inductive thematic analysis.

20.2.2.3 Inductive Thematic Analysis

Inductive thematic analysis was undertaken on the data in each of three high-level categories of schema, action, and world in order to uncover more detailed themes within this data. Inductive thematic analysis is the process by which the data are used to generate themes (Patton, 1990). In its purest form inductive analysis is a process of coding data without trying to fit it into a preexisting coding frame (Braun and Clarke, 2006). However, the data in this study were already classified into the three PCM codes and this therefore had some bearing on the nature of the themes generated in the inductive analysis process. The constant comparison technique was employed whereby each text segment was compared with previous items to see whether the same or a different phenomenon was described. The taxonomy was developed through an iterative process of review and refinement using the opinions and expertise of colleagues in the research group. The process of inductive analysis resulted in the identification of six schema subtypes, 11 action subtypes, and 11 world subtypes. These are presented in the PCM taxonomy in Appendix.

20.2.2.4 Relationship and Frequency Analysis

This research sought to explore the relationships between different elements of the perceptual cycle. To do this, for each CDM interview, each code was collated into a frequency table that captured "from-to" links between the different categories as they appeared in the coded transcripts. For example, a text segment coded as "action_decision action" (from), followed by a segment coded as "world_standard operating procedure" (to) were recorded in the frequency matrix. This was summed across the 20 interviews to create an amalgamated frequency count for each of the six phases and across the data set as a whole. This frequency count analysis was subjected to network analysis using the Agna™ software. This is a social network analysis (SNA) tool but is becoming an increasingly popular method for general network analysis. It provides a range of different metrics for analyzing networks and interested readers are directed to other texts for comprehensive descriptions of available metrics, including Houghton et al. (2006), Baber et al. (2013), and Stanton (2014).

Specifically, the metric of sociometric status (SMS) was of interest to define key information elements related to critical decision making. Sociometric status refers to the relative importance of a node (concept) within a network as its calculation is based on the connectedness (i.e., number of connections to other nodes) of a particular information element. The argument is that concepts with high sociometric status values represent key concepts as they are highly connected to other concepts within the critical decision-making network (Stanton, 2014). Here, the concepts (i.e., PCM subcategories) with a sociometric status value above the mean plus one standard deviation for the network were identified as primary concepts, those with a value higher than the mean but lower than the mean plus one standard deviation were identified as secondary concepts and those with a value lower than the mean were identified as tertiary concepts.

20.2.2.5 Reliability Analysis

It has previously been demonstrated that the original PCM coding scheme based on the three primary elements of schema, action, and world generated high levels of inter-rater (86%) and intra-rater (83%) reliability over a 4-week period (Plant and Stanton, 2013). To assess reliability, three additional coders were judged by the standard set by the expert coder in a blind condition, that is, raters were unaware of the expert's coding decisions. Reliability scores were calculated based on percentage agreement, that is, number of agreements divided by the number of times the coding was possible, multiplied by 100. This was in accordance with the literature that has suggested this is the most suitable way to calculate reliability scores with data of this nature (Boyatzis, 1998). There is general consensus that 80% agreement is the threshold for acceptable agreement (see Plant and Stanton, 2013) and this is used as the benchmark here for assessing reliability. The coders were presented with 200 text segments (10 from each interview) which represented 22% of the data. The text segments were selected using a random number generator. This randomly generated 10 numbers within the range of total number of text segments for each interview. Additionally, the original expert coder re-coded the 200 selected text segments. This occurred 13 months after the original coding had taken place. The results of the reliability assessment are presented in Table 20.1, all fall above the 80% threshold for agreement.

TABLE 20.1

Average Percentage Agreement from the Reliability Assessment of the PCM Taxonomy

	Inter-Rater Analysis (%)	Intra-Rater (%)
Schema subcategories	87	88
Action subcategories	84	82
World subcategories	82	88
Total (averaged across the three categories)	81	86

20.3 Results

Table 20.2 presents the primary, secondary, and tertiary concepts for the whole process of dealing with a critical incident (i.e., not phase of incident specific). These are listed in order of importance (highest to lowest) based on the sociometric status values. The sociometric status analysis was also conducted on each of the six phases of flight, these are described in relation to the primary and secondary concepts:

- *Pre-incident*: In this phase the pilots set the scene and described the antecedents to the incident. The most relevant concepts in this phase generally came from the world concepts. There were two primary concepts, both subtypes of world, being natural environmental conditions and operational context. Furthermore, location and physical cues were also defined as secondary concepts. Aviate and standard operating procedures were highlighted as the most relevant action concepts and declarative schema was the most important schema concept.

TABLE 20.2

Primary, Secondary, and Tertiary PCM Concepts When Dealing with a Critical Incident

Primary Concepts	Secondary Concepts	Tertiary Concepts
Aviate (A)	Operational context (W)	Standard operating procedure (A)
Decision action (A)	Physical cue (W)	Absent information (W)
Location (W)	Direct past experience (S)	Technological conditions (W)
Natural environment (W)	Trained past experience (S)	Insufficient schema (S)
Display indication (W)	Declarative schema (S)	Aircraft status (W)
	System monitoring (A)	System management (A)
	Communicate (A)	Concurrent diagnostics (A)
	Situation assessment (A)	Nonaction (A)
		Communicated information (W)
		Environment monitoring (A)
		Artifacts (W)
		Problem severity (W)
		Vicarious past experience (S)
		Navigate (A)
		Analogical schema (S)

Note: Letters depict whether the concept belongs to schema (S), action (A), or world (W) categories of the PCM.

- *Onset of problem phase*: This phase was characterized by the primary concept of physical cue. Technological conditions and display indication follow as the second and third most relevant world concepts. In this phase the most relevant action concept and third most relevant overall concept were aviate, the act of flying the aircraft. Other important action concepts included systems monitoring, concurrent diagnostics, and systems management. The most relevant schema concepts were direct past experience, trained past experience, and insufficient schema.

- *Immediate actions phase*: There were no primary concepts in this phase, but the most important concept was display indication (world concept) followed by trained past experience (schema concept). Action subtypes generally dominated this phase, with seven of the 14 secondary concepts coming from the action category, including aviation, systems monitoring, concurrent diagnostics, decision action, communicate, situation assessment, and standard operating procedures.

- *Decision-making phase*: In this phase, unsurprisingly, decision action was the primary concept. The remaining secondary concepts were evenly spread around the three elements of the PCM, with four world concepts (location, aircraft status, absent information, and display indication), four action concepts (situation assessment, communication, aviate, and standard operating procedure), and three schema concepts (direct past experience, trained past experience, and declarative schema).

- *Subsequent actions*: There are no primary concepts in this phase, but the top five most relevant secondary concepts are aviate, decision action, declarative schema, communicate, and display indication.

- *Incident containment*: In this phase, again, there are no primary concepts and only action and world subtypes appear as secondary concepts. Including in the action category; aviate, systems management, and communicate and in the world category; location, operational context, and aircraft status. Eleven of the 28 concepts do not feature in this phase of the incident, given that only 5% of the data are represented in this phase.

The primary and secondary PCM concepts were modeled into a network (see Figure 20.2). The links between each concept represent the directional flow of information and the strength of the links was informed by the Agna™ network data, the thicker the line, the stronger the link is between the concepts.

20.4 Discussion

Decision making can be one of the determining factors regarding whether or not normal situations turn into incidents, and incidents turn into accidents. Decision making in complex sociotechnical systems needs to be viewed through the lens of distributed cognition and the PCM achieves this by acknowledging the interaction between internal cognitive schemata held by the decision makers and the external environment in which decisions are made. However, the high level of description provided by the PCM has meant that only a limited level of detail is gained about the core areas of the model. This research applied a detailed PCM taxonomy to critical decision-making data and then analyzed the results using the SNA metric of SMS in order to determine the relative importance of PCM concepts during ACDM.

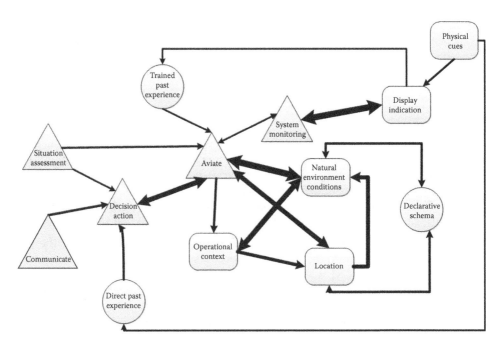

FIGURE 20.2
Network model of aeronautical critical decision making.

20.4.1 Summary of Results

The PCM taxonomy was developed to gain a more detailed understanding about the process of ACDM from the perspective of the PCM. The analysis discussed is a composite account summed across the 20 interviews and is considered both as a whole and by phase of incident. In both instances the PCM concepts were scored on sociometric status to determine their relative importance (the higher the status, the more important the concept). This resulted in the classification of primary, secondary, and tertiary concepts (only primary and secondary concepts are discussed as these are most pertinent to the ACDM process). The analysis showed that the most important concept for ACDM was the action concept of aviate; highlighted in Figure 20.2 as most of the other concepts are connected to aviate in some way. This is unsurprising given that aviate is the primary task management requirement in the "aviate–navigate–communicate–manage systems" strategy employed by all pilots when dealing with non-normal situations. It is of paramount importance that, regardless of what is happening around them, pilots continue to fly their aircraft. In a study of task management priorities, Schutte and Trujillo (1996) found that participant's prioritized aviate when dealing with non-normal situations. The next most important concept was "decision action." This is to be expected in data collected for decision-making research and increases confidence with the method for collecting decision-based data. The next three most relevant concepts were from the world category: location, natural environmental conditions, and display indications.

The absence of any schema subtypes in the primary concepts suggests that ACDM is predominately driven by bottom-up information processing, that is, the environmental information that is received drives the actions that are undertaken. This is supported by the ACDM model (Figure 20.2), which shows the strongest links between aviate and the world concepts of location and natural environment. This information processing strategy is akin to what Rasmussen (1983) described as skill-based behavior (SBB). SBB is the smooth, automatic, and execution of highly practiced physical actions. At this level of behavior world information and the pattern recognition, this induces maps directly onto physical actions. This develops through extensive training where operators acquire cued response patterns suited for specific situations and is therefore characteristic of expert performance. Rasmussen (1983) also described rule-based behavior (RBB) and knowledge-based behavior (KBB), the former requiring identification and recall of known rules which are stored in memory and the latter being effortful, conscious processing of unfamiliar situations. Through the process of training and learning, pilots will progress from KBB, through RBB, to SBB in the flight control task (aviating).

The phase specific analysis demonstrated that PCM concepts vary in importance according to phase of incident. For example, the "onset of problem" phase is characterized by the physical cues and technological conditions that generally alerted the pilot to the problem. Aviate is the most relevant action, in line with our previous discussion about the importance of maintaining adequate handling of the aircraft at all times. In the "immediate actions" phase, display indication was the primary concept and the next 11 secondary concepts were subtypes of the schema and action categories. This suggests that this phase was characterized by top-down information processing which is supported by the importance of trained past experience and direct past experience as secondary concepts. Insufficient schema also features as a secondary concept which was apparent when pilots talked about insufficient background knowledge to deal with the problem that was presented. It is unsurprising that in the "decision-making" phase decision action is the primary concept. Secondary concepts include a variety of all three subtypes as text segments in this phase related to the actions undertaken once the primary decision was made (e.g., situation assessment, communicate, and aviate) and how these actions were influenced by the world information (e.g., location, aircraft status, and absent information) and stored knowledge (e.g., direct past experience, trained past experience, and declarative schema).

20.4.2 Evaluation

The role of the perceptual cycle in Aeronautical Critical Decision Making (ACDM) has been previously identified (Plant and Stanton, 2012a, 2013, 2014a,b), but from this work the finer detail of the interactions was able to be explored. The results make intuitive sense, insomuch that concepts have the most relevance in the phases where they would expect to be found and are less relevant in phases where they are not expected. For example, decision action is the primary concept in the decision-making phase, whereas severity of problem and concurrent diagnostics do not feature in the pre-incident phase (because the incident had not happened yet). For exploratory studies, intuitive sense is important because it points toward the appropriateness of the data collection and analysis methods. The relative importance of concepts was objectively determined by the sociometric analysis function in Agna™ which suggests that the classification method employed with the PCM taxonomy was appropriate for gaining an increased understanding about ACDM.

A fundamental component of the PCM are schemata, these are akin to internal mental templates and therefore cannot be directly measured but only inferred through the manifestation of observable behavior or recalled information. As presented here, contemporary approaches for eliciting schema-based data and inferring perceptual cycle processes generally pair data collection methods such as interviews with qualitative data analysis including network modeling or thematic analysis. This is a form of CTA as they are approaches that determine the cognitive elements (i.e., mental processes and skills) required for task performance and the changes that occur as skills develop. The CDM is one of the most popular CTA methods. However, the CDM focuses on eliciting knowledge for behaviors classed as recognition-primed decisions (RPD), that is, decisions for which alternative actions are derived from the recognition of critical information and prior knowledge. The RPD model is the most popular and enduring model in the NDM domain. However, we have previously argued that it does not go far enough at acknowledging the interaction between schemata and environmental information and the modifying effect each can have on the other (Plant and Stanton, 2014a). Rather, the RPD model tends to focus on the decision-making processes that occur in the mind of the decision maker. However, this is only achieved at a generic level with questions such as "did your experience influence the decisions that you made?" The findings presented here suggest that schema data are not being captured as well as action and world data and therefore alternative methods are required that are more suited to extracting the three elements of the PCM.

20.4.3 Future Applications and Research

The potential exists to utilize the PCM taxonomy in any qualitative data analysis including data collected from interviews, communication transcripts, or verbal protocols. Stanton and Salmon (2009) stated that a valid taxonomy can be used either proactively to anticipate potential situations or retrospectively to classify and analyze situations after they have occurred. Both applications have associated practical implications, for example, in the identification of areas that may benefit from decision-making training or redesigning interfaces to facilitate the provision of information in the external environment. Salmon et al. (2012) argued that the use of taxonomies facilitates the aggregation of data across multiple cases, this is consistent with Fleishman and Quaintance's (1984) assertion that taxonomy development allows for the establishment of a base for conducting and reporting research studies to facilitate their comparison.

A comprehensive taxonomy should enable attributes that distinguish a category member from the general population to fall into a meaningful pattern, given knowledge of relevant literature, for example, contrasting expert and novice perspectives. With perceptual cycle processing, it is likely that experts would rely on schemata to select relevant environmental cues and therefore be more selective than novices in the information they attend to. Inexperienced and experienced pilots are known to have different accident types, the former being associated with loss of control accidents due to a lack of basic handling skills and the latter being associated with more cognitively-driven accidents, such as complacency (Civil Aviation Authority, 1997). Using the PCM taxonomy to highlight where and how differences in decision-making manifest between different demographic groups will increase its external validity and provide useful information about ACDM.

20.5 Conclusion

The PCM views decision making through the lens of distributed cognition, providing a process-orientated approach for understanding how internally held mental schemata interact with information perceived in the external environment to produce actions and behaviors. This study has demonstrated that different elements of the perceptual cycle differ in their importance depending on the phase of dealing with a critical incident. Understanding what information is utilized, when it is utilized, and how this interacts with actions undertaken is taking a step toward being able to develop decision-centered training aids, design solutions, or procedural strategies based on the principles of perceptual cycle information processing.

Acknowledgment

The authors thank the pilots who gave their time to participate in the interviews.

Appendix: PCM Taxonomy

Schema Subtype	Description	Examples
Vicarious past experience (VPE)	Statements relating to experiencing something in the imagination through the description by another person (e.g., hearing a colleague recall an incident they were involved with) or documentation (e.g., reading about a certain event in an industry magazine or incident/accident report)	*"I knew I had surged the engine…I had heard about surging…I hadn't experienced it but I knew that the engine was surging. It had been described to me, in books. You don't train for it. No one plays you a sound clip, it's more by reading documents I suppose"*
Direct past experience (DPE)	Statements relating to direct personal experience of similar events or situations in the past. This covers events experienced in live, operational contexts as opposed to those experienced through training	*"I have experienced levels of vibration on other aircraft and I know what is normal, what is abnormal, this exceeded it tenfold…"*
Trained past experience (TPE)	Statements relating to knowledge developed by direct personal experience of a specific task, event or situation, experienced within the confines of a training scenario (e.g., ground school training, simulator training, or training sorties)	*"The decision of what to do was in my experience because of training. I had seen this instance before in a simulator"*
Declarative schema (DS)	Statements relating to a schema that manifests as a descriptive knowledge of facts, usually as a product of the world information available	*"I knew it had just come out of maintenance, I was aware it could be a spurious event"*
Analogical schema (AS)	Statements relating to comparisons between things for the purpose of explanation and clarification. Typically these analogies will be structural analogies of physical objects or states of affairs in the world (akin to mental map or mental model)	*"How high am I, how fast am I, can't see a lot so having to make this picture in my head based on the information that I do know"*
Action Subtype	**Description**	**Examples**
Insufficient schema (IS)	Statements relating to inadequate or lacking knowledge, that is, a schema is not developed for a certain situation	*"It didn't fail like they do in training, so I wasn't instantly sure it was an engine failure"*
Aviate (Av)	Statements relating to direct manipulation (handling) of flight controls in order that the aircraft can be flown and safety is maintained	*"I attempted to roll the aircraft level"*

(Continued)

Action Subtype	Description	Examples
Navigate (Nav)	Statements relating to the process of accurately ascertaining position and planning and following a route or desired course	*"I followed the coast back"*
Communicate (Comm)	Statements relating to the sharing or exchange of information	*"I transmitted a non-standard mayday call"*
System management (Sys Man)	Statements relating to the processes of making an input into technological systems in order that the interaction or manipulation has an explicit output	*"I put in St. Albans head into the navigation system, so I typed in the three digit code which is St. Albans head"*
System monitoring (Sys Mon)	Statements relating to looking at (observing, checking) displays to gain an understanding of the situation	*"I did a complete scan of all the systems information"*
Environment monitoring (Env Mon)	Statements relating to observing or checking the internal or external physical environment in order to establish the current state-of-affairs	*"I was keeping eyes out for ground contact and searching for visual references"*
Concurrent diagnostic action (Conc Diag)	Statements relating to the process of determining, or attempting to determine, the cause or nature of a problem by examining the available information at the time the incident is occurring	*"We initially started looking for circuit breakers, to look if any had popped"*
Decision action (DA)	Statements relating to a conclusion or resolution that is reached after considering the available information	*"The first decision was to idle back the bad engine, rather than shut it down"*
Situation assessment (Sit Ass)	Statements relating to actions that relate to the evaluation and interpretation of available information	*"Trying to take into account the threats to you and the aircraft, that is, if I precede down a given path what is it likely to result in?"*
Nonaction (Non A)	Statements relating to actions that were not performed, either because the situation did not warrant a particular action or because equipment faults did not allow a particular action to be performed or because the pilot made an error or omission	*"I couldn't read any of the instruments or communicate"*
Standard operating procedure (SOP)	Statements relating to following the prescribed procedure that ought to be routinely followed in a given situation	*"I completed the pre-take off checks"*

World Subtype	Description	Examples
Natural environmental conditions (NEC)	Statements about natural environmental conditions (e.g., weather, light, temperature, noise)	*"Fortunately it was a clear day, nice sunny day"*
Technological conditions (Tech Cond)	Statements relating to the state of technological artifacts (e.g., with regard to appearance and working order)	*"…engines responded and all other stuff came back on"*
Communicated information (Comm info)	Statements relating to information available to the pilot from other people (e.g., other crew members, ATC, coastguard, etc.)	*"I received the cloud base report from Newquay"*
Location (Loc)	Statements relating to particular places or positions	*"…so now we were over the destination"*
Artifacts (Art)	Statements discussing physical objects, including written information, symbols, diagrams, or equipment	*"I had the flight reference cards"*
Display indications (Dis Ind)	Statements relating to the information elicited from the physical artifacts	*"Only thing identifiable was the high transmission oil temperature"*

(Continued)

World Subtype	Description	Examples
Operational context (Op Cont)	Statements relating to the routine functions or activities of the organization (e.g., search and rescue, police search, military training, etc.). This can include statements about the importance of being serviceable for the operational context or crew familiarity with the aircraft and how this effects decision making	*"the aircraft was relatively heavy because we were taking people back to Germany"*
Aircraft status (Air Stat)	Statements relating to the current status of the aircraft's integrity or performance (e.g., how good or bad it is flying, the current configuration of the aircraft, autopilot activation, etc.)	*"the aircraft was flying fine"*
Severity of problem (Prob Sev)	Statements relating to how bad (or otherwise) the critical incident is	*"we weren't in any immediate danger"*
Physical cues (Phys Cue)	Statements relating to external cues that provide information of conditions or states of being (e.g., noises, sounds, vibration, smells)	*"there was a loud bang, coughing and spluttering"*
Absent information (Abs Info)	Statements relating to information that was missing, not present or lacking. Reasons for this may include technical faults with equipment or nonexistent information	*"I didn't have a comprehensive map"*

References

Baber, C., Stanton, N.A., Atkinson, J., McMaster, R., and Houghton, R.J. 2013. Using social network analysis and agent-based modelling to explore information flow using common operational pictures for maritime search and rescue operations. *Ergonomics*, 56(6), 889–905.

Boyatzis, R.E. 1998. *Transforming Qualitative Information: Thematic Analysis and Code Development.* Thousand Oaks, California: Sage.

Braun, V. and Clarke, V. 2006. Using thematic analysis in psychology. *Qualitative Research in Psychology*, 3(2), 77–101.

Civil Aviation Authority. 1997. CAP 667: *Review of General Aviation Fatal Accidents 1985–1994.* London: Civil Aviation Authority.

Dekker, S.W.A. 2006. *The Field Guide to Understanding Human Error.* Aldershot: Ashgate.

Fleishman, E.A. and Quaintance, M.K. 1984. *Taxonomies of Human Performance. The Description of Human Tasks.* Academic Press: Orlando, Florida.

Houghton, R.T., Baber, C., McMaster, R., Stanton, N.A., Salmon, P.M., Stewart, R., and Walker, G. 2006. Command and control in emergency services operations: A social network analysis. *Ergonomics*, 49(12–13), 1204–1225.

Klein, G.A. 1998. *Sources of Power: How People Make Decisions.* Cambridge, Massachusetts: MIT Press.

Klein, G.A. and A. Armstrong. 2005. Critical decision method. In *Handbook of Human Factors and Ergonomics Methods*, edited by N. A. Stanton, A. Hedge, K. Brookhuis, E. Salas, and H. Hendrick, pp. 35.1–35.8. Boca Raton, Florida: CRC Press.

Klein, G.A., Calderwood, R., and Macgregor, D. 1989. Critical decision method for eliciting knowledge. *IEEE Transactions on Systems, Man and Cybernetics*, 19 (3), 462–472.

Neisser, U. 1976. *Cognition and Reality.* San Francisco: W.H. Freemand and Co.

Patton, M.Q. 1990. *Qualitative Evaluation and Research Methods*, 2nd ed. Newbury Park, Newbury Park, California: Sage.

Plant, K.L. and Stanton, N.A. 2012a. Why did the pilots shut down the wrong engine? Explaining errors in context using schema theory and the perceptual cycle model. *Safety Science*, 50, 300–315.

Plant, K.L. and Stanton, N.A. 2012b. "I did something against all regulations": Decision making in critical incidents. In *Proceedings of the 4th International Conference on Applied Human Factors and Ergonomics*, July 21–25, 2012, San Francisco. *Advances in Human Aspects of Aviation*. Landry, S (Ed), 2012 (Conference book). Boca Raton, Florida: CRC Press.

Plant, K.L. and Stanton, N.A. 2013. What's on your mind? Using the perceptual cycle model and critical decision method to understand the decision-making process in the cockpit. *Ergonomics*, 56(8), 1232–1250.

Plant, K.L. and Stanton, N.A. 2014a. All for one and one for all: Representing teams as a collection of individuals and an individual collective using a network perceptual cycle approach. *International Journal of Industrial Ergonomics*, 44, 777–792.

Plant, K.L. and Stanton, N.A. 2014b. The process of processing: Exploring the validity of Neisser's perceptual cycle with accounts from critical decision-making in the cockpit. *Ergonomics*, 58(6), 909–923. Doi: 10.1080/00140139.2014.991765.

Rasmussen, J. 1983. Skills, rules, and knowledge; signals, signs, and symbols, and other distinctions in human performance models. *IEEE Transactions on Systems, Man and Cybernetics*, 13(3), 257–266.

Salmon, P.M., Cornelissen, M., and Trotter, M.J. 2012. Systems-based accident analysis methods: A comparison of Accimap, HFACS, and STAMP. *Safety Science*, 50, 1158–1170.

Salmon, P.M., Read, G.J.M., Stanton, N.A., and Lenne, M.G. 2013. The crash at Kerang: Investigating systemic and psychological factors leading to unintentional non-compliance at rail level crossings. *Accident Analysis and Prevention*, 50, 1278–1288.

Schutte, P.C. and Trujillo, A.C. 1996. Flight crew task management in non-normal situations. In: *Proceedings of the 40th Annual Meeting of the Human Factors and Ergonomics Society*, pp. 244–248. Santa Monica, California: HFES.

Stanton, N.A. 2014. Representing distributed cognition in complex systems: How a submarine returns to periscope depth. *Ergonomics*, 57(3), 403–418.

Stanton, N.A. and Salmon, P.M. 2009. Human error taxonomies applied to driving: A generic driver error taxonomy and its implications for intelligent transport systems. *Safety Science*, 47, 227–237.

Stanton, N.A., Salmon, P.M., Rafferty, L.A., Walker, G.H., Baber, C., and Jenkins, D. 2013. *Human Factor Methods. A Practical Guide for Engineering and Design*, 2nd ed. Aldershot: Ashgate.

Stanton, N.A. and Walker, G.H. 2011. Exploring the psychological factors involved in the Ladbroke Grove rail accident. *Accident Analysis and Prevention*, 43(3), 1117–1127.

21

Exploring the Role of Culture in Helicopter Accidents

Helen Omole, Guy Walker, and Gina Netto

CONTENTS

21.1 Introduction

Despite making up only 12% of the total worldwide aviation fleet, helicopters account for 70% of accidents (General Aviation Manufacturer Association [GAMA], 2010). While the worldwide airline safety trend is improving, helicopter accident safety trends are not (IHST, 2012). Research shows that human error is implicated in around 70%–80% of aviation accidents (Yacavone, 1993; O'Hare et al., 1994; Sarter and Alexandar, 2000; Wiegmann and Shappell, 2001; Wiegmann and Shappell, 2003) and an important modifier of human behavior is the culture in which it takes place. Culture has been the topic of considerable previous study, especially in aviation domains (Helmreich, 1994; Orasanu et al., 1997; Soeters and Boer, 2000; Meshkati, 2002; and others). It has been studied at the level of organizations (Merrit and Helmreich, 1998), in maintenance activities (Soeters and Boer, 2000), design and manufacturing (Foushee, 1984), and pilots (Merrit and Helmreich, 1996a,b; Li et al. 2007; Strauch, 2010). Despite this, culture is still not a prominent component of accident analysis methods and neither is it studied in an integrated fashion across different layers of the system (Merrit and Helmreich, 1998). The aim of this chapter is fourfold. First, it is to link the extant knowledge-base on culture to the specific problem of helicopter safety and accident analysis. Second, to use this knowledge-base to propose a cultural framework linking elements of systems to features of culture and ultimately to human actions. Third, to perform a content analysis on real helicopter accident reports from two culturally distinct regions of the world to reveal the cultural factors actually in play. Fourth, to relate the discovered cultural factors back to the original framework in order to validate the components within it and their structure. The ultimate aim is to use this exercise to reveal the extent of the gap in current accident analysis approaches and provide a concrete way to gain traction on it.

21.2 Defining Culture

Merritt (1993) defines culture as the values/norms and practices shared with others which define a group, especially in relation to other groups. Kluckholn (1951) puts it that "culture consists in patterned ways of thinking, feeling and reacting, acquired and transmitted mainly by symbols, constituting the distinctive achievements of human groups, including their embodiments in artifacts; the essential core of culture consists of traditional ideas and especially their attached values" (Hofstede, 1997, p. 25). Culture defines the way of life of a particular group of people, and in turn, has wide ranging influences on their values and behaviors (Hofstede, 1997). Values are part of a process of socialization and represent principles or patterns of behavior which are held in high regard. Values influence people's actions by helping to sustain the social and economic structures in which they coexist (Kluckholn, 1951; Hofstede, 1997; Merritt, 1993). This can influence people to act in an individualistic or collective manner; how they relate to interpersonal power; how they will tolerate uncertainty, and the way they relate to members of the opposite gender (Hofstede, 1997, Chapters 4 and 5). These factors have been shown to play an important role in accident causation (Merrit and Helmreich, 1998, pp. 98–100).

This chapter focuses on national culture, defined as the traits which describe a group of people living in a particular region (Kluckholm, 1951). Value system dimensions are used to understand and measure the influence of national culture and have been identified by several authors such as Smith and Schwartz (1997), Schwartz (1999), and Hofstede (1980, 1997). They include "conceptions of the desirable that guides the way social actors (organizational leaders, policy makers, individual persons) select actions, evaluate people and events and explain their actions and evaluations" (Schwartz, 1999). It is "trans-situational" in that it guides a person's principle of living across a wide range of contexts (Kluckholm, 1951; Schwartz, 1994).

Organizational culture is a subculture of national culture, and professional culture (such as that which might exist among pilots or aircraft engineers) is a subculture of both, as shown in Figure 21.1. "The culture of a profession is manifested in its members by a sense of community and by the bonds of a common identity" (Goode, 1957). Within the aviation industry, for example, the professional culture of the flight crew is defined by its personality (used as predictors of performance); aptitude and intelligence; above average

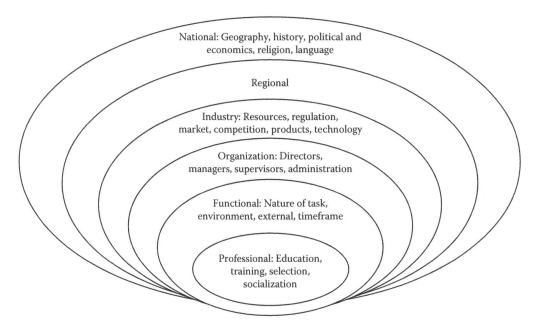

FIGURE 21.1
Interacting levels of cultural influence. (Adapted from Schneider, S. and Barsoux, J., *Managing Across Cultures*, Prentice-Hall, Harlow, 1997.)

motivation and interpersonal skills (Helmreich and Merritt, 1998). Critically, these sub-cultures interact in powerful ways with national culture (Helmreich and Davies, 2004; Strauch, 2010) to give rise to significant performance shaping effects. The extent to which these cultural interactions offer a predictor of performance lies partly in the notion of cultural context.

In high-cultural contexts (HCCs), the individual is subordinate to the group and the emphasis is on maintaining power relationships. HCCs manifest themselves in behaviors such as extreme politeness and careful judgment, verbal messages that are implicit in their content, the needs of individuals interfering with set objectives or rules, lengthy decision-making processes, and slow pace of change (Hall, 1974, 1959; Graham, 1985; Adler, 1991). It is difficult to generalize, but HCCs tend to be found in Africa, Arab nations, China, and Japan. Low-cultural context (LCC), on the other hand, tends to be the opposite. LCCs tend to foreground verbal messages that are explicit in context, impersonal negotiations, a focus on economic goals, quick and efficient decision making, and rapid change. It can be found in many Western countries such as North America and Europe (Hall, 1976; Nishimur et al., 2008; Neuliep, 2012). Neither HCC nor LCC is better than the other, rather, it depends on the context and setting. Indeed, according to Neuliep (2012), no national culture is exclusively HCC or LCC, instead falling along a continuum. The interest for the current chapter is how a national culture positioned at one end of the cultural context continuum interacts with organizational and professional subcultures positioned at the other. In aviation this is common. Aircraft and their associated operational, safety and maintenance procedures derive largely from LCC Western nations, but may find themselves embedded in HCC African or other nations. It would be surprising if there were not issues arising from this, but they are rarely addressed.

21.3 Culture in Aviation

An overriding feature of culture in aviation is that while the vast majority of aircraft are built by a small number of manufacturers in the United States and Western Europe, aviation business activities take place all over the world (Foushee, 1984). It is not merely the aircraft that are built in nonindigenous cultures, it is the associated maintenance, operational, and safety practices that go along with them. The key question is the extent to which this "alien" entity of people, technology, and processes translates into different cultural environments. A series of accidents have exposed the cultural dimension in aviation, including the Air Ontario Fokker F28 crash in March 1989 (MSSC, 1992), the 1992 Airbus A320 crash at Mount St-Odile (FMTT, 1993), and the Piper Navajo incident in June 1993 (AAIB, 1993). Accidents such as these, and numerous others, stimulate increased interest in the role of culture (e.g., Jing and Batteau, 2015) and the findings are revealing.

Merrit and Helmreich (1996a,b) studied the "cockpit management attitudes" between Asian and American flight crew and attendants using two of Hofstede's cultural dimensions; "power distance" and "individual-collective" (see Appendix). The report argues that the American crew were more independent, self-reliant (more individualistic), and exhibited a personal responsibility for their contribution to being an effective crew (low-power distance). This was in contrast to the Asian crew, who for similar cultural reasons were more likely to support the authority of a superior (high-power distance) and be satisfied with that role. Jing et al. (2001) compared accident outcomes in 59 countries and made similar observations of cultural differences, in particular Taiwanese and Chinese cultures scored highly in high-power distance and, with cockpit voice recordings indicating major communication difficulties between the crew in one accident, this supports Merrit and Helmreich (1996a,b) findings. The picture is not equivocal and highly dependent on context: there is no "one right" culture, as Merrit and Helmreich (1998) show in their study of 8000 pilots' responses to the statement: "written procedures are required for all in-flight situations."

The result revealed that in countries with low scores for uncertainty avoidance, and which had fewer rules and regulations, pilots were generally less stressed under emergency conditions and could respond naturally. In addition, pilots from these countries were more likely to act outside the rules using improvisation to solve problems independently (Hofstede, 1980, 1997). Soeters and Boer (2000, p. 126) compared NATO air forces accidents from the period 1991–1995 and, in contrast, to Merrit and Helmreich (1998) they reported that the stronger the level of uncertainty avoidance within the national (air force) culture the greater the chances of an accident. The more individualistic the national (air force) culture is, the lower the chances of a total loss accident.

Despite what seems like overwhelming evidence for the importance of culture's role in accident causation, it is still not fully accepted (Merrit and Helmreich, 1998; Maurino, 1998; Strauch, 2010). It is certainly more convenient to believe it is possible to have "culture free" work environments, yet it is clear there is high-cultural diversity within the industry which directly impacts on operations (Helmreich and Davies, 2004). The problem, as Maurino (1998) forcefully states, is that "cultural factors should routinely be considered during the safety investigation process, although this might be the toughest nut in the entire lot to crack, owing to the resilient conservatism of accident investigation" (p. xxiii). The industry is therefore faced with the challenge of developing a practical model to deal with the relationship between different cultures, and further research is needed to confront this challenge.

Recent attempts have focused on examining the relationship between the Western origin of the aviation industry and the diverse non-Western cultures into which it has been implemented. This research confirms the importance of the research area, but identifies a number of knowledge gaps that need to be addressed. Li et al. (2007) used the human factors analysis and classification system (HFACS; Wiegmann and Shappell, 2003) framework to compare accidents in China, the United States, and India to explore the role of cultural differences. The results show a significant difference in the accident outcome of the three countries in seven HFACS categories; however, certain causal factors could not be accounted for, such as cultural difference between the countries, the level of government oversight, and design/technology sophistication. This led Strauch (2010) to argue that the overall research methodology had flaws. To address gaps like this in existing methodologies what is needed is a systemic framework for cultural factors. Such a framework should take into account human failures, cultural beliefs, and the social context of the system. This framework should be developed for several reasons: first, it should identify the underlying cultural factors which led to the accident. Second, it should capture the contextual environment of the interactions. Third, it should examine the cultural dimensions contributing to those interactions. Fourth, it should highlight the benefits of acknowledging the sociocultural interaction in the accident investigation process and its outcomes. Such a framework is presented in the next section.

21.4 Proposed Multilevel Cultural Framework

The framework shown in Figure 21.2 is a synthesis of the wider background literature surveyed above, but refers more explicitly and directly to Hofsted's (1997) cultural dimensions, Hall and Hall's (1990) manifestations of culture in different contexts, Rasmussen's risk management framework (1997), and Reason's (2000) work on human error. Although structured differently to reflect the focus on cultural factors, the primary actors and elements from these models are all present. The multilevel framework of cultural factors consists of four levels. Between them they meet the objective of mapping out the sociocultural context of events leading to unsafe acts. The high-level cultural dimensions reviewed above (Hofstede, 2005) are found at the top of the framework, called Level 1. How these high-level cultural dimensions manifest themselves in different contexts (Hall and Hall, 1990) is represented at Level 2. Levels 1 and 2 are linked to Level 3, which contains the system elements to be found in Rasmussen's (1997) risk management framework. This, in turn, leads to Level 4 and the various points at which culturally influenced human actions occur (Reason, 1990). The sum total of these influences are safe or unsafe acts, and beyond that accidents or loss events. These events, of course, feedback into the model to influence culture (Level 1), cultural dimensions (Level 2), system parts (Level 3), and human actions (Level 4). In other words, consistent with sociotechnical systems theory (Walker et al., 2008), the framework is itself part of the environment and able to influence what happens in it.

This cultural framework (Figure 21.2) provides a way of decomposing cultural issues from a wider system of human, technical, and procedural interventions. Although the model was developed for the helicopter industry, its applications are more widespread. The framework is generic. It can be used alone as a single framework by virtue of its focus across all system levels, or in combination with other models. It also allows links

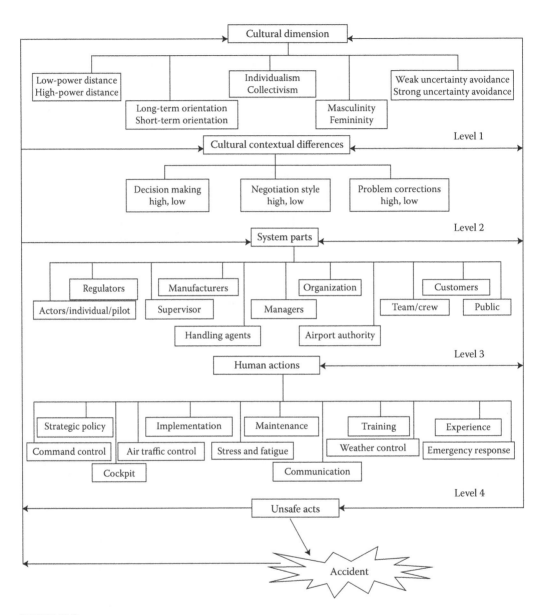

FIGURE 21.2
Proposed framework for the cultural analysis in accident causation.

between different system layers to be traced and for the role of culture to be more explicitly revealed.

To understand the interactions depicted in the model, a comparison between the United Kingdom and Nigeria was carried out. Both countries share common features of helicopter operation (as with a prominent off shore oil and gas industry) but differ in terms of cultural context. This provides an excellent manipulation of culture and enables the model to be explored fully. Content analysis was used to identify, explore, and understand the context and cultural characteristics of helicopter accidents occurring in both countries. This enables the following to be explored: (a) the links between culture and unsafe acts;

(b) the cultural dimensions shared by different parts of the system; and (c) the system roles and actions leading to unsafe acts. The next section describes the framework more fully, and how it goes on to directly inform a content analysis of accident investigation reports originating from Nigeria and the United Kingdom. The chapter then goes on to distil the results, highlighting culture and unsafe acts, with conclusions drawn on what this means for continued progress with accident analysis methodologies in the aviation sector and beyond.

21.5 Exploring and Validating the Framework

21.5.1 Design

Twelve formal air accident investigation reports were subjected to an in-depth theme-based content analysis in order to distil the role of culture in their causation. Six reports were sourced from the Nigerian Air Accident Department (NAAD) and six were from the UK Air Accident Investigation Branch (AAIB). The reports were selected on the basis of the accidents not being caused by purely technical and/or meteorological issues, instead focusing on human factors causes. The results of applying this criterion were a selection of both commercial and training flights from 1985 to 2011. A theme-based content analysis was selected as the means by which to extract cultural factors from the reports. The independent variable was the country of origin, with two levels (the United Kingdom or Nigeria). The dependent variables were a collection of 18 themes and 114 individual coding nodes driven from the cultural framework shown in Figure 21.2, and applied to the reports in a nonmutually exclusive manner (i.e., all applicable coding nodes could be applied to every thematic unit of text). The full list of themes and nodes appears in Appendix. Inter and intra-rater reliability was calculated to establish the reliability of the coding scheme.

21.5.2 Sampled Helicopter Accidents

The summary of the accident reports by country and accident characteristic(s) is shown in Table 21.1. The majority of the reports from both the United Kingdom and Nigeria were related to oil and gas operations. They each provided information about the accident in detail, including the sequences of events leading to and after the accidents, the immediate and causal factors, contributory and underlying causes, and the investigation recommendations. An interesting commentary on the further role of culture can be noted in relation to obtaining the Nigerian reports. These were constrained by availability and restricted to those released into the public domain by the NAAD. Extensive efforts were made both remotely and "in country" to assess further reports for possible inclusion, but this required a strongly inter-personal (HCC) approach and resources which far exceeded the time and budgetary allowances available to the project. The obtained reports do, however, provide a satisfactory working sample.

21.5.3 Procedure

The themes to be used as coding categories for the content analysis of the accident reports were extracted from the literature, and included items across all four levels of the cultural framework shown in Figure 21.2. The complete list of items appears in Appendix. As the coding categories were based on the existing literature, it meant that robust definitions for each term could be defined. This served as a further explicit aid for the coding scheme's reliability (Walker, 2005). Each report was taken in turn and in random order. The primary analyst scrutinized the report and divided the text into thematic portions (such as statements, sentences, and passages) and applied the coding scheme nonexclusively. As many viable coding items as applied to the given thematic portion were coded. This process was undertaken using the nVivo software tool.

Inter-rater reliability was performed in order to assess the reliability and repeatability of the coding scheme. Three individuals with no prior knowledge of the study were given six accident reports to code, three each from Nigeria and the United Kingdom. Training was provided prior to the coding process, and the raters made use of the same coding definitions as used by the primary analyst. One of the raters achieved a low 30% Cohen's Kappa coefficient, but upon closer analysis the rater had, for time and motivational reasons, significantly undercoded the reports. The remaining two raters performed a comprehensive coding of the reports and their Kappa coefficient reached a mean of 76.5%.

21.6 Results and Discussion

21.6.1 Analyzing the Components of the Cultural Framework

The sampled accident reports featured a total of 74 crew and passengers and 27 fatalities, covering the period 1985–2011. The content analysis of the accident reports resulted in a total of 2098 data points, a mean of 19.3 coded items for the UK reports and a mean of 11.6 coded items for the Nigerian reports. Within these coding points there are some very evident differences in how parts of the cultural framework are activated depending on whether the analyzed accident report described a UK or Nigerian helicopter accident. The following sections explore these cultural differences in more detail across the different layers of the framework.

21.6.1.1 Level 1: Cultural Dimensions

Level 1 of the cultural framework contains the high-level cultural dimensions derived mainly from Hofstede's prior work (2005). Table 21.2 shows that in the Nigerian accidents short-term orientation was linked to all six reports, weak uncertainty avoidance linked to five, collectivism linked to four, and high-power distance linked to three. In the UK accidents, strong uncertainty avoidance was linked to all six reports, while individualism was linked to five. The Nigerian accidents can be characterized, therefore, by greater tolerance for deviance from planned and required actions, and feeling relaxed about others' actions. The UK accidents can be characterized by stronger beliefs in conformity, stability, and formal operating principles. Table 21.3 provides exemplars from the coded transcripts.

21.6.1.2 Level 2: Cultural Context

Level 2 of the cultural model describes how the high-level cultural dimensions described in Level 1 manifest themselves in different contexts. Culturally, most Western countries

TABLE 21.1

Accident Report Characteristics

Country	Report	Operator	Aircraft Type	Crew Nationality Commander	Age	Gender	Place of Incident	Manu/ Inci Year	Type of Flight	Persons on Board	Fatalities	Primary Cause/Nature of Damage
UK	G-BKJD 12 1994	Bristow	Bell 214ST	British	37	M	Near the Petrojarl 1, East Shetland Basin	1982/1994	Passenger	17	0	External and operational/none
	G-REDL 2-2011	Bond	AS332 L2	British			11 nm NE of Peterhead, Scotland	2004/2009	Passenger	16	16	Technical/destroyed
	G-CHCF 02-09	CHC	AS332 L2	British	50	M	Aberdeen Airport, Scotland	2001/2007	Training	3	0	Technical/none
	G-REDG 04-12	CHC	AS365N3	British	37	M	Norwich Airport	2010/2011	Passenger	7	0	Operational/none
	G-BLUN 7-2008	CHC	SA365N	British	51	M	Near the North Morecambe gas platform	1985/2006	Passenger	7	6	Operational/destroyed
	G-SEWP 06-11	PSNI	AS355F2	British	42	M	31 nm south of Belfast Aldergrove Airport, Northern Ireland	1991/2010	Passenger	4	0	Operational/destroyed
Nig	2008-07-22-F	NCAT	TAMPICO Club 9	Nigerian	41	M	Zaria Aerodrome	1997/2008	Training	1	0	Operational/destroyed
	2006-10-10-F	NCAT	TAMPICO Club 9	Nigerian	49	F	At Fanfulani village, Zaria, Kaduna State	1998/2007	Training	4	0	Technical and operational/damaged
	2008-03-24-F	AERO	AS 365 N2	Nigerian	49	M	Bonny Airstrip, Bayelsa State	1991/2008	Training	2	0	Operational/damaged
	2007-08-03-F	Bristow	Bell 412	Australian/British	47	M	Qua Iboe Terminal Akwa Ibom State	2005/2007	Passenger	1	1	Operational/destroyed
	1985 06 12 cia119	Aero	ALOUETTE III	British	34	M	MEREN 24 OFF-SHORE LANDING PAD (JACKET)	*/1985	Passenger	1	0	Operational/damaged
	1991 24 02 cia161	Bristow	Bell-212	British	54	M	Eket Off-shore Near BOP Oil Rig	*/1991	Passenger	13	9	Operational/destroyed

Note: * represents year of Aircraft Manufacture absent from the report.

TABLE 21.2

Activation of Level 1 Cultural Dimensions in UK and Nigerian Accident Reports

		Cultural Dimension								
		Low-Power Distance	High-Power Distance	Individualism	Collectivism	Weak Uncertainty Avoidance	Strong Uncertainty Avoidance	Short-term Orientation	Femininity	Muscularity
UK	G-BKJD	X	X	X	X	X	X	X		X
	G-REDL			X			X			
	G-CHCF						X			
	G-REDG			X	X	X	X			X
	G-BLUN		X	X		X	X	X		X
	G-SEWP			X		X	X			X
	Totals	**1**	**2**	**5**	**2**	**4**	**6**	**2**	**0**	**4**
Nig	5N-CAV				X	X		X		
	5N-CBF		X		X	X		X	X	
	5N-BJF					X		X		
	5N-BIQ		X	X	X	X	X	X		X
	5N-ALD				X	X		X		
	5N-AJY		X	X			X	X		
	Totals	**0**	**3**	**2**	**4**	**5**	**2**	**6**	**1**	**1**

Note: Bold numerals represent total number of nodes in each group for Nigeria and UK, respectively.

TABLE 21.3

Exemplars of Level 1 Cultural Dimensions in UK and Nigerian Accident Reports

Example Cultural Dimension	UK; Report G-BKJD 12 1994	Nig; Report 2007-08-03F
Weak uncertainty avoidance	*Let the future happen* "The information available to the crew did not alert them to the potential influence of the cumulo-nimbus adjacent to the Petrojarl 1" "The procedures for transition from the hover outside ground effect into climbing forward flight by sole reference to the flight instruments were inadequately defined" *Relaxed about others* "The company did not keep records of pilots' subsequent deck landing experience and there was no requirement, either company or regulatory, for pilots to keep such records in their personal flying logbooks"	*Deviance is tolerated* "The heliport is unlicensed and operated by Mobil Producing Nigeria (MPN)" *Let the future happen* "QIT heliport has no air traffic control services" *Relaxed about others* "The organizational safety management system did not identify, intervene and mitigate stress and crisis that developed in the circumstances of the pilot, days before the accident"
Strong uncertainty avoidance	*Beliefs in conformity, stability and principles* "At about 1250 hrs the crew carried out a brief air test which was satisfactory and they then shut down the aircraft, completed the necessary post air test work and prepared for the forthcoming flight to Sumburgh" *Dislike of risk and ambiguity* "Both pilots felt uncomfortable in the turbulence. The co-pilot remembered monitoring the torque and both pilots were aware that the airspeed was lower than normal in the climb" "The successive delays during the morning must have been frustrating for the crew but frequent changes of plan and task are the norm when flying helicopters to offshore oil installations"	*Beliefs in conformity, stability and principles,* *Intolerance toward deviant persons and ideas* "He was regarded as a dedicated, hardworking, highly professional and meticulous pilot, who constantly craved for excellence. Many believed he knew more about emergency procedures than any other pilot in the fleet" *Dislike of risk and ambiguity* "Investigation revealed the pilot's fear and anxiety and subsequent directives to his wife on security and safety of the family. The pilot was constantly disturbed and worried about the possibility of being kidnapped by the Niger Delta militants"

tend to show LCC characteristics while the reverse (HCC) has been identified for non-Western countries (Hall, 1974). These characteristics are clearly manifest in Table 21.4, with marked differences in the decision making, negotiation style, and problem correction approaches across the sampled Nigerian and the UK reports. In the United Kingdom, features of LCC were evident in all six reports, with the data inverting quite strongly for the Nigerian reports, all of which showed features of HCC. Table 21.5 provides exemplars from the coded transcripts.

21.6.1.3 Level 3: System Parts

Levels 1 (cultural dimension) and 2 (cultural context) are linked in the cultural model to Level 3, which contains the system elements to be found in Rasmussen's (1997) risk management framework. These system elements differ among the Nigeria and UK accidents

TABLE 21.4

Activation of Level 2 Cultural Dimensions in UK and Nigerian Accident Reports

		Cultural Context Differences							
		Low-Contextual Differences	Low Decision Making	Low Negotiation Style	Low Problem Correction	High-Contextual Differences	High Decision Making	High Negotiation Style	High Problem Correction
UK	G-BKJD	X	X	X	X	X	X		X
	G-REDL	X	X	X	X	X	X		X
	G-CHCF	X	X	X	X				
	G-REDG	X	X	X	X	X	X		
	G-BLUN	X	X	X	X	X	X		X
	G-SEWP	X	X	X	X				
	Totals	**6**	**6**	**6**	**6**	**4**	**4**	**0**	**3**
Nig	5N-CAV					X	X	X	X
	5N-CBF					X	X	X	X
	5N-BJF					X	X		X
	5N-BIQ	X	X	X	X	X	X	X	X
	5N-ALD					X	X	X	X
	5N-AJY	X	X	X	X	X	X	X	X
	Totals	**2**	**2**	**2**	**2**	**6**	**6**	**5**	**6**

Note: Bold numerals represent total number of nodes in each group for Nigeria and UK, respectively.

TABLE 21.5

Exemplars of Level 2 Cultural Dimensions in UK and Nigerian Accident Reports

Example Cultural Context Differences	UK—G-REDL		Nig—CAA 161
Low decision making	*Decisions are reached quickly* "From his previous experience, he did not think that this discovery was unusual due to the conical housing/rotor head replacement, which had been completed on 1 March 2009"	High decision making	*Decisions take time* "This investigator is yet to understand why a large proportion of officers in these Companies volunteered that the accident was anticipated and yet did very little to prevent its happening"
Low negotiation style	*Focus on economic goals* "Despite being a mandatory fit for G-REDL, HUMS is regarded as a maintenance advisory tool by industry, and is not considered by the manufacturer as the primary method of detecting gearbox degradation. As such, HUMS is provided by the helicopter manufacturer as an option and, at 1 April 2009, 44 of the global fleet of 82 AS332 L2 helicopters were fitted with HUMS"	High negotiation style	*Disagreement avoided because it is personally threatening* "…it is obvious that the displacement of the loose hatches is one day bound to happen. It was also gathered during the interviews that the pilots were very much aware and had experienced the hatches giving way under them"
Low problem correction	*Change is fast* "Immediately, on being notified of the accident, they quarantined all the documentation and records relating to the operation and maintenance of G-REDL"	High problem correction	*Change is slow* "There was no designated Air Traffic Control in the area…" "There was no formal fire traffic control" "There were no formal meteorological data available for the off-shore operations." "The jackets which are very limited in their total landing area, have no facilities whatsoever…"

in terms of cultural effects, although to a lesser extent than other coding categories. The regulators were linked to four of the six Nigeria reports, while in the UK the regulators and manufacturers were linked to five of the six reports. In general, more system parts (n = 46) were implicated in cultural issues in the United Kingdom compared to the Nigerian reports (n = 36). Table 21.6 shows the key differences in coding, while Table 21.7 presents exemplars from the transcripts.

21.6.1.4 Level 4: Human Actions

Level 4 of the cultural framework shows the various points at which culturally influenced human actions occur (Reason, 1990). A large number of human activities were coded, specifically those relating to safety concerns, emergency response, aircraft and helipad design and management, cost priority, and lack of regulatory oversight within the system. In the Nigeria reports, strategic policy, communication, and air traffic control were human actions that were linked to five of the six reports. In the UK reports, maintenance,

TABLE 21.6

Activation of Level 3 Cultural Dimensions in UK and Nigerian Accident Reports

					System Parts							
		Team	Supervisor	Regulators	Public	Organization	Manufacturer	Manager	Customers	Handling Agent	Airport Authority	Actor
UK	G-BKJD	X	X	X		X	X	X	X	X		X
	G-REDL	X	X	X		X	X		X			X
	G-CHCF			X	X	X	X					X
	G-REDG	X	X	X		X	X	X		X	X	X
	G-BLUN	X	X	X	X	X	X	X	X	X		X
	G-SEWP	X	X		X	X			X			X
	Total	**5**	**5**	**5**	**3**	**6**	**5**	**3**	**4**	**3**	**1**	**6**
Nig	5N-CAV		X		X	X		X				X
	5N-CBF	X	X	X		X		X				X
	5N-BJF	X	X			X						X
	5N-BIQ	X	X	X	X	X		X	X			X
	5N-ALD			X		X	X		X			X
	5N-AJY	X		X	X	X	X	X	X			X
	Total	**4**	**4**	**4**	**3**	**6**	**2**	**4**	**3**	**0**	**0**	**6**

Note: Bold numerals represent total number of nodes in each group for Nigeria and UK, respectively.

TABLE 21.7

Exemplars of Level 3 Cultural Dimensions in UK and Nigerian Accident Reports

Example System Parts	UK—G-BLUN 7-2008	Nig—2008-07-22-F
Organization	"The recovered recordings do not provide information on all system selections or indications. The helicopter was not equipped with image recorders and none of the avionic systems fitted were designed to record data; in particular the GPS data was not recorded." "The CAA met with the operator to discuss the results of the audit, which raised concerns about the Company's management organization, training and accident prevention and flight safety programme."	"There is no evidence to show that above components of the emergency plan are in place or tested every two years or with a similar exercise every year as requested by ICAO, neither was the emergency plan activated on the day of the accident". "Zaria aerodrome had an Emergency Response Plan (ERP) in place but there was no evidence that it had been tested in accordance with ICAO annex 14 recommendations."
Actor	"The co-pilot was flying an approach to the North Morecambe platform at night, in poor weather conditions, he lost control of the helicopter and requested assistance from the commander." "Since the co-pilot was in his first year with the operator, the company required that he complete a 6-monthly line check until he had successfully completed three such checks."— No evidence on this	"Inappropriate use of carburettor heat by the student" "There was no second opinion on the performance of the student before she was released for the solo flight." "From the available records, the student did not satisfactorily complete the required fourteen exercises before being released for the first solo flight."

communication, and command control were linked to five of the six reports. Table 21.8 presents the outcomes of the content analysis in detail, with Table 21.9 providing exemplars directly from the reports.

21.6.2 Analyzing the Interconnections within the Cultural Framework

Having examined the role of different coding nodes in the UK and Nigerian accidents, and mapped these to the different levels of the cultural framework, it is also important to see how these nodes connect. An analysis was performed in nVivo based on cluster analysis. This was deployed in an exploratory manner to by grouping nodes that shared similar themes using an inbuilt hierarchical clustering algorithm. Node similarity was quantified using the Pearson correlation coefficient. R-values of less than 0.35 denote a low or weak level of nodal interconnection, 0.36–0.67 modest or moderate interconnection, 0.68–0.89 strong or high interconnection, with r coefficients greater than 0.90 representing very high interconnection (Weber and Lamb, 1970; Mason et al., 1983). Discovering these correlations between nodes enables the links in the proposed cultural framework to be explored and quantified. Table 21.10 presents a rank ordered list of the inter-node correlations with an r-value greater than 0.70, a conservative criterion in which only those interconnections regarded as "strong" or "very strong" are admitted. This is a strict criterion applied in order to minimize the over-stating of relationships that may not be present, and to focus the analysis on critical links within the framework.

Some key cross national features from Table 21.10 can be highlighted. The UK reports revealed a high positive relationship between LCC decision making and weak uncertainty avoidance, with an r-value of 0.74. Quick decision making in the LCC is associated with an avoidance of risk taking or having to face an ambiguous situation. In the United Kingdom,

TABLE 21.8

Activation of Level 4 Cultural Dimensions in UK and Nigerian Accident Reports

		Human Actions														
		Training	Stress and Fatigue	Strategic Policy	Security	Politics	Maintenance	Media	Experience	Emergency Response	Communication	Command Control	Cockpit	Air Traffic Control	Weather Control	Implementation
UK	G-BKJD	×	×	×			×		×	×	×	×	×	×	×	×
	G-REDL			×			×				×					×
	G-CHCF	×					×					×				×
	G-REDG			×			×			×	×	×	×	×		×
	G-BLUN	×	×	×			×		×	×	×	×	×	×	×	×
	G-SEWP	×	×							×	×	×			×	×
	Total	**4**	**3**	**4**	**0**	**0**	**5**	**0**	**2**	**4**	**5**	**5**	**3**	**3**	**3**	**6**
Nig	5N-CAV	×		×			×		×	×	×	×		×		×
	5N-CBF			×			×				×	×				×
	5N-BJF	×							×		×	×		×		×
	5N-BIQ		×	×	×	×	×			×	×	×		×		×
	5N-ALD			×			×		×		×	×		×		×
	5N-AJY	×		×			×		×	×		×	×	×		×
	Total	**3**	**1**	**5**	**1**	**1**	**5**	**0**	**4**	**3**	**5**	**6**	**1**	**5**	**0**	**6**

Note: Bold numerals represent total number of nodes in each group for Nigeria and UK, respectively.

TABLE 21.9

Exemplars of Level 4 Cultural Dimensions in UK and Nigerian Accident Reports

Example Human Actions	UK—G-REDG	Nig—CAA 191
Communication	"As there was no dedicated means for ground staff to inform ATC of the incident, in order to alert the crew." "The base manager attempted to contact the crew but was unsuccessful." "He had also tried to contact the crew on the company frequency but with no success." "The co-pilot admitted he had not recognized the hand signal for 'fire'…" "…there had been issues with the quality of the audio"	"Communications and control of helicopter movements were carried out by the use of radio sets located inland at Escravos." "Off-shore meteorological services should be provided to cover helicopter operations. A forecast station can be located on one of the centrally located heliports, for example, Meren 1."
Implementation	"There was an additional delay due to his inability to attract the crew's attention, followed by confusion over the hand signals used." "The investigation estimated that over three minutes elapsed between the engineer initially observing the flames and the DATCO alerting the RFFS."	"Gulf Oil Company personnel that they are all so engrossed in oil production that subsidiary supportive roles were cast in the shadows. This investigator is yet to understand why a large proportion of officers in these Companies volunteered that the accident was anticipated and yet did very little to prevent its happening. …"

TABLE 21.10

Inter-Node and Inter-Level Correlations (r)

Node	Level	Node	Level	R-Value
Command control	4	Actors UK	3	0.84
Individualism	1	SP Actors Nig	3	0.77
Communication	1	Actors UK	3	0.77
Individualism	1	NG high-power distance	1	0.76
Individualism	1	UK high-power distance	1	0.76
Training	1	HCD negotiating styles Nig	2	0.75
Individualism	1	SP Actors UK	3	0.75
Weak uncertainty avoidance	1	LCD decision-making UK	2	0.74
Weak uncertainty avoidance	1	LCD decision-making Nig	2	0.73
LCD decision making	2	SP Actors UK	3	0.72
LCD negotiation styles	2	SP Actors UK	3	0.72
Manufacturers	3	UK maintenance	4	0.72
Cockpit	4	Actors UK	3	0.72
Implementation	4	LCD decision making	2	0.72
LCD decision making	2	Communication UK	4	0.71
LCD negotiation styles	2	Communication UK	4	0.71
LCD negotiation styles	2	UK individualism	1	0.70
Individualism	1	LCD decision-making UK	2	0.70

there is also a strong positive relationship between LCC and human actions categorized as communication (r-value = 0.71) and implementation (r-value = 0.72). In the UK reports, therefore, a feature that emerges is the role of an LCC, whereas the Nigerian reports tend to show the opposite. The Nigerian reports also showed a strong correlation (r = 0.76) between individualism and high-power distance. There seems to be evidence of seniority (perhaps even superiority) in the relationships between the individuals involved in the sampled accidents, and the tendency for them to feel they can act with autonomy (r = 0.77). These are all features the wider literature would lead us to expect, and they have been discovered independently here via the content analysis process. This provides some reassurance in terms of construct validity.

21.6.3 Driving the Cultural Framework with Data

Performing the content analysis on the UK and Nigerian air accident reports enables us to validate the model. This is achieved in two ways. The first is to take the component-level analysis and highlight what components of the model were more strongly coded in one nation compared to another. Figure 21.3 shows this by color coding the "active" nodes in the model based on how many times (as a proportion of the total number of coding instances) a particular node was coded. The bigger/bolder the box the more popular the node. The number in the box shows the actual proportion.

The second way the model is validated is to take the inter-node correlations presented above in Table 21.10 and use those to ascribe a strength to a particular link in the model. For example, the proposed links in the model are all present as dotted lines. If a correlation between nodes at different levels of the model is actually present in the current dataset, the link is emboldened. The thickness of the emboldened line is in proportion to the r-value shown in Table 21.10. The actual r-value is also shown.

In practical terms, the darker boxes shown in Figure 21.3 reveal the concentration of the coded nodes for each country. For instance, individualism, short-term orientation, and weak uncertainty avoidance are the main coded nodes in the Nigerian context on the left. On the other hand, low-power distance, individualism, short-term orientation, and weak uncertainty avoidance dominated the UK accident reports on the right.

21.7 Conclusions

This chapter set out with four explicit aims. The first was to link the extant knowledge-base on culture to the specific problem of helicopter safety and accident analysis. This has been achieved by a review of the literature which enables us to alight on a definition of culture, understand what links culture to actual behaviors, and where different cultural features reside within a wider sociotechnical system. The second aim was to use this knowledge-base to propose a cultural framework linking elements of systems to features of culture and ultimately to human actions. This too has been achieved. The multilevel model refers to Hofsted's (1997) cultural dimensions, Hall and Hall's (1990) manifestations of culture in different contexts, Rasmussen's risk management framework (1997), and Reason's (2000) work on human error. It is an integrative model that seeks to bring these different elements, often studied in different domains, into a common representation. The third aim of the chapter was to explore the extent to which the features highlighted in the model were actually present in helicopter accident reports, and to see if activation of those features was contingent on the culture in which the accident took place. The use of a sample of reports from Nigeria and the United Kingdom provides an excellent cultural manipulation and key differences did emerge. In particular, uncertainty avoidance and cultural context seem key. The fourth and final aim was to relate the discovered cultural factors back to the original framework in order to validate the components within it and their structure. This aim was also achieved. What arises, then, is a situation whereby cultural factors are clearly evident in the sampled accident reports, yet culture is currently not well represented in accident analysis methods. This is a potentially significant omission.

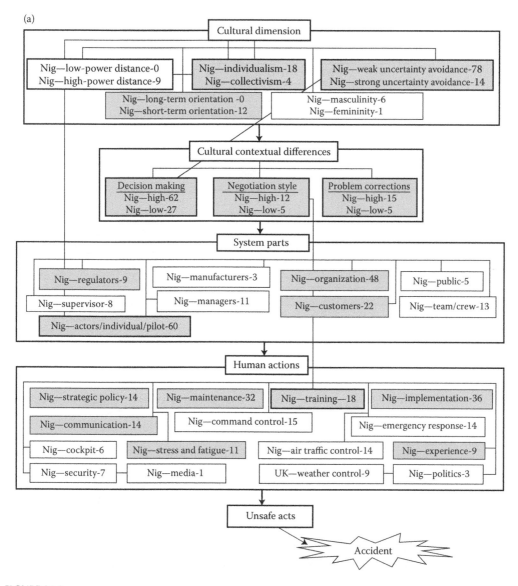

FIGURE 21.3
Validation of the cultural framework and a comparison between (a) Nigeria and (b) the United Kingdom. The relative size of the boxes, and the thickness of the lines, is driven from the content analysis and shows the activation of these model components based on the accident reports. *(Continued)*

(b)

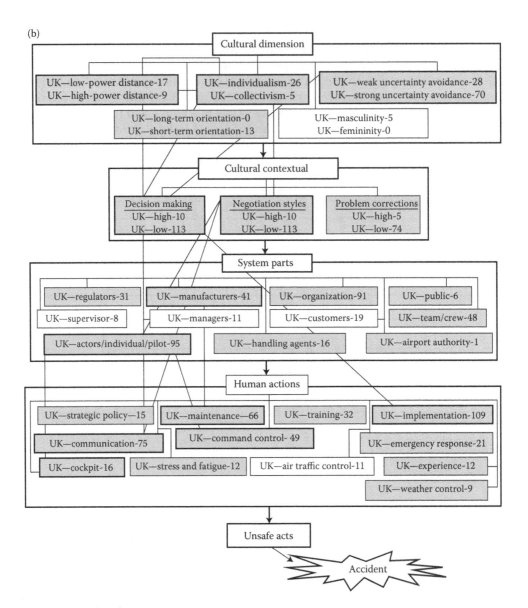

FIGURE 21.3 (Continued)
Validation of the cultural framework and a comparison between (a) Nigeria and (b) the United Kingdom. The relative size of the boxes, and the thickness of the lines, is driven from the content analysis and shows the activation of these model components based on the accident reports.

This study is one of the few that relates national cultural differences to the rising incidence of helicopter accidents, a key area of risk in the worldwide aviation sector. The promise that work on culture holds is the ability to make progress on air accidents that are currently classified simply as human error (a component view) and reassess them in terms of wider cultural influences on behavior (a systems view). The cultural framework developed and tested in this study contributes to our understanding of how and why culture influences individual or group action in a work environment. Accordingly, when these traits are adopted in an investigation process, the reasons why actions are taken can be more fully understood. Future research is being targeted at accident investigation methodologies, and using the cultural framework is a way of helping them capture this increasingly vital aspect of human and system performance.

Appendix: The Concepts Used for the Nodes and the Coding Consistency

This is the coding scheme used for the content analysis. Coding is theme based (phrases and other meaningful units of text) and mutually inclusive (more than one category can apply to a theme). The nodes are the themes used in the analysis, for instance, in level 1 as shown below the node is individualism, while the subnodes are the examples.

Cultural Dimension Nodes (Level 1)

Node	Individualism "I"	Collectivism "We"
Subnodes	Loosely knit social framework	Tightly knit social framework
	High individual autonomy—self-centered	Low individual autonomy—we centered
	Take care of selves and immediate family only	Will be taken care of when needed
		Loyalty to family, group, clan
	Weak Uncertainty Avoidance	**Strong Uncertainty Avoidance**
Subnodes	Let the future happen	Control the future
	Relaxed about others	Do not like risk and ambiguity
	Practice more important than codes for belief and behavior	Beliefs in conformity, stability, and principles
	Deviance is tolerated	Intolerance toward deviant persons and ideas
	Low-Power Distance	**High-Power Distance**
Subnodes	No acceptance of inequalities	Acceptance of inequalities
	Strive for power equalization	Power is distributed unequally
	Differences must be justified	Acceptance of hierarchies
	Little acceptance of hierarchies	Everybody has his/her place
	Masculinity (Performance/Achievement)	**Femininity (Welfare/Relations)**
Subnodes	Winner take all	Welfare for all
	Preference for achievement, heroism, assertiveness, and material success	Preference for relationships
		Modesty, caring for the weak, quality of life
	Maximum social differentiation between the sexes	Minimum social differentiation
	Performance societies	Focus on peoples' welfare
	Long-Term Orientation (Future Rewards)	**Short-Term Orientation (Past and Present)**
Subnodes	Emphasis on persistence	Emphasis on quick results
	Relationships ordered by status, personal adaptability	Status not ordered by relationships
	Face considerations common but seen as a weakness	Personal steadfastness and stability
	Leisure time not too important. Save, be thrifty, invest in real estate	Protection of one's face is important
	Relationships and market position important	Leisure time important
	Good or evil depends on circumstances	Spend, invest in mutual funds
		Bottom line important
		Belief in absolutes about good and evil

Contextual Differences (Level 2)

	Decision Making	
Node	**In Lower-Context Cultures**	**In Higher-Context Cultures**
Subnodes	Decisions reach quickly and efficiently	Details are important
	Reaching an agreement on main/facts points	Decisions take time
		Trust comes first
	Details worked out later by others	Verbal message is implicit in context
	Verbal message is explicit	Time is not easily scheduled
	Tasks scheduled to be done at particular times	Needs of people may interfere with keeping to a set time/SOP
	One thing at a time	Important is that activity gets done
	Important is that activity is done efficiently	

	Negotiation Style	
	Low-Context Cultures	**High-Context Cultures**
	Negotiations impersonally	Emphasize relationships
	Focus on economic goals	Sociable atmosphere when negotiating
	Precise words and intend them to be taken literally	Personally face-to-face communication avoided because it is personally threatening
	Directly	Indirect and personally
	Low use of nonverbal elements	High use of nonverbal elements

	Problem Correction	
	Low-Context Cultures	**High-Context Cultures**
Subnodes	Encourage open disagreement	Avoid confrontation and debate
	Straight forward, concise, and efficient for action expected	Flowery language
		Humility due to power distance
	Change is fast	Elaborate apologies are typical
	Change and see immediate results	Change is slow
		Things are rooted in the past; slow to change, and stable

System Parts (Level 3)

Node	Actors/Individual/Pilot
Subnodes	Team/crew
	Supervisors
	Managers
	Regulators
	Manufacturers
	Organization
	Public
	Media
	Customers
	Heli-port operators

Human Actions (Level 4)

Node	Humans in the System Are Performing/ Undertaking/Experiencing
Subnodes	Strategic policy
	Command control
	Implementation
	Communication
	Air Traffic control
	Maintenance
	Emergency response
	Media
	Experience
	Stress and fatigue
	Politics
	Security
	Weather control

References

AAIB, 1993. *Report on the accident to Piper PA-31-325 C/R Navajo*, G-BMGB, 4 nm south east of King's Lynn, Norfolk, Aircraft accident report 6/94, Air accident investigation breach, Department of transport.

Adler, N. 1991. *International Dimensions of Organisational Behaviour*, Boston: PWS-Kent.

Boer, P.C. 1997. Beleidsbeslissingen en bedrijfsveiligheid in de krijgsmacht [Policy decisions and operating safety in the armed forces], *Carré*, 1, 10–13 in Soeters, J.L., and Boer, P.C. 2000. Culture and flight safety in military aviation. *International Journal of Aviation Psychology*, 10 (2), 111–133.

FMTT, 1993. *Official report of the commission of investigation into the account on the 20 January 1992 near Mont Sainte Odile (Bas-rhin) of the Airbus A.320 registered F-GGED operated by Air Inter.*

Foushee, H.C. 1984. Dyads and triads at 35,000 feets: Factors affecting group process and aircrew performance, *American Psychologist*, 39, 886–93.

General Aviation Manufacturer Association (GAMA), 2010. *General Aviation Statistical Databook and Industry Outlook*, GAMA, accessed November 11, 2012.

Goode, W.J. 1957. Community within a community: The Professions, *American Sociological Review*, 22, 194–200.

Graham, J. 1985. The influence of culture on the process of business negotiation: An exploratory story, *Journal of International Business Studies*, Spring, 79–94.

Hall, E. and M. Hall. 1990. *Understanding Cultural differences: German, French and Americans*, Yarmouth: Intercultural Press.

Hall, E.T. 1959. *The Silent Language*, Greenwich, Connecticut: Fawcett, 39pp.

Hall, E.T. 1974. *Beyond Culture*, New York: Doubleday.

Hall, E.T. 1976. *Beyond Culture*, Garden City, New York: Anchor Press/Doubleday, 79pp.

Helmreich, R.L. 1994. Anatomy of a system accident: The crash of Avianca Flight 052, *International Journal of Aviation Psychology*, 4(3), 265–84.

Helmreich, R.L. and J.M. Davies. 2004. Culture, threat, and error: Lessons from aviation, *Canadian Journal of Anesthesia*, 51, R1–R4.

Helmreich, R.L. and A.C. Merritt, 1998. *Culture at Work in Aviation and Medicine: National, Organizational and Professional Influences*, Brookfield, Aldershot: Ashgate Publishing Limited.

Hofstede, G. 1980. *Culture's Consequences: International Differences in Work Related Values*, Beverly Hills, California: Sage.

Hofstede, G. 1997. *Culture and Organizations: Software of the Mind*, New York: McGraw-Hill, 279pp.

International Helicopter Safety Team, IHST. 2012. *Update 2011 overview*, www.ihst.org, accessed 08/20/12.

Jing, H.-H., J. Lu, and S.L. Peng. 2001. Culture, authoritarianism and commercial aircraft accidents, *Human Factors and Aerospace Safety*, 1(4), 341–359.

Jing, H.S. and A. Batteau. 2015. *The Dragon in the Cockpit: How Western Aviation Concepts Conflict with Chinese Value Systems*. Farnham: Ashgate.

Kluckholn, C. 1951. *The* study of culture, in Lerner, D. and Lasswell, H.D (eds), *The Policy Sciences*, Stanford, California: Stanford University Press, in Hofstede, G., 1980. *Culture's Consequences: International Differences in Work Related Values*, Beverly Hills, California: Sage, 25pp.

Li, W., D. Harris, and A. Chen. 2007. Eastern mind in Western cockpits: Meta-analysis of human factors mishaps from three nations, *Aviation, Space and Environmental Medicine*, 78, 420–425.

Mason, R., D. Lind, and W. Marchal. 1983. *Statistics: An Introduction*, New York: Harcourt Brace Jovanovich, Inc, pp. 368–383.

Maurino, D. 1998. *Culture at Work in Aviation and Medicine: National, Organizational and Professional Influences*, Brookfield, Aldershot: Ashgate Publishing Limited.

Merritt, A. 1993. The influence of national and organization culture on human performance, *The CRM Advocate*, 93–1, 5–6.

Merritt, A.C. and R.L. Helmreich. 1996a. Human factors on the flight deck: The influence of national culture, *Journal of Cross-Cultural Psychology*, 27(1), 5–24.

Merritt, A.C. and R.L. Helmreich. 1996b. Culture in the cockpit: A multi-airline study of pilot attitudes and values, in *Proceedings of the Eighth International Symposium on Aviation Psychology*, April 24–27, 1995, pp. 676–681. Columbus, Ohio: Ohio State University.

Meshkati, N. 2002. Macroergonomics and aviation safety: The importance of cultural factors in technology transfer, in Hendrick, H. and Kleiner, B. (eds.), *Macroergonomics Theory, Methods, and Applications*, pp. 323–330, Mahwah, New Jersey: Lawrence Erlbaum.

MSSC, 1992. *Commission of inquiry into the Air Ontario Crash at Dryden, Ontario (Canada)*, Final report, 1992, http://epe.lac-bac.gc.ca/100/200/301/pco-bcp/commissions-ef/moshansky1992-eng/moshansky1992-eng.htm, accessed 23/09/2014.

Neuliep, J.W. 2012. *Intercultural Communication: A Contextual Approach* (5th edition), Thousand Oaks, CA: Sage.

Nishimura, S., A. Nevgi, and S. Tella. 2008. *Communication Style and Cultural Features in High/Low Context Communication Cultures: A Case Study of Finland, Japan and India*, pp. 783–796, retrieved on July 11, 2013.

O'Hare, D., M. Wiggins, R. Batt, and D. Morrison. 1994. Cognitive failure analysis for aircraft accident investigation, *Ergonomics*, 37(11), 1855–1869.

Orasanu, J., U. Fischer, and J. Davison. 1997. Cross-cultural barriers to effective communication in aviation, in S. Oskamp and C. Gransone (eds.), *Cross-Cultural Work Groups: The Claremont Symposium on Applied Social Psychology*, pp. 134–160. Thousand Oaks, California: Sage.

Rasmussen, J. 1997. Risk management in a dynamic society: A modelling problem, *Safety Science*, 27, 183–213.

Reason, J. 1990. *Human Error*, New York: Cambridge University Press.

Reason, J. 2000. Human error: Models and management, *British Medical Journal* 320(7237), 768–770.

Sarter, N. and H. Alexander. 2000. Error types and related error detection mechanisms in the aviation domain: An analysis of aviation safety reporting system incident reports, *The International Journal of Aviation Psychology*, 10(2), 189–206.

Schneider, S. and J. Barsoux. 1997. *Managing Across Cultures*, Harlow: Prentice-Hall, 1997.

Schwartz, S.H. 1994. Beyond individualism/collectivism: New cultural dimension of values, In: Kim, C., Triandis, H.C. Kagitcibasi, C. Choi, C., and G. Yoon (eds.) *Individualism and Collectivism: Theory, Method, and Applications*, pp. 85–119, Thousand Oaks, California, Sage.

Schwartz, S.H. 1999. Cultural value differences: Some implication for work, *Applied Psychology: An International Review*, 48, 23–47.

Shappell, S. and D. Wiegmann. 2004. *HFACS Analysis of Military and Civilian Aviation Accidents: A North America Comparison*, Gold Coast Australia: ISASI.

Smith, P.B. and S.H. Schwartz. 1997. Values in Berry, in Poortinga, W.B., Segall, Y.H., Dasen, M.H. (eds.) (2002) (2nd edition) *Cross-Cultural Psychology: Research and Applications*, Cambridge, UK: Cambridge University Press.

Soeters, J.L. and P.C. Boer. 2000. Culture and flight safety in military aviation, *International Journal of Aviation Psychology*, 10(2), 111–133.

Strauch, B. 2010. Can cultural differences lead to accidents? Team cultural differences and sociotechnical system operations human factors: *The Journal of the Human Factors and Ergonomics Society*, 52, 246.

Walker, G.H. 2005. Verbal protocol analysis, In: Stanton, N. A., Hedge, A., Brookhuis, K., Salas, E., and Hendrick, H. (eds.) *The Handbook of Human Factors and Ergonomics Methods*. Boca Raton, Florida: CRC Press, pp. 30-1–30-9.

Walker, G.H., N.A. Stanton, P.M. Salmon, and D.P Jenkins. 2008. A review of sociotechnical systems theory: A classic concept for new command and control paradigms. *Theoretical Issues in Ergonomics Science*, 9(6), 479–499.

Weber, J.C. and D.R. Lamb. 1970. *Statistics and Research in Physical Education*, St. Louis: CV Mosby Co, pp 59–64, 222.

Wiegmann, D. and S. Shappell. 2001. Human error perspectives in aviation. *The International Journal of Aviation Psychology*, 11(4), 341–357.

Wiegmann, D. and S. Shappell. 2003. *A Human Error Approach to Aviation Accident Analysis*, Aldershot, England: Ashgate.

Yacavone, D. 1993. Mishap trends and cause factors in Naval aviation: A review of Naval Safety Center data, 1986–90. *Aviation, Space and Environmental Medicine*, 64, 392–395.

22

Pilot–Controller Communication Problems and an Initial Exploration of Language-Engineering Technologies as a Potential Solution

Bettina Bajaj and Arnab Majumdar

CONTENTS

22.1 Introduction

Despite considerable research into communication problems between pilots and controllers, conducted from a variety of angles, such problems continue to exist. While Barshi and Farris have recently stated that "misunderstandings occur with an alarming frequency" (Barshi and Farris 2013: 15), the data on which this quote is based stem from research published almost 35 years ago. With no current studies available that analyze the current levels of communication issues, this study therefore presents a small-scale study consisting of three independent data analyses with the aim of determining the extent to which such issues still present a problem.

Following the review of related work, the subsequent three sections will present and discuss this study. It is then explored how an additional communication system based on language-engineering technologies may work and reduce communication issues while having a positive effect on workload and situational awareness.

22.2 Related Work

Since the early 1980s pilot–controller communication problems have been the subject of considerable research.* One of the earliest collection of research studies was compiled by Billings and Cheaney in 1981. In one of these studies, Billings and Reynard (1981) analyzed a total of 12,373 Aviation Safety Reporting System (ASRS) reports, using a typological approach, as part of which they classified information transfer issues into several groups. Billings and Reynard's research is of particular relevance to the present study as they quantified the number of miscommunications between pilots and controllers and their results are still quoted in current research. According to them, over 73.3% of the reports have revealed information transfer problems (Billings and Reynard 1981: 11). They argue that because so many voice communications are impeded, verbal communication is not the best medium for effective information transfer (Billings and Reynard 1981: 12). It is interesting to note that as far back as 1981, Billings and Reynard proposed that one way to improve such issues would be to use data link methods, although they surmised that these may cause different issues, for example, misread numbers on displays (Billings and Reynard 1981: 12–13). With regard to the current state of pilot–controller communication, the remarks with which Billings and Cheaney conclude their overall report are worth highlighting since they are still valid, as will be shown later in this chapter.

> These and previous studies lead us to conclude that there is a real and present need for better information transfer […]. […]. We conclude that there is insufficient awareness of the pervasive nature of the information transfer problem in its various manifestations, and that this lack of awareness may be in part responsible for nonstandard and inadequate communications practices on the part of both controllers and pilots. (Billings and Cheaney 1981: 92–93)

Subsequently, numerous studies have applied similar typological approaches (e.g., Monan 1983; Morrison and Wright 1989), although the objects of these investigations differed. For instance, ATC (air traffic control) audio and/or video tapes were also analyzed (e.g., Cardosi and Boole 1991) and human-in-the-loop simulations conducted (e.g., Kanki and Foushee 1989). A systematic review of these is provided in Prinzo and Britton's (1993) comprehensive overview of pilot–controller communication literature, which also lists studies based on other approaches, such as acoustical and cognitive-psycholinguistic approaches. By citing the draft of the final report of the *Work Group of Human Factors Relating to Controller and Pilot Errors* (1992: 9), Prinzo and Britton (1993: 4) highlight that all these studies are largely descriptive without providing systematic research into the causes of communication problems. The literature does, however, record various studies which have not only done this, but have also looked at the causes from various perspectives. For instance, Goguen et al. (1985) have taken a communicative approach, whereas Cushing (1994) sees generic communication problems as the cause for pilot–controller

---

* Communication problems due to medical reasons, e.g., noise-induced hearing loss, are not considered here.

miscommunications. More recently, a cognitive-linguistic approach was applied by Barshi and Farris (2013). A number of linguistic studies have led to vital recommendations for improvements, for example, suggestions have been made regarding the length and complexity of ATC messages and phraseology wordings. Other studies have put forward new methods and tools for testing and training in ATC communications (e.g., Elliot 1997; Alderson 2011). Nonlinguistic studies have addressed such issues as frequency congestion, and noise reduction, and, as suggested by Billings and Cheaney (1981), have examined alternative methods of conveying ATC messages using Controller Pilot Data Link Communications (CPDLC) (e.g., Schneider et al. 2011).

Since Billings and Reynard's research, there appear to have been no studies that examine whether communication problems still amount to 73.3%. Barshi and Farris (2013), for instance, quote this figure without questioning its validity, and in other recent literature we can only get a vague idea. For example, in a study by EUROCONTROL* (2006), 535 communication occurrences (reported between October 2004 and March 2005) and 344 pilot and controller surveys were analyzed. While the aim of this study was not to determine an overall percentage of communication issues, the fact that 535 occurrences were reported within 6 months nevertheless appears to be indicative of a problematic situation. A more recent study by the International Air Transport Association, the International Federation of Air Line Pilots' Associations, and the International Federation of Air Traffic Controllers' Associations (IATA, IFALPA, and IFATCA 2011) has examined the nonuse of standard phraseology and related language issues using questionnaires. The results show, for example, that 44% of the 2070 participating pilots encounter nonstandard phrases *at least once per flight* (IATA, IFALPA, and IFATCA 2011: 13) and 52% of the 568 controllers report that they come across these *at least daily* (IATA, IFALPA, and IFATCA 2011: 45). Other results of interest include the fact that 48% of the pilots operated to airports where ICAO (International Civil Aviation Organization) standard phraseology is not used (IATA, IFALPA, and IFATCA 2011: 17–18) and 11% of the pilots cited the use of local languages as a factor in decreasing situational awareness (IATA, IFALPA, and IFATCA 2011: 28–30). We will see later that nonstandard phrases and local languages are indeed a considerable problem for pilots.

22.3 Method

The methodology consists of three data analyses of (1) reports from CHIRP[†] (UK Confidential Human Factors Incident Reporting Programme), (2) aircraft accident/incident reports from the UK Air Accident Investigation Branch (AAIB), and (3) communication flight logs completed by British airline pilots. Only British data sources were used because, as far as could be ascertained, there are no studies that deal only with British pilot–controller communication issues (cf. also*) and also to achieve homogenous sampling. The reason for using two types of report and flight logs is twofold. First, we wanted to compare the number of issues mentioned in the reports and on the flights to those without such issues. Second, we were interested in the difference in number of miscommunications found in the two

---

* The results of many studies into pilot–controller miscommunications might not be fully applicable everywhere as most studies relate to North America. In contrast, there are few studies on this topic in Europe or other parts of the world.

† CHIRP reports were chosen because, unlike in the United States, ASRS reports in the United Kingdom are submitted to airlines and are confidential, hence requiring special permission for access.

report types and wanted to know what communication issues occur during actual flights, the basis of which is the recent comment by IATA, IFALPA, and IFATCA who noted that

> the use of non-standard phraseology, local accents, and the use of local languages in radio communication are infrequently reported as contributing factors to incidents and accidents. However, the vast majority of the survey's respondents stated that these factors were a concern and routinely caused misunderstanding (IATA, IFALPA and IFATCA 2011: 7–8).

22.3.1 CHIRP Reports

These reports are published in CHIRP feedback online publications dealing with air transport. Each publication contains confidentially submitted reports from controllers, pilots, engineers, relevant organizations, and cabin crew. CHIRP publications also include editorials and comments from readers in response to reports published in preceding publications. At the time of analysis, 86 air transport publications were available, of which the most recent 43 publications (from 2015 to 2000) were selected for analysis in this study. The 43 publications were then analyzed with a view to identifying any reports relevant to this study. For reports to be considered relevant, they had to (i) concern commercial flights and (ii) they had to be submitted by pilots or controllers, resulting in a total of 531 reports relevant to the study. All other reports were omitted, including reports in the form of comments since these refer to reports in preceding publications and would have led to duplication (or more) of reports. Each report was subsequently screened for citations of communication problems by means of an online search using the following search words and truncations*: *callsign, call sign, comm, congest, discipl, English, language, local, native, non-stand,* and *phrase*. The extracted problems were then allocated to the categories below, which were compiled specifically for the purposes of this study on the basis of the various communication issues found in the literature:

1. The *use of local languages*
2. *English language issues* (e.g., variations in English pronunciation/enunciation as well as in prosody, that is, tempo, rhythm, pitch and loudness, by both native and nonnative speakers)
3. *Message issues* (e.g., incomplete, misunderstood, omitted transmissions, and so on)
4. *Phraseology issues* (e.g., use of nonstandard phrases; lack of global standardization/ harmonization of ATC phraseology, resulting in various standards being in use in addition to the ICAO standard)
5. *Frequency issues* (e.g., blocked, congested frequencies, and so on)

The aim of analyzing the CHIRP reports was twofold: to establish the proportion of reports submitted that mention communication problems and the exact nature of these problems.

22.3.2 AAIB Aircraft Accident/Incident Reports

The same number of air accident/incident reports were analyzed, that is, 531, which are available in a searchable database on the AAIB website. As before, any reports considered

---

* Truncated word forms were used to identify any occurrences which contained the specific string of characters.

relevant had to involve commercial flights. The AAIB database offers a variety of report types relating to an accident/incident, including full formal reports, summaries, and addenda (additional material relating to a report), and to avoid counting a report more than once only one of these report types was included in the total number of relevant reports. The reports were selected in reverse chronological order from 2015 to 2004 and the analysis of these 531 reports was conducted in the same manner as for the CHIRP reports, with the extracted communication issues allocated using the same categories. The aim of analyzing the AAIB reports was to determine how many of the total number of accident/incidents examined in the reports involved communication problems and their exact nature.

22.3.3 Communication Flight Logs

A total of 30 pilots working for several UK airlines were asked at random to complete 4-week flight logs for the purposes of recording any communication issues during their flights. Each of the 4-week flight logs consisted of four consecutive weeks, and all the 4-week periods recorded were also consecutive. A total of eight pilots responded* (26.6%), including a female pilot (12.5%),† and of these there were four captains (including two training captains) and four senior first officers. Six of the respondents fly long-haul routes and two are on short haul. The pilots were provided with a prepared flight log template accessible on their work-issued iPads. Five pilots were able to provide three 4-week periods, while three recorded two 4-week periods due to standby and leave, resulting in a total of 21 flight logs. Instructions included that the total number of sectors flown in each 4-week period needed to be recorded in a table, in which the pilots also had to highlight those sectors with communication issues. For each sector with such problems, they also needed to give a brief description of the exact nature of the problems. The completed flight logs did not need to be subjected to the same analysis as the above reports since no filtering was necessary. The reported miscommunications were then allocated to the same categories as above. In addition, the data were also analyzed per pilot in order to answer the following questions: (i) Do short-haul and long-haul routes produce different results? (ii) How many communication issues are there per pilot? (iii) Do all the pilots experience the same types of problem? The goal of analyzing real-life flight logs was both to obtain an idea of the extent to which pilots are exposed to communication problems during their day-to-day flying schedule and the nature of these problems.

22.4 Results

The data were analyzed using descriptive statistics. Note that due to the small size of the samples and the fact that they refer to the United Kingdom, the results do not lend themselves to generalization using inferential statistics.

---

* Although the response rate appears low, it fares well in view of the comment made in CHIRP Issue 72: "In the survey of flight crew, ATCOs and Licensed Engineers, we received 1,790 completed responses, around 6% of the total number of forms sent out. This percentage return is within the range expected for a survey of this kind" (2004: 1).

† 212 (5.5%) out of 3866 pilots employed by British Airways are women (Pilot 1 (R.B.), personal communication, July 25, 2015).

22.4.1 Results of the Analysis of CHIRP Reports

Out of the 531 CHIRP reports that were analyzed, 117 reports were found to contain communication issues (22%). Since some reports revealed multiple communication problems, a total of 170 individual communication issues were counted. These 170 issues were subsequently allocated to the five categories of communication problems that could be identified in the relevant literature (see Section 22.3), but as the results in Table 22.1 show, four further categories needed to be added in order to accommodate the types of problem found. As can be seen, the largest number of problems belong to the message/transmission category (43%), closely followed by phraseology and terminology problems (21.8%). In third place, with 17.1%, we find frequency issues.

22.4.2 Results of the Analysis of Aircraft Accident/Incident Reports

Thirty three out of the 531 accident/incident reports contained communication issues (6.2%). The subsequent analysis of these revealed a total of 58 individual communication issues. As was the case with the CHIRP reports, the number of categories had to be extended by a further four. In Table 22.1, it can be seen that, as with the case of the CHIRP reports, the largest number of problems belongs to the message/transmission category (34.5%) and the second largest category with 17.2% concerns phraseology and terminology. Also in second place is the category to which unspecified communication issues were allocated, for example, if it was impossible to identify why there was an unclear communication situation. In third place, issues with radio communication units were found in 12.1% of the cases, which is a much higher percentage than in the case of the CHIRP reports (2.4%). This was closely followed by issues regarding the quality of English (10.3%), which were cited twice as often as in the CHIRP reports (5.3%). It should be noted that no mention of local languages was made and that frequency issues play a lesser role than in the CHIRP reports but, as will be seen, they are much more frequently cited in the flight logs.

22.4.3 Results of the Analysis of Communication Flight Logs

A total of 21 4-week flight logs were produced by eight pilots. In turn, these flight logs consist of a total of 240 sectors: 125 long haul and 115 short haul. The pilots reported communication problems in 85 out of the 240 sectors (35.4%). On short haul, 26 out of the 115 sectors produced problems (22.6%), while on long haul 59 out of 125 sectors generated issues (47.2%). For the 85 sectors with communication issues, a total of 232 individual problems were logged. Table 22.1 shows that the message/transmission category is ranked highest (37.5%), followed by frequency issues with 26.3%. In third place, there is the quality of English category with 22%, and the use of local languages came fourth (6.9%). It can be seen that most problems also fall into the message/transmission issues category, but that, unlike in the two report types, phraseology/terminology issues were rarely reported. However, the quality of English and local languages categories were logged much more often than in the reports. It is noteworthy that on three occasions the pilots encountered an unwillingness to communicate by ATC, which was cited just once in the CHIRPs, but was absent in the accident/incident reports.

Table 22.2 shows the results per pilot, which indicate that within the long-haul and short-haul groups the percentage of communication issues is somewhat similar. On long haul, most pilots reported problems on 37%–54% of their sectors flown, with the exception of pilots 6 and 7 who logged problems in 93.8% and 21.1% of the sectors, respectively. The two

TABLE 22.1

Results of the Three Data Analyses

Categories of Communication Issues	CHIRP Reports		Aircraft Accident and Incident Reports		Communication Flight Logs	
	N = 170	f/N (100%)	N = 58	f/N (100%)	N = 232	f/N (100%)
1. Unspecified communication issues (lack of, ineffective, unclear, low quality of communications)	f = 9	5.3%	f = 10	17.2%	f = 10	4.3%
2. Problems with radio communication units (complete/temporary loss)	f = 4	2.4%	f = 7	12.1%	f = 1	0.4%
3. Quality of English issues (heavily accented English by nonnative speakers, by native speakers, using English colloquialisms)	f = 9	5.3%	f = 6	10.3%	f = 51	22%
4. Use of local languages	f = 6	3.5%	N/A	N/A	f = 16	6.9%
5. Phraseology and terminology issues (ambiguous, nonstandard, unclear, lack of international harmonization)	f = 37	21.8%	f = 10	17.2%	f = 3	1.3%
6. Frequency issues (busy, congested, blocked, unreadable, static interference, mis-set, closed, confusion, unauthorized frequency changes, lack of frequency change instructions, frequency "black holes," HF communications of inadequate quality)	f = 29	17.1%	f = 3	5.2%	f = 61	26.3%
7. Message/transmission issues (incomplete, incorrect, missed, misheard, misconstrued, unclear, conflicting, unrealistic, omission of, too complex, spoken too fast, split messages, unauthorized, at inappropriate times, numerals incorrectly pronounced, repeated several times over, call sign issues)	f = 73	43%	f = 20	34.5%	f = 87	37.5%
8. Unwillingness-to-communicate issues (deliberately ignoring transmissions, not clarifying instructions)	f = 1	0.6%	N/A	N/A	f = 3	1.3%
9. Other communication issues (due to O₂ masks, high ambient cockpit noise, side tone in controller headsets causing confusion, unreliable new RT systems for ATC)	f = 2	1.2%	f = 2	3.4%	N/A	N/A

TABLE 22.2

Results by Pilot

Pilot	Gender	Rank[a]	Fleet	Flight Length[b]	Number of 4-Week Periods	Number of Sectors	Number of Sectors with Communication Issues	f/N (100%)
1	M	TrngCPT	B747-400	LH, MH	3	24	f = 13	54.2%
2	M	CPT	B747-400	LH	3	20	f = 9	45%
3	M	TrngCPT	B747-400	LH, MH	2	24	f = 9	37.5%
4	M	SFO	B747-400	LH, MH	3	22	f = 9	40.9%
5	M	SFO	A321/20/19	SH, MH	2	91	f = 19	20.9%
6	M	CPT	A380	ULH, LH	3	16	f = 15	93.8%
7	F	SFO	B747-400	ULH, LH, MH	2	19	f = 4	21.1%
8	M	SFO	A320/19	SH, MH	3	24	f = 7	29.2%
Total					21	240	N = 85	35.4%

[a] TrngCPT = Training captain; CPT = captain; SFO = senior first officer.
[b] ULH = Ultra long-haul sector, LH = long-haul sector, MH = medium-haul sector, SH = short-haul sector.

short-haul pilots produced much lower percentages, that is, 20.9% by pilot 5 and 29.2% by pilot 8. In summary, the flight logs revealed a much higher percentage of communication problems (overall 35.4%; long haul 47.2%; short haul 22.6%) than in both the CHIRP (22%) and aircraft accident/incident reports (6.2%).

22.5 Discussion

The results from the three analyses* are clearly not at a similar level to the information transfer issues reported by Billings and Reynard (1981), who, as will be recalled, reported that 73.3% of the analyzed ASRS reports included such issues. However, the lower number of reported issues is not unexpected as since 1981 many improvements have been made regarding ICAO phraseology, new training, and testing methods in aviation English for nonnative speakers were introduced, and human factors training intensified. Also, since 1981 fewer high frequencies (HFs) have been in use, which has improved the quality of ATC transmissions. However, in view of the results from the flight logs, in which still relatively high numbers of communication issues have been reported, this explanation may not give us the full picture. It is necessary therefore to investigate the factors which may have influenced the three data analyses in this study.

A likely reason behind such a discrepancy between Billings and Reynard's 73.3% and the CHIRP result of 22% could be the number of ASRS reports they analyzed, that is, 12,373, all of which were submitted in a period of 2 years. In contrast, in the United Kingdom considerably fewer CHIRP reports were filed during a 2-year period as only 531 relevant reports (the number of omitted reports is roughly the same) were counted from

---

* Limitations of this study can be seen in the small sample sizes and that only pilot flight logs but no ATC communication logs could be obtained. The accident/incident and CHIRP reports though include inputs from both pilots and controllers.

2000 to 2015. The reason for this could be fourfold. First, far more flights take place per day in the United States compared to the United Kingdom. Second, while ASRS reports are anonymous, CHIRPs are confidential but not anonymous, which means that pilots and controllers may be reluctant to file reports for fear of repercussions. Third, the low number of reported issues to CHIRP may also be due to a lack of interest in reporting them. Finally, the legal implications of ASRS reporting in the United States cannot be discounted as an explanation for the high level of reporting.

The low figure of 6.2% of cited communication problems in the aircraft accident/incident reports ties in with the observation by IATA, IFALPA, and IFATCA that nonstandard phrases, local languages, and accents are rarely cited as being contributory to accidents, but that for the large majority of pilots and controllers these issues are a cause of concern and regularly result in misunderstandings (IATA, IFALPA, and IFATCA 2011: 7–8). They fail to provide any explanation for this, but it may be speculated that, perhaps, the specific personality traits of pilots play a role here. Commercial pilots are a fairly homogenous group in terms of stress resistance, multitasking abilities, ability to find solutions fast, willingness and ability to adapt, to cope with problems under pressure, and they have a "can-do" attitude when faced with solving problems (cf. Green et al. 1996; The Air Pilot's Manual 2013). Hence, in view of the relatively high number of problems reported in the flight logs, in particular by pilot 6 who cited issues in 93.8% of his sectors, it may indeed be the personality traits of pilots which prevent communication issues from developing into an accident. The same could perhaps be said of the personality traits of controllers which appear to be similar.* This is an area that could warrant further investigation. However, it can be argued that this reliance from regulators and airlines on pilots and controllers being able to cope with and resolve communication problems on a continual basis means that this is an accident waiting to happen. Concerns are also expressed by controllers: "What did he say? Asking 'Say again,' often leaves us none the wiser. Whilst we manage on a day-to-day basis, we are concerned that we would not understand them in the event of an emergency" (CHIRP Issue 80 2006: 3).

The pilots reported in their flight logs that the most frequent communication issues belong to the message/transmission category, in which they, for example, listed incomplete, incorrect, missed, misheard, and too complex messages. This was closely followed by frequency issues, where the major problems reported are congested, blocked, and busy frequencies, and also the inadequate quality of HFs over Africa and South-East Asia, which means that they regularly fly long distances without any ATC contact. Both categories overlap, however, inasmuch as some message/transmission issues are brought on by frequency issues, for example, by low-quality HF, by blocked, or congested frequencies. The pilots often reported that they had to spend a lot of time discussing with each other what the controller had said, and they also mentioned how often messages needed to be clarified, for example, pilot 1 reported that one message had to be repeated four to six times. The third most cited problem was the quality of English category, which also overlaps with the message/transmission category as instructions given in heavily accented English often need to be repeated several times. Another frequent problem was the use of local languages, which the pilots unanimously reported as causing a decrease in situational awareness. A surprising issue was logged by pilot 7 who said that in a particular airspace, ATC would not speak to her because, in her opinion, she was female.[†] Other

---

* For instance, cf. https://www.eurocontrol.int/articles/skills-required-be-air-traffic-controller. Courtesy of EUROCONTROL.
[†] Pilot 7 also reported that she had similar experiences as a short-haul pilot.

reported refusals to communicate stem from political differences between two states. Due to the small sample, we cannot know whether such refusals, particularly in the case of the female pilot, are just isolated occurrences or whether this could be indicative of a bigger problem. The discrepancy between short-haul and long-haul results can be explained by the different countries and regions that are flown to or over. The problems on long haul are much more characterized by frequency issues, the quality of English spoken, and the use of local languages than is the case for the predominantly European short-haul flights.

In terms of mental workload and situational awareness, it was in particular the long-haul pilots that were adamant that communication issues had a negative effect. Since situational awareness is part of a pilot's decision-making process, which in turn is affected by high stress/workload situations, it is clear that communication issues leading to such a situation are undesirable since the loss of situational awareness could lead to accidents. Hence, although the percentages of 35.4% overall and 47.2% long-haul problems were much lower than Billings and Reynard's 73.3%, these figures can nevertheless be considered to be too high. It is thus surprising that despite continuous human factors training, improved training and testing methods for aviation English, the ongoing process by ICAO to improve ATC phraseology, and the decrease in operational HFs, such a high number of problems should still exist in today's flight decks and ATC workstations.

Therefore, it is pertinent to ask why such communication issues have not been reduced even further. There may be two obvious answers and one less obvious one to this question. First, the human factor will always be present in interpersonal communications and this makes an eradication of errors unlikely. Second, technology has not advanced enough yet to solve frequency problems once and for all, and the planned expanded CPDLC via satellites will only solve communication issues to some extent but cannot solve the use of local languages and the quality of English. Third, and less obviously, it may be argued that in pilot–controller communications, there is an inconspicuous concept involved that so far seems to have resisted most human factors efforts inasmuch as that many pilots and controllers are still not sufficiently mindful of what we may call *communication awareness.** This often manifests itself in *"us-and-them"* communication situations instead of *"we"* situations. Indeed, as far back as in 1981, Billings and Cheaney surmised that a lack of awareness by pilots and controllers of the exact nature of communication could partly be responsible for causing communication issues. Hence, it can be argued that while some issues have been reduced, too many still occur and pilot–controller communication awareness has not yet improved sufficiently.

22.6 Potential Solution

Since the results of the communication flight logs have shown that voice communications remain problematic and routinely cause misunderstandings and since there still

---

* Taking Billings and Cheaney's quote (cf. Section 22.2) as the basis, we propose the following working definition for the concept of communication awareness: It is vital to know exactly the nature of the communication situation we are in, to know what needs to be communicated, how it should be communicated, and when exactly. It also means that we are as fully informed as possible about the other person's communicative environment and that we are *willing* to communicate appropriately. For example, it is important to communicate slowly with the awareness that the other person may take longer to understand the information due to the environment and situation they are in.

appears to be a considerable lack of communication awareness among pilots and controllers, it is vital that such communication issues be minimized. Below, we explore the idea of developing an additional communication system using language-engineering technologies in automatic speech recognition (ASR), machine translation (MT), and term extraction (TE); however, due to limitations of space, the individual technologies will only be outlined briefly.

22.6.1 ASR, MT, and TE

The goal of ASR is to transfer speech to text on the basis of speech recognition algorithms which transform a sequence of acoustic waves into a sequence of written text (Jurafsky and Martin 2009). The problem of speech being automatically recognized irrespective of any surrounding conditions poses problems, but recently substantial performance improvements have been achieved in the application of deep neural networks (DNN) into ASR technology (e.g., Maas et al. 2012). Performance levels can also be raised if ASR is applied to *controlled languages* (characterized by standardized phrases, reduced/disambiguated terms, limited syntax, and repetitive information) and ATC phraseology fulfils the requirements of such a language. Like ASR, the quality of MT depends on many problems inherent in language and speech. MT performance also increases the more a language is reduced in terms of its grammatical structures, linguistic devices, and if a domain is delimited. For example, high-quality results have been achieved in the domain of meteorology (Gotti et al. 2013), which is a field characterized by standardized weather reports, reduced and disambiguated terminology, controlled language (e.g., limited use of syntax), and repetitive information (cf. Jurafsky and Martin 2009). The state-of-the-art technology used in MT systems is hybrid, which brings together rule-based as well as statistical and example-based methods (e.g., Forcada 2010). The main advantage of this combined approach is that within a delimited domain, the quality of translations is likely to be high. As part of the MT process, TE often takes place simultaneously (cf. Vivaldi and Rodríguez 2007) and automatically identifies term candidates from spoken or written specialist texts. Approaches to TE include linguistic and statistical methods, but more recent tools are hybrid. Like MT, performance levels are likely to be high if used in delimited domains with controlled languages.

22.6.2 Proposed Advanced Intelligent Communication System (AICSys)

ASR systems for use in aircraft have been investigated for several years with a view to applying ASR in voice input systems in cockpits (e.g., Baber and Noyes 1996; Lennertz et al. 2012) and for training controllers (e.g., Cordero et al. 2012). For example, direct voice input by pilots is already used in the Eurofighter Typhoon (2014) and in the F-35 Lightning II Joint Strike Fighter (Schutte 2007). Similar to our study, Geacăr (2010) investigates speech input as a back-up solution to voice communications, which means that any voice message is transcribed into text and transmitted as such via data link. In contrast, Lennertz et al. (2012) study data link messages that are accompanied by synthetic speech outputs in addition to voice communications with a view to reducing head-down time during single-pilot operations. It is clear, however, that the above approaches would be unable to deal with local-language use as well as English accents and dialects. Geacăr's suggestion to use ASR for transcribing all ATC radio messages and to transmit these using data link seems to correspond to the ASR phase described in our system (see below), but he does not mention how local-language messages or messages spoken with accents or in dialects would be dealt with. Although Lennertz et al. claim that adding synthetic speech commands to textual data link messages did not "introduce additional complications" (Lennertz et al. 2012: 31), it does not seem to be well justified. Since their experiment involved single-pilot operations, the combined synthetic speech-text method may have its benefits in such a situation. It remains to be seen, however, how synthetic speech outputs of data link text messages, in addition to the voice communications, will fare in two-pilot operations, which are more complex in terms of pilot tasks and aircraft systems. Moreover, in two-crew operations, crew resource management (CRM) requires the crew to communicate with each other to a large extent and interruptions from synthetic speech outputs could disrupt workflow resulting in increased workload.

Key findings in the areas of ASR and MT show that the performance levels of both technologies depend on the application of the latest artificial intelligence techniques and on how issues such as dialects, accents, prosody, channel, noise, vocabulary size, domain delimitation, grammar differences, etc. are dealt with. The task of creating an intelligent communication system which produces results of the highest quality will be challenging. However, as we have seen, all three technologies should perform well with the controlled language of ATC phraseology. Consequently, the proposed system should be able to reduce many of the communication issues mentioned in this study as follows:

- During the first phase, ASR technology would convert spoken English messages into text in real time. This allows pilots and controllers to see the message on a display, including any nonstandard phrases, which could then be queried with ATC if necessary. Since state-of-the-art ASR systems are trainable at source by individual speakers, this should also address pronunciation, enunciation, and prosody issues for both native and nonnative speakers.
- The simultaneously running TE phase would highlight safety-critical terminology as high-priority information.
- If local languages are used, in the second, though simultaneous, phase MT technology is triggered into action and would enable translations into English in real time.
- Since we propose to use data link methods, frequency issues should not affect the transcribed/translated messages.

22.6.3 AICSys on the Flight Deck and in ATC Workstations

The system is intended as an additional communication system to voice communications, thus contributing to the level of system redundancy in ATC workstations and in particular in flight decks, where thus far voice radio communications are without redundancy.

The proposed system would provide pilots and controllers with text versions of voice messages. For example, for every voice message to an aircraft, a transcribed text version would be sent via data link and be shown in real time on a display on the flight deck. Simultaneously, any other aircraft in the same airspace which has the system installed would also receive the text version of that particular message on their display, so that pilots would also be able to read messages intended for other aircraft. If a local language is used on the radio, the MT system would get activated and pilots and controllers would receive English translations of the foreign-language message on their respective displays in real time. Pilots would therefore always be able to understand messages in a local language and hence be in the picture of what nearby aircraft are doing and their intent.

It would, however, be up to the pilots to decide whether they need to look at messages on the displays or not. For instance, as long as the pilot dealing with ATC communications hears transmissions in English using standard phrases, s/he would probably have no need to look at the display. However, if the message is garbled or if local languages are used, or if the aircraft is close to any runways, the pilot would want to know what has been said and look at the display.

Similarly, during times of high mental workload, controllers may filter messages from aircraft (Airbus 2014), hence not all communications addressed to them may be listened to or acknowledged if they are, for example, talking to another aircraft. The additional transcripts/translations on a suitable display in the workstation would therefore give them the chance to read a message should the need arise. Consequently, the role of the AICSys in avoiding miscommunications on the ATC side is evident. Using such a system, it should be possible to avoid ambiguity in critical information regarding, for example, clearances, and pilots and controllers might be prevented from mishearing call signs, information about flight levels or speed, irrespective of dialects, etc. Given the fact that controllers deal with more than one aircraft at a time, design and implementation considerations will be crucial in terms of how to separate the various transcripts/translations from individual aircraft in such a way that they are easily distinguishable (possibly separated from each other in space and by color).

22.6.4 Human Factors Considerations

It is vital during the high workload phases of flight that communications are accurate and not time consuming. The proposed system may therefore have the potential to reduce high workload since pilots would have more time to aviate and navigate rather than having to spend time double-checking messages with ATC or by figuring out among each other what was said. When workload increases, communication tends to disintegrate (cf. Barshi and Farris 2013), for example, messages decrease in length under high workload (e.g., Raby and Wickens 1994) as a result of restrictions on a human being's time-sharing abilities that underlie complex performance (Jennings and Chiles 1977). How would this time-sharing ability be affected by the additional transcripts/translations in the flight deck? According to Wickens' four-dimensional multiple resource model (Wickens 2008), the perception of auditory and visual information occurs in different parts of the brain, which means that time-sharing between tasks that use different resources should be less conflicted. However, he points out that if both the auditory and the visual tasks need processing at a higher level, which is what the comprehension of spoken words and written text would require, then this "will still compete for common perceptual resources (and may also compete for common code-defined resources [...])" (Wickens 2008: 450). We nevertheless argue that the provision of transcripts/translations on a display would be an improved situation for pilots since in phases of high workload the pilot-not-flying knows that if she/he has not been able to hear or understand a message over the radio, it will be available as text on the display. Through the help of the displayed text, it is argued that mental resources in the pilots' brains are freed.

For controllers, considerations of mental workload, situational awareness, and attention allocation are equally as important. The use of the proposed communication system could provide improved situational awareness for controllers as the lack of ambiguity in communications will greatly enhance the accuracy of the information provided. This being the case, then the certainty of information will in addition enable controllers to manage their workload in a fashion considerably more enhanced than at present.

22.7 Conclusions

This impetus behind this work was twofold. The first aim was to establish the current level of communication problems between pilots and controllers. The results from the data analyses showed that, although the number of issues has fallen considerably since Billings and Reynard's seminal research in 1981, problems during pilot–controller communications are still numerous as was shown in particular by the analysis of the flight logs compiled by British commercial pilots. As a result of the ongoing problems, the second aim was to propose and explore a potential solution in the form of an additional communication system in order to minimize such problems. This study presented a description of how such a system may improve pilot–controller workload and situational awareness. Further research will focus on creating a prototype of the system, while gaining a better understanding of what factors directly affect pilot–controller communications, as part of which the concept of communication awareness will be developed further.

References

Airbus. 2014. Flight operations briefing notes: Human performance—Effective pilot/controller communications. http://www.airbus.com/fileadmin/media_gallery/files/safety_library_items/Air busSafetyLib_-FLT_OPS-HUM_PER-SEQ04.pdf (accessed July 14, 2015).

Alderson, J. C. 2011. The politics of aviation English testing. *Language Assessment Quarterly* 8(4): 386–403.

Baber, C. and J. Noyes. 1996. Automatic speech recognition in adverse environments. *Human Factors* 38(1): 142–155.

Barshi, I. and C. Farris. 2013. *Misunderstandings in ATC Communication*. Farnham: Ashgate Publishing.

Billings, C. E. and E. S. Cheaney. 1981. The information transfer problem: Summary and comments. In *Information Transfer Problems in the Aviation System*, eds. C. E. Billings and E. S. Cheaney, 85–93. NASA Technical Paper 1875. Moffett Field, California: NASA Ames Research Center.

Billings, C. E. and W. D. Reynard. 1981. Dimensions of the information transfer problem. In *Information Transfer Problems in the Aviation System*, eds. C. E. Billings and E. S. Cheaney, 9–14. NASA Technical Paper 1875. Moffett Field, California: NASA Ames Research Center.

Cardosi, K. M. and P. W. Boole. 1991. *Analysis of Pilot Response Time to Time-Critical Air Traffic Control Calls*. Cambridge, Massachusetts: National Transportation Systems Center.

CHIRP Issue 72. 2004. An Air Transport Safety Newsletter from the Confidential Human Factors Incident Reporting Programme CHIRP, Farnborough, UK. Available from www.chirp.co.uk

CHIRP Issue 80. 2006. An Air Transport Safety Newsletter from the Confidential Human Factors Incident Reporting Programme CHIRP, Farnborough, UK. Available from www.chirp.co.uk

Cordero, J. M., M. Dorado, and J. M. de Pablo. 2012. Automated speech recognition in ATC environment, ATACCS'2012, Research Papers, 46–53. www.hala-sesar.net/sites/default/files/documents/p46-cordero.pdf (accessed July 14, 2015).

Cushing, S. 1994. *Fatal Words: Communication Clashes and Aircraft Crashes*. Chicago: University of Chicago Press.

Elliot, G. 1997. English in aviation safety: Testing and training solutions. In *Selected Proceedings of the 1997 Symposium "Aviation Communication,"* April 9–11, 1997, eds. P. Quigley and P. McElwain, 21–23. Prescott, Arizona: Embry-Riddle Aeronautical University.

EUROCONTROL. 2006. *Air-Ground Communication Safety Study Causes and Recommendations*, Edition 1.1. http://www.skybrary.aero/bookshelf/books/162.pdf (accessed July 14, 2015).

Eurofighter Typhoon. 2014. Technology: Cockpit. http://typhoon.starstreak.net/Eurofighter/cockpit.html (accessed: July 14, 2015).

Forcada, M. L. 2010. Machine translation today. In: *Handbook of Translation Studies*, Vol. 1, eds. Y. Gambier and L. Van Doorslaer, pp. 215–223. Amsterdam: John Benjamins Publishing.

Geacăr, C.-M. 2010. Reducing pilot/ATC communication errors using voice recognition. In *Proceedings of the 27th International Congress of the Aeronautical Sciences, ICAS 2010*. www.icas.org/ICAS_ARCHIVE/ICAS2010/PAPERS/441.pdf (accessed July 14, 2015).

Goguen, J. A., C. Linde, and M. Murphy. 1985. *Crew Communication as a Factor in Aviation Accidents*. NASA Technical Memorandum 88254. Moffett Field, California: NASA Ames Research Center.

Gotti, F., P. Langlais, and G. Lapalme. 2013. Designing a machine translation system for Canadian weather warnings: A case study. *Natural Language Engineering* 1(1): 1–36.

Green, R. G., H. Muir, M. James, D. Gradwell, and R. L. Green. 1996. *Human Factors for Pilots*, 2nd edition. Aldershot: Avebury Aviation.

IATA, IFALPA, and IFATCA. 2011. *Pilots and Air Traffic Controllers Phraseology Study*. Montreal: International Air Transport Association.

Jennings, A. E. and W. D. Chiles. 1977. An investigation of time-sharing ability as a factor in complex performance. *Human Factors* 19: 535–547.

Jurafsky, D. and J. M. Martin. 2009. *Speech and Language Processing: An Introduction to Natural Language Processing, Computational Linguistics, and Speech Recognition*, 2nd international edition. Upper Saddle River, New Jersey: Pearson Prentice Hall.

Kanki, B. G. and H. C. Foushee. 1989. Communication as group process mediator of aircrew performance. *Aviation, Space and Environmental Medicine* 60(5): 402–410.

Lennertz, T., J. Bürki-Cohen, A. L. Sparko, et al. 2012. NextGen flight deck data comm: Auxiliary synthetic speech—Phase I. In *Proceedings of the Human Factors and Ergonomics Society* 56: 31–35.

Maas, A. L., Q. V. Le, T. M. O'Neil, et al. 2012. Recurrent neural networks for noise reduction in robust ASR. http://www1.icsi.berkeley.edu/~vinyals/Files/rnn_denoise_2012.pdf (accessed July 14, 2015).

Monan, W. P. 1983. Addressee errors in ATC communication: The call sign problem. *NASA Contractor Report 166462*. Moffett Field, California: NASA Ames Research Center.

Morrison, R. and R. H. Wright. 1989. ATC control and communication problems: An overview of recent ASRS data. In *Proceedings of the Fifth International Symposium of Aviation Psychology*, Vol. 2, ed. R. S. Jensen, 902–907. Columbus, Ohio: Ohio State University.

Prinzo, O. V. and T. W. Britton. 1993. *ATC/Pilot Voice Communications—A Survey of the Literature*. Oklahoma City: FAA Civil Aeromedical Institute.

Raby, M. and C. D. Wickens. 1994. Strategic workload management and decision biases in aviation. *The International Journal of Aviation Psychology* 4: 211–240.

Schneider, V. I., A. F. Healy, I. Barshi, and J. A. Kole. 2011. Following navigation instructions presented verbally or spatially: Effects on training, retention, and transfer. *Applied Cognitive Psychology* 25: 53–56.

Schutte, J. 2007. Researchers fine-tune F-35 pilot-aircraft speech system. Human Effectiveness Directorate. http://www.afmc.af.mil/news/story.asp?id=123071564 (accessed July 14, 2015).

The Air Pilot's Manual. 2013. *Human Factors & Pilot Performance*, Vol. 6. Shoreham Airport: Pooleys-Air Pilot Publishing.

Vivaldi, J. and H. Rodríguez. 2007. Evaluation of terms and term extraction systems: A practical approach. *Terminology* 13(2): 225–248.

Wickens, C. D. 2008. Multiple resources and mental workload. *Human Factors* 50(3): 449–455.

Work Group of Human Factors Relating to Controller and Pilot Errors. 1992. Final Report.

23

Experimental Study for the Empirical Risk Analysis of Sociotechnical Systems in ATM

Lothar Meyer, Katja Gaunitz, and Hartmut Fricke

CONTENTS

23.1 Introduction

The current methods for estimating the risk of sociotechnical systems in air traffic management (ATM) mostly rely on accident and incident reports, expert judgment, or model-based approaches. The predictive risk estimation of novel systems, in particular, is traditionally performed by the subjective adaptation of the expert's operational experiences to the expected operation after the hypothetical start-up of the target system. In this regard, the term risk complies with the definition: *"Risk is defined as the probability that an accident occurs during a stated period of time"* (Blom et al. 2003).

The most promising model-based approaches offer the advantage of coping with enormous sample spaces, by providing objective data and the statistical power to prove even very little probabilities of the accident event, for example, the target level of safety in ATM with a maximum of 1.55E-8 accidents per operating hour (Blom et al. 2001). An exhausting validation of all modeled *a priori* assumptions regarding the safety effects on a new design in realistic operating conditions is extremely challenging as there are usually no means of obtaining and transferring a direct evidence from the current systems and operations: *"errors are likely to be made when designers apply error modeling techniques"* (Johnson 1999). This might impair the external validity of the model for unknown or unexpected cases.

For the problem described above, human-in-the-loop simulations (HITLSs) offer an empirical approach that is often used for estimating the performance of sociotechnical systems in a predictive way, for example, by means of workload measures. HITLS has also been successfully used to accompany failure modes and effects analysis (FMEA) studies, namely to quantify isolated probabilities in the interaction between the operator and the working environment as well as human error probabilities (HEPs) that can be used for the quantification of model parameters (Stroeve et al. 2013). In contrast, a pure HITLS approach is rarely used only for risk analysis. This is due to the enormous efforts needed to obtain valid data as well as to the limited sample spaces that can be achieved in real-time simulation (Shorrock et al. 2001). Studies that involve operational experts perform a few hundred hours of simulation time at best (Stroeve et al. 2013), providing insufficient statistical power for a reliable elimination of rare and risk-inducing events. This is expressed by the ATM-safety iceberg in Blom et al. (2001). When applying statistical testing, the type-I error rate would be unacceptably large, in the case that an unsafe system is assumed as a null hypothesis. This error can be explained by the *Weak Law of Large Numbers*, also known as *convergence in probability* or, more specifically, Bernoulli's theorem. It describes a decreasing difference between an observable frequency and the true probability with increasing sample spaces. The difference is a quantifiable metric of the type-I error rate and can be estimated by *Chebyshev's inequality*. When assuming one operating hour as a basic unit of the population that has the end-state *accident* or *no accident*, with an underlying *binomial distribution*, the error could be estimated as

$$P(|X - \mu| \geq k) \leq \frac{n \cdot p(1-p)}{k^2}, \tag{23.1}$$

with the random variable X, the mean μ, the variance of the distribution $\sigma^2 = n \cdot p \cdot (1 - p)$, and the confidence tolerance level k. Even with a sample space of 1.0E9 hours and a target safety of one accident per 1.0E9 operational hours, there is still a 13.6% probability to declare an unsafe system as safe when no accident has been detected in the experimental time. For instance, 3.0E9 operational hours are needed for gaining 95% confidence. Thus, the empirical approaches to cope with such rare events suffer from practicability to prove the novel system by means of HITLS.

Our proof-of-concept study is based on an approach by which insufficient sample spaces are compensated by intensifying the probability to detect safety indicators and by which, therefore, the power is increased with samples that are held constant. Hence, it addresses a problem definition of Swain: *"the problem remains of how raw data from training simulators can be modified to reflect real-world performance"* (Swain 1990).

As a possible solution to this problem, we developed a concept called *accelerated risk analysis (AccSis)*, which describes a methodology to gain the desired acceleration effect needed for intensifying the probability of safety-relevant occurrences and the related safety metrics. This acceleration effect shall practically be reached by the induction of a calibrated time pressure that stimulates the occurrence of human error. Concerning the time-pressure induction, we developed a procedure following the *time-budget (TB)* principles (Bubb and Jastrzebska-Fraczek 1999, Bubb 2005) named *competitive performance (ComPerf)*. It puts the test person under the impression of not having sufficient time to solve the problem (Chang and Mosleh 2007). This approach is motivated by the *accelerated-life-testing* (ALT) methodology, which forward the *mean time to failure* into the experimental period by means of an accelerated and calibrated stress induction during the experiment (Nelson 2009). Hence, AccSis explicitly addresses the occurrence of *right censoring* (Cox 1972).

Referring to this reasoning, this chapter presents the *time-pressure–risk model* and the related conceptual framework, named *AccSis* in Section 23.2. The primary subject of investigation is the problem of how to adapt the stochastic methods from ALT to the risk analysis of sociotechnical systems in ATM, considering the stochastic human behavior instead of stochastic processes of product aging. A HITLS experimental design is presented for the evaluation of AccSis and *ComPerf* following an innovative advanced surface movement guidance and control system (A-SMGCS) for air traffic control in the scope of a proof-of-concept study in Chapter 3. Three test persons were trained and qualified for conducting the experiments, whose individual results served for evaluating the hypothesis of the *time-pressure–risk model*. This chapter discusses the findings identified in the results of the HITLS and delivers insights in the effects of the stress-induction procedure indicated by means of the detected runway incursion (RI) frequencies and reaction times, all of which are outlined in Section 23.4.

23.2 Methodology

23.2.1 The Concept of AccSis

This conceptual framework has the objective of estimating the compliance of sociotechnical systems with a given target probability of an accepted safety metric (e.g., the accident), expressed as the alternative hypothesis $p < p_{target}$, by means of HITLS-based empirical data. Facing the problem of mitigating the statistical type-I error starts with analyzing *Chebyshev's inequality*. The mitigation can proceed as follows:

1. By increasing the number of generated samples n.
2. By modifying the simulated working conditions in the experimental design that rescales the probability by an acceleration factor a. A symmetric and linear rescaling of the target safety p_{target} and the true probability of the system p by the acceleration factor leads to $\hat{p}_{target} = a \cdot p_{target}$ and $\hat{p} = a \cdot p$ in which the alternative hypothesis is maintained. Applying the rescaling to *Chebyshev's inequality*, an effective mitigation of the type-I error can be determined as follows:

$$\frac{n \cdot \hat{p}(1-\hat{p})}{\hat{k}^2} = \frac{\hat{p}(1-\hat{p})}{n \cdot \hat{p}_{target}^2} = \frac{1}{a} \cdot \frac{p(1-p \cdot a)}{n \cdot p_{target}^2} \approx \frac{p}{a \cdot n \cdot p_{target}^2},$$

with $\hat{k} = n \cdot \hat{p}_{target}$. When defining $p \ll 1$, one can approximately assume $p(1 - p \cdot a) \approx p$. The mitigation effect of the error can be quantified to a^{-1} and affects a virtual accumulation of the samples generated, described as $a \cdot n$.

The second approach thus constitutes an approach to face the safety-iceberg problem by describing a procedure that accelerates the convergence of the type-I error by modifying the boundary conditions of the HITLS. This affects a calibrated rescaling of the target and the system probability for safety-relevant events.

In reliability testing, the acceleration effect is practically achieved by a stress induction of, for example, thermic or mechanic stress that forwards the targeted failure event into the experimental time. In this way, the problem of *right censoring* is addressed, which describes the problem of measuring the time of an event that lies beyond the experimental time (Nelson 2009). The approach of ALT can be split into two tasks:

1. Failure stimulation—The experiment is to be conducted under varying gradations of stress, which deflect the load from design stress to accelerated stress. Three gradations of load are practically recommended for capturing sufficient samples of failure events of the product in ALT.

2. Regression analysis—The failure distributions of each load level are fitted to analytic or nonparametric distribution models. A regression model (life-stress model) is to be applied that extrapolates the trend of the distribution shape to design stress (see Figure 23.1).

The idea to adapt this concept to accelerate the occurrence of safety-relevant events in HITLS is severely impaired by the fact that human performance is a complex field that suffers from nonlinearity and nonreplicability compared to the functionality of technical products. For this reason, we identified the systematic differences between the analysis of product failure events and the commitment of errors by operators, when acting in a sociotechnical system.

- The most-significant difference is the stochastic that contrasts accident events of sociotechnical systems and technical failure events. The product lifetime is temporally limited as a result of the progress of aging that is attributed by a dependent stochastic distribution (lifetime distribution). In contrast, we assume the accident event in aviation to be the result of a failed operator's decision, which, hence, is regarded as an independent event with a limited temporal relation to the preceding operational actions and in which a distribution cannot be modeled over time when assuming a Bernoulli distribution for accidents.

- The second difference, which is the fact that the stress-inducing procedures of ALT are completely incompatible with sociotechnical systems, is related to the first one.

- The third difference is the missing accident-stress model for human behavior for the regression analysis, since state-of-the-art models, although describing the relationship between human error and stress, fail to deliver a domain-specific model curve.

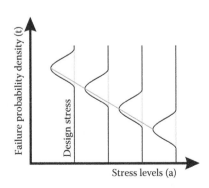

FIGURE 23.1
Stress-life relation according to the ALT concept.

This chapter considers AccSis to be the subject of a long-term validation strategy due to the reasons given above. Therefore, our current research follows a stepwise validation strategy to overcome the mentioned differences, in which finally a full compliance of AccSis with the requirements of the risk analysis of sociotechnical systems shall be achieved. On the basis of this consideration, we chose the first step to be a proof-of-concept study: the controlled acceleration of safety-relevant events by intensifying human error. The first objective is, hence, to gain an understanding of the principles of applied-stress induction while being constrained by existing procedures of the working environment in ATM.

To explain our choice of human errors as the key factor, we refer to the integrated risk picture (IRP), which describes the contribution of human errors to accidents in the combination of causal factors by means of a fault-tree model (Spouge and Perrin 2006). For a sociotechnical system, the IRP can be regarded as a significant fingermark of risk, in which branches of failure catenation form the resulting accident probability. One has to bear in mind that only branches affected by the acceleration effect are taken into account for this study.

When considering causal factors in the context of AccSis, organizational, technical, and human errors can be distinguished as the principal accident causes. This complies with Reason's "a trajectory of accident opportunity" that models the human error propagation in the presence of corresponding hazards as unsafe acts (Reason 1990). Human error has been identified as the most-frequent contribution to accidents and incidents in aviation with a share of 60%–80% (Shapell and Wiegmann 1996) or 75%, respectively (Müller 2004). The focus on human error thus addresses a causal key factor of sociotechnical systems: the major contribution of human error to risk. A vast amount of causal branches must hence be covered by acceleration. Following the ceteris-paribus principles, procedures, tasks, and other boundary conditions are to be held constant during HITLS that implies a major requirement on seemingly unimportant contextual conditions of the simulation.

23.2.2 The Role of Time Pressure for the Stimulation of Human Error

Besides uncertainty, time pressure seems to be of particular relevance when considering human decision-making processes (Rastegary and Landy 1993). Rastegary defines time pressure "*... as the difference between the amount of available time and the amount of time required to resolve a decision task.*"

According empirical findings, time pressure is known to significantly affect human performance (Freedman et al. 1988). This relation points to the vital impact of time pressure on human performance, that is, on acting correctly according to the procedures. This influence can be explained by the fact that the performance of cognitive information processing is a function of time pressure that affects a minimization of cognitive effort in a cost/benefit frame of reference. It is reported that an increased selectivity of information is observable. Under time pressure, more pieces of information are used but in a shallower way (Edland and Svenson 1993).

Time pressure contributes more to HEPs than additional tasks when performing time-critical tasks (Bubb and Jastrzebska 1999). Therefore, a TB was defined, which puts the time-available t_a into relation to a time needed for decision t_n, as follows:

$$TB = \frac{t_n}{t_a}. \tag{23.2}$$

An increased error probability was measured by a factor of 14 under the condition of time pressure. This observation corresponds to the assumptions of the Human Reliability Assessment THERP, which considers a factor of 10 under stress conditions (Swain and Guttmann 1983).

Time pressure and human error are causally linked and can be transferred to a continuous quality metric for human actions that is ultimately classifiable as acceptable or not acceptable. Specifically, the deflection of actions below a minimum quality can be regarded as not acceptable or, in line with conventional theories, human error. Continuing, quality is linked to performance as follows:

$$P = \frac{Q}{t},\tag{23.3}$$

with the human performance P, the quality of human action Q, and the time given t (Bubb and Jastrzebska-Fraczek 1999). Thus, time pressure affects Q, divided by time. We identified the definition of TB as an inherent advantage for the stimulation of human error for two reasons:

1. It induces a calibrated time pressure by setting t_a
2. Human performance is sensitive to time pressure

To summarize, the concept of accelerating the occurrence of accidents unifies many theories about accident causation and human error to a comprising causal catenation, as shown in Figure 23.2, with each of the links being already empirically validated by the elementary findings (Freedman et al. 1988, Reason 1990, Bubb and Jastrzebska-Fraczek 1999).

The introduced concept for utilizing the TB *principle* to stimulate time pressure and hence human errors to thus intensify the probability of accidents is a summative generic

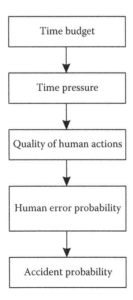

FIGURE 23.2
Causal relationship between TB and the accident probability.

description of the effect mechanism. It is necessarily a domain-specific challenge to develop a procedure that produces a *calibrated* time pressure by means of this principle.

23.2.3 Competitive Performance

Most ideas for the implementation of an induced time pressure aim at setting boundary conditions to effectively shorten the available time. Secondary tasks might, for example, shorten t_a by forcing the operator to organize task sharing and prioritization according to the time constraint. This sharing will as well change the pattern of activities and impact the IRP picture without any control of the deflection from the design stress. The same holds true for the conventional means of HITLS calibration, namely the intensification of the task load, for example, the traffic volume.

As time pressure is transformed from an objective condition to a subjective feeling, we decided to choose the approach of a "competitive arousal." Following this approach, time pressure is generated by providing a competitive environment that triggers the desire of the operator to win (Kersholt 1994, Malhotra 2010). Our concept establishes a "competitive arousal" by forcing the operator to compete with a "calibrated reference operator" that operates under the same contextual conditions (cloned worlds) and is capable of acting according to a calibrated performance (see Figure 23.3), named ComPerf. When the human operator acts, his or her performance metric, for example, the throughput of the system, is measured and fed back for instant comparison. The headstart is the quantified indicator for the performance of the human operator compared to the reference operator. The reference operator in this instance is a model-based software agent that supports the gradations of performance.

If the lead of the human operator shrinks below a given threshold, a hard penalty applies to challenge the test person to compete as hard as possible. In this instance, the effort needed to finish the scenario successfully was increased by generating additional tasks or enlarging time constraints, such as the scenario's finish time. As expressed before, the implementation must carefully compensate for the changed boundary conditions stemming from applied penalties, to achieve constant and comparable contextual conditions. The advantage of controlling the available time t_a by varying the performance of the

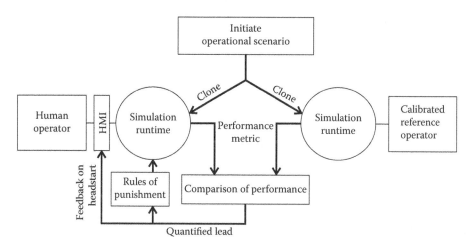

FIGURE 23.3
The concept of ComPerf.

reference operator and, therefore, by establishing the TB principles in relation to the decision times of the human operator t_n, is the inherent feedback loop that is highly suited for automatic tuning of the perceived time pressure.

23.3 Empirical Study

The introduction to the conceptual methodology of risk analysis by means of *AccSis* and the approach regarding time-pressure induction with the help of *ComPerf* were both deduced to an experimental design, in which the plausibility of the risk model, as shown in Figure 23.2, should be the subject of investigation. The controller working place (CWP) of the air traffic controller (ATCo) has been chosen as an exemplary safety-critical working environment within air traffic control. The related task is to control traffic at the airport in the function of a tower controller according to the procedures defined by ICAO PANS-ATM Doc. 4444. The principal tasks of the ATCo are defined in section 7.1.1.1 as follows: *"Aerodrome control towers shall issue information and clearances to aircraft under their control to achieve a safe, orderly and expeditious flow of air traffic on and in the vicinity of an aerodrome with the object of preventing collision(s). ..."*

The hypotheses were formulated as follows:

- The time needed t_n is sensitive to the target load set by *ComPerf*
- The relative frequency of safety-relevant events is sensitive to the target load set by *ComPerf*

These hypotheses set the focus on two major causal relationships of the risk model (see Figure 23.2).

We decided to choose the RI as the target safety-relevant event instead of an accident event. In the present context of aerodrome traffic control, the RI is a precursor of an accident event and is as such selected as a risk-indicating event, defined by ICAO Doc. 4444 as the following: *"Any occurrence at an aerodrome involving the incorrect presence of an aircraft vehicle or person on the protected area of a surface designated for the landing and take-off of aircraft."*

The notion of RIs as precursors of accident events is backed by safety management principles and the statistical understanding that the occurrence of collision accidents relates to RIs in a ratio of 1:100, which would, by the way, imply a runway collision accident rate of one every 3.7 years (Birenheide 2010b).

23.3.1 Experimental Tasks and Simulation Scenarios

The chosen HITLS consists of test persons that operate a Surface Manager HMI as the primary working device (Figure 23.4). The device complies with the Eurocontrol A-SMGCS Implementation level 3 (Birenheide 2010a), with the functional exception of a missing device that prevents RI (runway incursion prevention and alerting systems, RIPASs) automatically. The tasks to be performed by the test persons are defined by ICAO Annex 11 and ICAO PANS-ATM Doc. 4444 for tower and ground-control services. The Surface Manager HMI allows for the selection of a target aircraft by pen strokes, as well as granting pushback, taxi, lineup, or take-off clearances on an airport surface surveillance radar screen presenting the entire traffic situation at Frankfurt airport (ICAO code: EDDF).

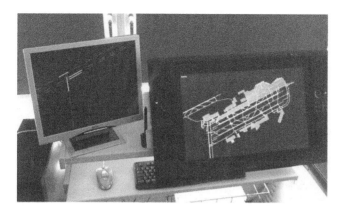

FIGURE 23.4
The Surface Movement Manager HMI consists of a ground surveillance of the airport and a secondary surveillance radar of the vicinity of the airport.

The generated traffic consists of inbound and outbound a/c traffic movements at Frankfurt airport on the three active runways (RWYs) in direction 25, operating 25L as a landing-only RWY, 18 as take-off-only RWY, and 25R in a mixed mode. This complies with the former operational concept before RWY north started operating. RWY dependencies are given for RWY 18 and RWY 25R, as well as for RWY 18 and RWY 25L. The dependency between 25R and 25L was considered according to the reduced RWY separation and semi-mixed parallel RWY operations. The random traffic generator initially distributes 160 movements over 240 simulated minutes per execution run according to a given set of stochastic parameters with uniformly distributed destination routes or departure gates (including north and south area stands) and RWYs. We accelerated the simulation speed by a factor of 2. The routes of the ground movements are initialized by the *Floyd and Warshall algorithm*, which optimizes routes according to a given operational concept and ensures a similar task load for all experimental executions. The software aircraft/pilot agents are capable of self-separating on taxiways and to solve taxi obstruction and crossing conflicts autonomously according to the rules laid out in ICAO Annex 2—Rules of the Air. The execution scenario demands that the test persons work on both ground and tower positions parallelly, that they control the whole airport at a severely increased task load (160 movements at doubled real time).

The concept of *ComPerf* was adapted to the experiment by the application of a simple controller agent, who is capable of acting as an ATCo. The evaluation of the agent's decisions by a traffic-movements predictor affects the resulting operation to be conflict-free to a verifiable degree. No prioritization is implemented, since the agent handles all the movements simultaneously and independently. The agent is configurable by a reaction time t_r per clearance, which calibrates the performance concerning the number of aircraft handled per time. By setting t_r, the decision making of the controller agent allows the human operator a controllable advantage in the context of the performance comparison of *ComPerf*. The human operators' time necessary for decision-making t_n is hence set into competition with t_r, by which the TB principles are established when defining $t_r = t_a$. Setting a desired rapidness, t_r of decision making can consequently be assumed as a target load for the human operator.

The absolute number of traffic movements, which depart from the simulated airport or reach their designated stand, has been chosen to be the key performance metric for

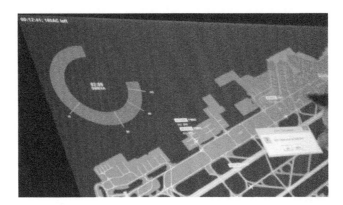

FIGURE 23.5
The clock on the ground surveillance display feedbacks the lead to the human operator.

ComPerf. Leaving the system is defined by the moment of (1) granting the clearance for takeoff for outbound movements or (2) granting the last taxi clearance before entering the aircraft stand. The comparison calculates the performance lead of the human operator by comparing these metrics to the autonomous software competitor. Presuming that the test person would not take any action, a time can be calculated for which the lead becomes zero. This can be regarded as a quantified headstart, calculated on the basis of a fast-time simulation of the controller-agent's world that establishes the complete timeline of the agent, including timestamps of all the operational events, in very little time.

The countdown was visually and acoustically fed back to the human operator by the visualization of a clock on the ground surveillance display (Figure 23.5) and by an alarm noise. The noise indicated the lead time, graded from 300 to 180, 30, and 10 simulated seconds, accompanied by an increasing playback volume. A lead of zero was accompanied by an unpleasant alarm noise, indicating the time-error (TE) condition that results in penalty. The visualization of the headstart consisted of a circle-like clock that covered 6 min as a full circle with a logarithmic time axis.

The penalty was implemented as an increase of the aircraft queue by two additional movements. This consequently increased the duration of the experiment indirectly by the time necessary for handling and finalizing the movements. As the simulated world of the agent is synchronized with the test person's world, the duration of the experiment effectively lies in the test person's hand. This mechanism is regarded as a sufficient measure of motivation for winning the competition, since we presume that all test persons are not only motivated to successfully compete with the controller agent but, moreover, to finish the simulation in time (and be done).

23.3.2 Test Persons and Training

For the empirical study, we acquired three students of the study program "Transport Engineering" in the 4th year of their diploma to act as novice test persons. We educated them according to the tasks described above and trained them by means of the test setup. Every test person successfully completed a training consisting of 10 h and final tests that indicated whether the rules of RWY separation could be mastered according to the trained procedures.

TABLE 23.1

Target Load Parameters

ComPerf A/B	t_r (s)
A	30
B	20

23.3.3 Measurements

The measurements consisted of three metrics, namely, the necessary time, the frequency of RI, and the frequency of TEs, which fulfilled our requirements to capture reactions to the gradations of load according to our hypotheses.

Firstly, we recorded RI events as the principal safety metric during the experiment. RIs were automatically detected as soon as rules of the reduced RWY separation minima and parallel RWY operations described in ICAO PANS-ATM Doc. 4444 were violated. Secondly, the necessary time t_n is regarded as an indicator of the cognitive decision time and is the measured time period from the request of clearance by the aircraft until the clearance is granted by the human operator. Third, the frequency of TE was recorded, quantifying the number of penalties applied when the lead was zero and the TB, therefore, equaled >1.

23.3.4 Calibration of Target Load

The calibration procedures were performed prior to the experiment and consisted of a trajectory that varies t_r over time through a predefined bandwidth between 0 and 150 s. The calibration procedure is explained in more detail in Meyer et al. (2014). Two target load levels were quantified as parameters for the controller agent, which defined two experimental configurations (Table 23.1).

23.3.5 Executive Planning

The experiments were conducted according to a sequence plan that varied the configurations and its target load in a systematic order. It is assumed that the quality of the novice person's decisions is continuously increasing due to the experience gained during the series of experiments. To prevent the measurements from being affected by the training effect, the sequence plan follows an alternating order of the configurations.

23.4 Results

23.4.1 The Necessary Time

According to the stated hypothesis for correlation, it was expected that the decreasing reaction times of the controller agent (increasing the target load) would effect an accelerated working speed of the human operator (hence, a decreased time needed t_n). With respect to this expectation, the three test persons showed unclear reactions in the time needed to grant clearances. This is indicated by the measurements ($n > 1000$), illustrated

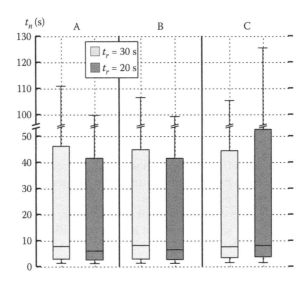

FIGURE 23.6
Reaction times of t_n over the varying target loads of t_r.

TABLE 23.2

U Test of Reaction Times t_n

Test Person	p-Value (%)
A	1.11
B	9.89
C	5.09

in Figure 23.6, which contrasts t_n as box plots according to the selected target loads for each test person. The measurements of test persons A and B indicate the tendencies of an accelerated working speed. In contrast, test person C shows a tendency to maintain his or her working speed.

For testing these observations objectively, the Mann–Whitney–U test provides a probability (p-value) for two independent nonparametric samples on its central tendency to belong to the same population.

The test results (Table 23.2) show no clear rejection of the null hypothesis for all test persons. Only the distribution of test person A exhibits a significant increase in reaction time, indicated by a value of $p < 5\%$. Test person B shows the same tendency. The reaction of test person C is contrary to A and B.

23.4.2 Runway Incursion

RIs were measured as an absolute frequency per target load and test person. The frequency was divided by the number of take-off clearances granted by the human operator. This should compensate for those varying periods of the execution scenarios due to the extensions by the applied penalties. A common tendency can be found through all the measurements (Figure 23.7). This confirms a sensitivity of the target load to the resulting

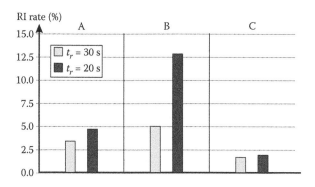

FIGURE 23.7
RI rate.

frequency of safety-relevant events. Therein, test person B shows the largest increase and C the lowest, indicating a decrease in the quality of decision making.

The measured frequency of RI was subject to a learning curve, indicating an increase in competence and therefore also in quality over the course of the experiment.

23.4.3 TE and TB

The absolute frequency of TE indicates the compliance of the working speed with the given target load. For this reason, TE is a metric that shows the ability of the human operator to respond to the induced time pressure. The test persons show a two-track reaction on the increasing load (Figure 23.8). Test persons A and C showed less reactions to the increased target load than test person B, while test person C shows a smaller overall susceptibility to the induction procedure. From the view point of the human operator, permitting a higher frequency of TEs might be an attractive means to effectively extend the available time t_a while accepting that the experimental period is extended by the penalty. Thus, a correlation between the frequency of TE and the mean *TB* (Table 23.3) can be expected.

Dividing the samples of t_n by the reaction time t_r delivers the TB samples whose mean values are summarized in Table 23.3.

The Spearman-correlation rank coefficient was 60% (*p*-value: 20.8%). Even when no significance can be proven, it provides an indication of a strong relation between *TE* and *TB*.

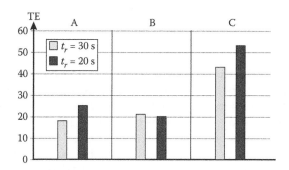

FIGURE 23.8
Frequency of TEs.

TABLE 23.3

Mean TBs and TEs

	A		B		C	
$t_r(s)$	30	20	30	20	30	20
Mean TB	1.30	1.82	1.29	1.76	1.37	2.20
Sum TE	18	25	21	20	43	53

23.5 Conclusion

In summary, the results clearly show the reactions of the test persons to increased stress, as well as a lowered quality of work for all test persons, at the same time. However, the data gathered from this small test group are not sufficient to validate the risk model (Figure 23.2), since the number of samples does not provide the required power for statistical testing. Nevertheless, the findings serve satisfactorily for giving clues on the success of the induction procedure and the plausibility of the hypothesis in the scope of the proof-of-concept study.

The tendency observed indicates an increase of the probability of safety-relevant events and human error when increasing the target load (Figure 23.7). Thus, an increased uncertainty during decision making can be concluded from the data gathered from the three test persons. The high variance of the amplitude of the *RI rate* may be explained by considering the frequency of *TE* as a crucial influencing factor of uncertainty. This consideration manifests in all measurable actions of the human operator, forcing him or her to balance between the quality and working speed in a subjective speed–accuracy trade-off (SAT) that is in line with the principles described by *Fitts' law* (Fitts 1954). This trade-off is illustrated as an example in Figure 23.8, which shows the individual operating points on the RI-rate–TE chart and the balance of the available performance between the two claims of the task definition defined in section 7.1.1.1, ICAO PANS-ATM Doc. 4444.

A plausible analytical description of the trade-off can be derived from the relation between quality, performance, and time needed, introduced in Section 23.2.2. Assuming the *RI rate* as a reciprocal metric of quality and the frequency of *TE* as a valid measure of the TB (cf. correlation test), the relationship can be expressed as

$$\frac{t_n}{Q} = \frac{1}{P} \Leftrightarrow RI_{rate} \cdot TB \cdot t_r = \frac{1}{P} \Leftrightarrow \boxed{RI_{rate} \cdot TE = \frac{1}{P \cdot t_r}}. \tag{23.4}$$

This term describes the product of t_r and P as the reciprocal function of a surface, with the measured factors forming the dimensions of the related rectangle. With a view on Figure 23.8, the shift of the operating points of the test persons A and C approximately follows this relationship, leading to a measure of performance that takes the target load t_r into account, while maintaining constant proportions of the trade-off. Figure 23.9 illustrates this relation.

Test person B obviously switched the strategic and tactical priorities, which led to a shift in the trade-off's proportions in favor of the *RI rate*. The cause of this rebalancing might be explained by using Hollnagel's *Contextual Control Model* (Hollnagel 1993). In this model, test person B's behavior could refer to a *scrambled control* mode, indicating an overload situation or insufficient training of the actions that do not follow a stable action pattern.

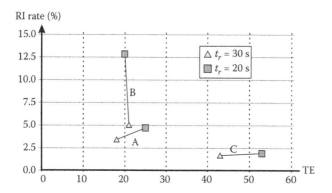

FIGURE 23.9
RI-rate–TE chart of stress reactions and the trade-off.

In contrast, the test persons A and C show stress reactions that indicate the systematic relations between the *RI rate* and *TE*. This relation might explain the effective acceleration of safety-relevant events that occur under the present time constraints.

The lowest degree of sensitivity most clearly show the reactions of test person C. Being insensitive to time-pressure arousal is an essential mark of quality air traffic control. Permitting *TE* fully complies with the training, by which the test person is qualified to act correspondingly to their role model. An alternative conclusion can be deduced from the reactions of test persons A and C, whose proportions of the trade-off remained stable, while time pressure increased. This can be best observed in the time needed by test person C (Figure 23.7), who showed the robustness against stress induction by maintaining the working speed in the best way. This observation complies with the findings of Rastegary (Rastegary and Landy 1993), by which test person C can be classified as a time-urgent individual, whose sensitivity to time pressure provides for a constant performance over a larger interval.

As a conclusion, it can be said that the risk of a sociotechnical system under the conditions of design stress might best be represented by an individual operating point on the RI-rate–TE plane. The estimation of this point can then be achieved by using stress reactions of varying accelerated-stress conditions for the regression analysis. Since, currently, there is no valid regression model known that can analytically relate to time pressure, *TE*, and the safety metrics, the results of the stress reaction, illustrated in Figure 23.8, can give clear indications of the qualities needed for the model, which are

- Basis performance: any individual provides a specific basis performance for design-stress conditions, consisting of specific proportions of the *RI rate* and *TE*.
- Variance of the proportions: as the reactions of test person B show, the proportions of the trade-off depend on the present time pressure. The Contextual Control Model might provide qualitative indications of the curve progression of the proportions of the trade-off over an increased time pressure.

By way of modeling the relations analytically for use in a regression analysis, the findings on the field of *speed-versus-accuracy* studies might be helpful. Furthermore, enhancing the related SAT curves by using ATM-related safety metrics instead of human error events might lead to a successful adaptation of the models known to use in the current context.

References

Birenheide, M. 2010a. Definition of A-SMGCS Implementation Levels. Article. EUROCONTROL Headquarters. https://www.eurocontrol.int/articles/advanced-surface-movement-guidance-and-control-systems-smgcs (accessed August 7, 2015).

Birenheide, M. 2010b. Generic cost–benefit analysis of A-SMGCS levels 1 and 2. Article. EUROCONTROL Headquarters. https://www.eurocontrol.int/articles/advanced-surface-movement-guidance-and-control-systems-smgcs (accessed August 7, 2015).

Blom, H. A. P., G. J. Bakker, P. J. G. Blanker et al. 2001. Accident risk assessment for advanced air traffic management. *Progress in Astronautics and Aeronautics* 193: 463–80.

Blom, H. A. P., G. J. Bakker, M. H. C Everdij et al. 2003. Stochastic analysis background of accident risk assessment for air traffic management. Hybridge Report D2.2. National Aerospace Laboratory NLR. http://hybridge.nlr.nl (accessed August 7, 2015).

Bubb, H. 2005. Human reliability: A key to improved quality in manufacturing. *Human Factors and Ergonomics in Manufacturing and Service Industries* 15.4: 353–68.

Bubb, H., and I. Jastrzebska-Fraczek. 1999. Human error probability depending on time pressure and difficulty of sequential tasks. *Safety and Reliability—Proceedings of ESREL 99*, 1: 681–86.

Chang, Y. H. J., and A. Mosleh. 2007. Cognitive modeling and dynamic probabilistic simulation of operating crew response to complex system accidents. Part 2: IDAC performance influencing factors model. *Reliability Engineering and System Safety* 92.8: 1014–40.

Cox, D. R. 1972. Regression models and life-tables. *Journal of the Royal Statistical Society. Series B (Methodological)* 34.2: 187–220.

Edland, A., and O. Svenson. 1993. Judgment and decision making under time pressure—Studies and findings. In *Time Pressure and Stress in Human Judgment and Decision Making*, eds. O. Svenson, and A. J. Maule, 27–40, New York and London: Plenum Press.

Fitts, P. M. 1954. The information capacity of the human motor system in controlling the amplitude of movement. *Journal of Experimental Psychology* 47.6: 262–69.

Freedman, J. L., D. R. Edwards, and J. E. McGrath. 1988. Time pressure, task performance, and enjoyment. *The Social Psychology of Time: New Perspectives* 91: 113–33.

Hollnagel, E. 1993. *Human Reliability Analysis: Context and Control.* London: Academic Press.

Johnson, C. 1999. Why human error modeling has failed to help systems development. *Interacting with Computers* 11.5: 517–524.

Kerstholt, J. H. 1994. The effect of time pressure on decision-making behaviour in a dynamic task environment. *Acta Psychologica* 86.1: 89–104.

Malhotra, D. 2010. The desire to win: The effects of competitive arousal on motivation and behavior. *Organizational Behavior and Human Decision Processes* 111.2: 139–46.

Meyer, L., K. Gaunitz, and H. Fricke. 2014. Investigating time pressure for the empirical risk analysis of socio-technical systems in ATM. *Advances in Human Aspects of Transportation: Part I* 1: 33–45.

Müller, M. 2004. Risk and risk management in aviation. *Zeitschrift für ärztliche Fortbildung und Qualitätssicherung* 98.7: 559–65.

Nelson, B. N. 2009. *Accelerated Testing: Statistical Method, Test Plans, and Data Analysis.* New Jersey: John Wiley & Sons.

Rastegary, H., and F. J. Landy. 1993. The interactions among time urgency, uncertainty, and time pressure. In *Time Pressure and Stress in Human Judgment and Decision Making*, eds. O. Svenson, and A. J. Maule, 217–39, New York and London: Plenum Press.

Reason, J. 1990. *Human Error.* New York, USA: Cambridge University Press.

Shapell, S., and D. Wiegmann. 1996. U.S. naval aviation mishaps 1977–92: Differences between single- and dual-piloted aircraft. *Aviation, Space, and Environmental Medicine* 67(1): 65–9.

Shorrock, S. T., B. Kirwan, H. MacKendrick, and R. Kennedy. 2001. Assessing human error in air traffic management systems design: Methodological issues. *Le travail humain* 64.3: 269–89.

Spouge, J., and E. Perrin. 2006. 2005/2012 integrated risk picture for air traffic management in Europe. Eurocontrol EEC Note 2006/05, https://www.eurocontrol.int/eec/public/standard_page/DOC_Report_2006_009.html (accessed August 7, 2015).

Stroeve, S. H., H. A. P. Blom, and G. J. Bakker. 2013. Contrasting safety assessments of a runway incursion scenario: Event sequence analysis versus multi-agent dynamic risk modelling. *Reliability Engineering and System Safety* 109: 133–49.

Swain, A. D. 1990. Human reliability analysis: Need, status, trends and limitations. *Reliability Engineering and System Safety* 29.3: 301–13.

Swain, A. D., and H. E. Guttman. 1983. *Handbook of Human Reliability Analysis with Emphasis on Nuclear Power Plant Applications*. NUREG/CR-1278. Albuquerque: Sandia National Laboratories.

24

What about the Next Generation? Assessing Experts' Judgments of Human Abilities Required for Working in a Future ATC Environment

Dirk Schulze Kissing

CONTENTS

24.1 Introduction

The future roles of air-traffic controllers (ATCOs) are supposed to change significantly especially in the terminal maneuvering areas (TMAs) of large airports. Due to a system-wide information management, automated planning systems will be able to utilize complex optimization algorithms to keep flight plans synchronized with constraints from network plans and airport processes on a tactical level.

Although ATCOs are conceived to remain a central element within future air-traffic management (ATM) (cf., Single European Sky ATM Research Program [SESAR], 2012), it still has to be clarified how the difficulties for human operators to coordinate agents whose intentions are hard to comprehend (Sarter et al., 1997), because their behavior is based on decisions which result from complex optimization algorithms, can be overcome and automation biases (Cummings, 2004) prevented. The contribution of personnel selection is to proactively assess not only how such evolutions of the human role may affect the human operator, but also how abilities required to fit into the new job profile will change, so selection profiles for ATCOs will be modified accordingly in time.

This chapter focuses on analyzing the job requirements and their corresponding abilities in a potential future approach control setting. A simulation experiment is conducted with different levels of ATCOs' assistance up to the point where the humans have to supervise a self-executing automation (cf., Willems, 2002). The roles under scrutiny all refer to the tactical level. The ATCOs are focused on traffic situations in a terminal maneuvering area (TMA) as displayed at the controller working position (CWP). With monitoring skills getting an even more central requirement in future air traffic control (ATC) tasks due to automation (Broach, 2013), and empirical indication that this is accompanied by changes in monitoring behavior (cf., Voller and Low, 2004), a focus is set on the assessment of the ATCOs' eye-gaze behavior to explore the way information is scanned contingent with role changes.

24.2 Theory

24.2.1 Requirements of Controlling Air Traffic in TMA

ATCOs in a high density TMA often have to handle 10 and more aircraft (a/c) at a time (Freed and Johnston, 1995). Certain cognitive mechanisms supposedly accomplish the coordination and supervision of many a/c, that is, the control of air traffic.

The key for enabling an ATCO (air-traffic controller) to handle so many is that the task for any a/c usually is entire routine (Freed and Johnston, 1995). The inbound traffic into the TMA is guided over only a few sector inbound fixes, and is normally kept on a standard arrival route (STAR) before handed over to the referring tower at a final approach fix. This allows the ATCO to rely on well-trained routines for handling the individual a/c and consolidating these into an overall sector-control plan. In a well-trained routine the ATCO identifies an inbound a/c on the radar screen when it calls over radio telephony (r/t), naming the call sign and the current level. Acknowledging radar contact via r/t by then the ATCO formally accepts responsibility for the a/c. The ATCO then checks for the correct destination on the flight progress strip, then selects a path, routinely the STAR, and instructs the pilot to follow it or alternatively, vectors to the nearest fix on the flight plan are given. The pilot acknowledges all clearances. The ATCO observes when the a/c is approaching a cleared fix and then clears the plane to an altitude that is required for the descent to the airport, and then vectors it to the final approach. The pilot again acknowledges all clearances. When the ATCO observes the a/c approaching the final approach fix, she/he provides the pilot with a clearance for an ILS (instrument landing system) approach, and hands-off responsibility for the a/c to the tower ATCO by instructing the pilot to change to tower frequency, and contacting the tower when passing the final approach fix, which the pilot acknowledges. With the acknowledgment of the a/c the ATCO intends to act according to the plan routine, namely to provide adequate clearances in time. With its moving along to the sector center, that is, the area of the final approach the a/c receives higher priority.

In order to have a high level of situation awareness, ATCOs have to continually sample the airspace on the radar display to ensure that no separation conflicts occur which may require them to take actions, check for flight progress strips and other information displays, as well as information obtained through radio communications and phone lines. The ATCO therefore regularly scans the sector from inner to outer areas to frequently update the information on location, speed, heading, and altitude of these a/c, and to continuously combine this information with other context information, like weather, into a coherent mental representation of the current situation (Sheridan and Parasuraman, 2005; Loft et al., 2007). The ATCO also regularly scans relatively consistent spatial locations (cf., Wickens et al., 2001, 2003) for critical events. An example for such a spatial hot spot would be TMA fixes where down-winding traffic streams on different standard arrival routes (STARs) are merging. This task is also called monitoring and comprises of deciding on a moment-to-moment basis which information is most important to attend for the updating of the mental model (Redding, 1992; Seamster et al., 1993).

Based on this mental model of the situation the ATCO builds up a sector-control plan and decides about clearances given to pilots on speed, heading, and altitude for the different a/c to maintain separation, and thus to guide the a/c safely through their sector (Loft et al., 2007). ATCOs also have to delay intentions and regularly check how far the plan has proceeded, if conditions for the action implementation are met, like providing a clearance for an altitude necessary to descend to the airport only after a following a/c with a higher speed on the level below has passed, or turning a northbound a/c to final only after a preceding south-inbound a/c is established on the final approach.

As the mental model is supposed to provide the structure for an efficient retrieval of sector-control plan routines (Seamster et al., 1993), it can be assumed for the level of cognitive processes that once a routine is retrieved from long-term memory it gets activated as an intention and is thus transferred into working memory (cf., Cowan, 1999) to await the appropriate set of conditions so that it can become selected to control action (Norman and Shallice, 1986). As long as the conditions are not met the routine remains active as an intention until its execution is completed (cf., Miller et al., 1960). The active routines, or intentions, respectively, have special ways of being remembered that are necessary for coordinating the parts of the sector-control plan under execution (cf., Miller et al., 1960). Triggers allow suitable activated schemas to be initiated at the precise time required (cf., Norman and Shallice, 1986). Conscious prioritization provides additional activation to the intentions according to their subjectively experienced importance, so there is a higher probability they receive attention first when it is distributed. Mechanisms of time-based prospective remembering (cf., Harris and Wilkins, 1982; Ellis, 1996; Block and Zakay, 2006) supposedly also come into play to control attention to frequent and timely flow back to the different intended tasks. And finally, an overall metacognitive management of the sector-control plan, with assessments on how far the plan has progressed, which action is of utmost priority, or if contextual changes make plan revisions necessary, shall be executed by deliberate, conscious control processes (for further reading, see Norman and Shallice, 1986). Once the mental model and the sector-control plan are set up, their revisions are closely coupled as changes in the situation often require changes in the sector plan (Redding, 1992; Seamster et al., 1993).

As illustrated by this the current ATC in a TMA can be conceived as a visually and cognitively demanding task with the core requirement to build-up and maintain situation awareness (Redding, 1992; Seamster et al., 1993).

24.2.2 Measuring Scanning Behavior as a Means to Assess Monitoring Abilities

Broach (2013) presumes that with the introduction of advanced automation, scanning of visual sources will be one of the abilities which will become more important for future ATC job performance. The appraisal that virtually nothing is known of how ATCOs scan visual sources (Stein, 1992) does no longer fully hold true today as considerable work has been done on this topic since the early 1990s (e.g., Wickens, 2000; Wickens et al., 2001, 2003; Willems, 2002; Parasuraman et al., 2008; Moore and Gugerty, 2010; Rovira and Parasuraman, 2010). However, because of the technological progress on the accuracy and robustness of eye-tracking measurement systems in recent years, there now seem to be new opportunities to push the topic of ATCOs' visual information sampling further.

The position of the eye is supposed to reflect the current direction of attention, with spontaneous and task relevant looking as two different types of eye movement (Kahneman, 1973). Occurring independently from intentions, spontaneous looking is regarded to be a bottom-up process driven by salient features of events, like movement or novelty, and serves the function of information seeking (cf., Kahneman, 1973). Task relevant looking on the other hand is related to top-down decisions in which an area of interest (AoI) on the radar screen is likely to be richest in relevant information. These decisions require a quick and unconscious weighing of expectancies, effort, and value (cf., Senders, 1964). The sequential allocation of glances called visual scanning is mediated by both bottom-up and top-down processes, and thus comprises both spontaneous and task relevant looking. According to the model of scanning behavior proposed by Wickens et al. (2001), the four factors that influence the frequency of scanning are the physical salience of an event, the physical distance between two consecutively attended AoI, the expectancy for a change of information, and the value of processing or the cost of not-processing that information. The model predicts that an AoI, like a flight data block (FDB), will be visually scanned to the extent that it is expected to contain new information, but even in the absence of high expectations will be sampled more frequently when the AoI is related to a high priority task. Finally, visual sampling may be inhibited by excessive effort requirements, which may cause a complacency bias in scanning behavior (cf., Parasuraman and Manzey, 2010).

An alteration of attention between phases of focused and distributed attention is a characteristic of skilled scanning (Moore and Gugerty, 2010). Even while ATCOs are focusing on a specific area of the radar screen, they must continue to be aware of what is occurring around their sector. Consequently, ATCOs should not neglect to utilize both focused and distributed attention strategies to achieve and maintain situation awareness.

Scanning behavior is assumed to change with ATCOs assigned to the supervisory control (SC) role with high requirements on system monitoring. Parasuraman et al. (2008) suppose that operators scan the raw information sources less frequently when tasks are delegated to an advanced automation, and Rovira and Parasuraman (2010) assume that ATCOs will invest the unbound cognitive resources to focus on secondary tasks, indicated by fewer fixations exhibited to the radar display compared with manual control (MC) conditions. According to Voller and Low (2004) ATCOs reported to be aware of changes in their scanning behavior under simulated future task conditions, without being able to specify this further. Willems (2002) reports on a reduced visual scan structure under conditions of higher levels of automation.

It is an open question what the main characteristics of visual scanning behavior within a TMA task setting are, and in which ways these may change with the introduction of advanced automation. Before moving on to this question, a short introduction into the topic of automation in ATC is provided.

24.2.3 Perspectives on Automation in ATC

An increased level of automation (LoA) support for ATCOs appears to be mandatory to maintain safety and efficiency in a future time-based environment (cf., Kirwan, 2001). As technology tends to push to automate tasks as fully as possible (Miller and Parasuraman, 2007), the future locus of control in ATM is expected to switch from human to automation (Smith et al., 2001). ATCOs supposedly will become traffic managers responsible for resolving exceptional situations instead of controlling every a/c individually like it is today (Dekker and Woods, 1999). Arnaldo et al. (2012) discuss the question whether humans will remain within the future trajectory management processes and will perform some strategic functions where their skills come into play, or if tactical decisions in the future may imply autonomous and fully automated processes. In some far-reaching concepts, it is supposed that in the long-term future automation will take over (Truman and de Graaf, 2007). Truman and de Graaff (2007) also envisage that human intervention and control will be at an absolute minimum in such a self-monitoring and self-controlling future ATM system, with the ATCOs in the role of supervisory controller.

On the other hand there are good reasons to keep the human ATCO within the decision loop and not let automation take over the whole task set of trajectory management within the TMA. Cummings (2004) warns that in time critical environments like ATC, with many external and changing constraints, higher levels of automation are not advisable because automation today, and maybe also in the future does not show the adaptability, flexibility, and problem-solving skills the human operators show to cope with these changing circumstances. Mogford (1997) pointed out that removing the mental picture of the situation from the human domain into automation by keeping the human out of the decision loop would disable ATCOs to bring their problem-solving capabilities into play (see also Bainbridge, 1983). High levels of decision-making automation (LoA) reduces the operators awareness of system dynamics (Miller and Parasuraman, 2007). Humans tend to be less aware of changes when those changes are under the control of another agent. Expert panels therefore recommend that automation efforts in ATC should stop at the level of suggesting decision or action alternatives (Wickens et al., 1998).

When thinking of future ATCOs as monitors of a system it has to be considered that the mechanisms of goal-directed behavior to generate and maintain a sector-control plan, as explained above, would no longer take effect when the ATCO is out of the decision loop, with all the drawbacks this might have on the monitoring behavior and the referring situation awareness.

The current study addresses the following problems: What do experts assess to be the main changes in job requirements when the locus of control in TMA air-traffic control gradually switches from human to automation?

What is guiding ATCOs' scanning behavior when no underlying plan to activate intentions can be assumed to create expectations? What are the observable differences in monitoring behavior in different LoA task settings? Are there gaze-behavioral correlates indicating out of the loop performance?

24.3 Method

24.3.1 Prospective Job Analysis

With the effort that is put into the deployment of advanced automation into ATC, the prognosis of abilities for future ATCOs is co-emerging as a related human factors topic. Alexander (Ammerman et al., 1987; Alexander et al., 1989) conducted a job analysis to assess how future ATCO work would change after the introduction of advanced automation into ATC. Nickels et al. (1995) developed a list of future ATC-work requirements which was the foundation of a selection system for the hiring of ATCOs. At EUROCONTROL Voller and Low (2004) developed a technique named *Solutions for Human Automation Partnership in ATM* (SHAPE) to predict how new automated systems might affect ATCOs' skill requirements. Manning and Broach (1992) at the FAA (Federal Aviation Administration) conducted a strategic-job analysis to assess the changes of the ATCO job as it is predicted to exist in the future as a result of increasing automation. After an analysis of the present job, they gathered information about the kinds of issues, like technological and organizational developments which may affect the job in the future. Broach (2013) conducted a strategic-job analysis for the tower CWP. He used the overall profile of abilities derived by Nickels et al. (1995) to compare current with future ability requirements. He performed an extensive review of current ability profiles and of assessable documentation on technology changes which are expected to be introduced until the year 2018. Based on the current profile and the expected technology changes, he infers a future ability profile. Broach resumes that scanning of visual sources will be one of the abilities which will become more important for future performance.

Eissfeldt et al. (2009) used a technique for future-oriented job analysis first introduced by Schneider and Konz (1989). They held several workshops with subject matter experts (SMEs) to discuss future issues of ATCOs' and pilots' jobs. After being primed with the discussion outputs, the SME filled out the Fleishman Job Analysis Survey (F-JAS, Fleishman and Reilly, 1995) once rating the current job requirements and once the requirements for the job as it is considered for the future. Eissfeldt et al. (1999) used a simulation-based approach to assess future job requirements. The simulation features were data-link communication, a stripless system, planning tools, and automatic conflict detection aids. In a 1 h exercise four brief ATC simulation scenarios were performed by the SMEs. Before and after a session, they filled out the F-JAS. Eissfeldt et al. (1999) found a general decrease of cognitive requirements for future ATC, with the strongest effects occurring for oral expression and oral comprehension, written expression, and number facility. These decreases were attributed to the effect of data-link communication.

In the current experiment ATCOs' judgments about future job requirements are assessed after they experienced working in some future ATCO role settings. To allow a prognosis of what changes are up to come, the key features in the European ATM Master Plan (SESAR, 2012) were taken into consideration. Time-based operations are the key concept to be introduced with SESAR Step 1 (interval-based metering). In a time-based environment the initial synchronization of the traffic arriving in high traffic density TMA will be generally ensured via the allocation of a controlled time of arrival (CTA) on initial approach fixes (IAFs). The introduction of a/c capable of meeting a CTA with appropriate accuracy improves the performance and reliability of an Arrival Manager (AMAN) system. In the time-based environment proposed in the SESAR ATM target concept (SESAR Consortium, 2007), automation is assumed as a means to expand the ATCOs functional envelope, for example, by calculating a/c trajectories with respect to the a/c landing times and providing the ATCO with tactical guidance advisories accordingly. Another SESAR concept envisages the ATCO in the role of a central decision maker who still manages air traffic, but delegates the execution of certain spacing and positioning tasks to the pilots of ASAS (airborne separation assurance system) equipped a/c. Such a change in roles and responsibilities is envisaged in the SESAR concept of self-merging and -spacing, where pilots now take the responsibility for identifying target a/c and establishing separation based on instructions from the ground.

With the delegation of specific tasks to other agents, it is assumed that ATCOs will be relieved from the task load so they are able to adjust to a more strategic work style (cf., Willems, 2002; SESAR Consortium, 2012) in trajectory management, as it supposedly is required within a time-based environment.

The interaction of ATCOs with the simulation environment is assessed by tracking their eye movements to gain further objective indication for changes in their operational monitoring behavior. As the most important information for an ATCO is related to the moving targets on a situational display, the analysis focuses on attentional switches (transitions) between dynamic area of interests (dAoI) which are defined by the moving zone each a/c blip with related FDB covers on the radar screen (cf., Willems, 2002; Gross et al., 2010). The analysis of fixation transitions within experimental runs is supposed to reflect the monitoring activities to build and retain situation awareness. It is claimed that transitions between a/c indicate the activation of meaningful relations within this overall traffic representation, with the total transition number reflecting the priority of that constellation (but see Ellis and Stark, 1986 for a critical discussion of this assumption).

Research Question 1: What do experts assess to be the main changes in job abilities required for their working position after performing new roles in scenarios with advanced automation within their familiar TMA environment?

Research Question 2: Is there objective evidence for fundamental changes in monitoring (measurable by behavioral correlates) when the new roles imply that central task of TMA management is delegated to cockpit and/or to automation?

24.3.1.1 Hypotheses

H1: For multiagent conditions, the ATCOs experience higher job requirements with regard to monitoring capabilities. Accordingly, significantly higher ratings on the situation awareness job-ability scale (regarded as the ability to perceive the current automation–traffic interaction, understand the reasons behind it, and project future development; cf., Fleishman and Reilly, 1995) are expected.

H2: For multiagent conditions, the ATCOs delegate tasks and use released resources to switch to prolonged phases of distributed attention (cf., Moore and Gugerty, 2010), that is, focusing their visual scanning to what is occurring around their sector. Accordingly, more transitions of fixations between distant a/c on the radar screen are expected. A more strategic work style should also be reflected in a comparably higher level of situation awareness. The release of resources should also be indicated in lower levels of experienced workload.

24.3.2 Experimental Design

AMANs normally are a means to avoid overcrowded TMA by controlling the flow en route before reaching the IAFs, where a/c are handed over from area control centers (ACCs) to TMA. However, the DLR AMAN research prototype 4D-CARMA (four-dimensional cooperative arrival manager) (Ehr and Uebbing-Rumke, 2013), which is classified as a strong executing system which is compliant to SESAR requirements, is also calculating trajectories with respect to the landing time on ground and accordingly generates tactical guidance advisories. These advisories are shown in timely and precise manner to the approach controllers at all approach phases on the radar screen.

Several kinds of future scenarios are constructed with 4D-CARMA as the central element to create different role settings for ATCOs. 4D-CARMA is configured to various levels of automation (Sheridan and Verplank, 1978; Endsley, 1987; Endsley and Kaber, 1999). On the five-level LoA hierarchy proposed by Endsley (1987), the AMAN is configured to the LoAs "manual control," "decision support," and "monitored AI" (these are the experimental conditions also selected by Willems, 2002). This corresponds to the LoAs "manual control" (MC), "decision support" (DS), and "supervisory control" (SC) as the second highest LoA (cf., Endsley and Kaber, 1999). The taxonomy proposed by in the RHEA (Role of the Human in the Evolution of ATM systems; Nijhuis, 2000) project is used to assign the ATCOs' roles accordingly:

- Experimental run 1 (baseline): ATCO handles traffic according to the first-come-first-serve principle.
- Experimental run 2: ATCO delegates responsibility of certain tasks to suitably-equipped a/c and aircrew.
- Experimental run 3: 4D-CARMA predicts what needs to be done for certain tasks and makes suggestions to the ATCO.
- Experimental run 4: 4D-CARMA controls—ATCO monitors only, and overrides when necessary (management by exception).
- Experimental run 5: Automation controls—ATCO monitors only, and overrides 4D-CARMA when necessary, with both working together in the control loop during an emergency event.

Experimental run 1 (baseline) resembled current operations in which ATCOs had standard systems support. 4D-CARMA was running during the baseline scenario because it is a standard tool in all Langen ACC sectors. Experimental run 2 (ATCO delegates control to a/c) also resembled current operations from the point of system support, but the concept of a/c controlled merging and spacing was introduced for the three piloted a/c. From the pilots' perspective, the new requirements during in-flight procedures were to (1) meet the target time of overflow at the IAF (initial approach fix), calculated on ground by 4D-CARMA and displayed in their CDTI, and (2) to follow the merging and spacing instructions given by ATC, which comprises to maintain distance and time to their target a/c ahead. In experimental run 3 (ATCO controls—4D-CARMA suggests) 4D-CARMA supported the ATCO by recommending level, speed, and turn-to-base advisories and their timing to meet time-based criteria for the incoming flights. The 4D-CARMA advisories for ATCOs were displayed within the corresponding flight data blocks (FDB) on the radar screen. From the cockpits perspective run 3 functioned as a repeated measurement with the same requirements as in experimental run 2. In experimental run 4 (4D-CARMA controls—ATCO monitors) all of the ATCO's tasks, that is, determining speeds, headings, and levels to the a/c and building an arrival sequence, are delegated to 4D-CARMA. The work of the ATCO here moves toward SC. From the pilots' perspective in comparison to the baseline requirements change in receiving ATC advisories from ground-based automation via data link and reading it back via r/t to keep the supervising controller informed about their intent. The ATCO monitored the r/t and always was able to intervene via r/t. In experimental run 5 (4D-CARMA and ATCO in control loop) again, the ATCO was in the supervisory role but had to intervene due to a simulated emergency. From the pilots' perspective the requirements were analogous to experimental run 4.

The ATCOs tasks were to receive incoming traffic at the IAF, coordinate these a/c according to the traffic flow and provide sufficient lateral and vertical separation between a/c, as well as guiding them to the final approach. In the baseline scenario 4D-CARMA was active, but just displaying a planned sequence on a time ladder. To the subsequent experimental runs the CTA and merge-&-space operation, 4D-CARMA, and ATCO–pilot data-link communications (CPDLC) were selectively introduced. For the first future condition (experimental run 2) 4D-CARMA was active, but not in a strong executing mode. The job includes all core tasks of the baseline condition, added to by the tasks related to the merge and space procedures, such as instructing the pilots and monitoring the on-going operations. For the next, more advanced future condition (experimental run 3) 4D-CARMA was set into strong executing mode. In this setting the ATCO is advised to generally instruct the pilots with the 4D-CARMA clearances, if no safety concerns stand against them. Also the core tasks to determine headings, level, and speeds according to the intended approach sequence where suspended as 4D-CARMA made recommendations to follow. However, the list again was expanded by the new core task to evaluate AMAN advisories before communicating them via r/t. For the most advanced future condition (experimental run 4) 4D-CARMA was set into a self-executing mode. Almost all core tasks are now performed by 4D-CARMA via data-link communication with the pilots. In the role of supervisory controller, the new core tasks are to monitor 4D-CARMA performance on the referring display, monitor a/c behavior on the radar screen, monitor r/t communication, and to intervene to take MC in case of an exception, such as an emergency. This most advanced future scenario was performed a second time with shuffled a/c signs and a provoked emergency event that forced the ATCO to perform tactical interventions (experimental run 5).

In the experimental runs only incoming traffic was simulated. All conditions were simulated within the structure of the Frankfurt approach sector, which is the sector in the TMA for all incoming a/c with destination Frankfurt, in its configuration valid until October 2011 (i.e., no third landing runway integrated). These flights approach the area via four different (STARs) and IAFs. The latter are coded as OSMAX (western sector inbound), ROLIS (northern sector inbound), GEDERN (eastern sector inbound), and PSA (southern sector inbound). The arriving a/c are lined up on the downwind and then "fed" onto the final approach. The sector controls all levels from surface up to flight-level 115. The arrival routes are designed in a way so as not to interfere with each other. In the experimental trials, only a sector configuration for using runways 25 is applied.

In all experimental task settings the participants were advised to control or monitor the incoming traffic for the whole TMA. Generally the participating ATCOs were advised to guide the down-winding traffic via the STARs. All ATCOs reported being very familiar with this assignment, as this is a standard setting in real-life during periods of lower traffic load. The participants also worked in a one-man setting without the assistance of a coordinate ATCO. The experimental baseline scenario (experimental run 1) is oriented at the actual method of operations for approach control.

24.3.3 Simulator Preparation

The human-in-the-loop simulations were performed at DLR's integrated air-ground simulator AviaSim (Schulze Kissing et al., 2010; Bruder et al., 2013). AviaSim comprises one ATCO working position and three pilot WPs, two pseudo-pilots, and one pseudo-ATCO console, as well as an experimental-control console. The ATC working position is equipped with a SmartEye® Pro remote eye-tracking system using a five-camera layout. The ATCO is seated in front of four displays with the radar screen as the main window in front. On the left the electronic flight strip bay is located. On the right the 4D-CARMA generated timeline for the arriving a/c is shown. The timeline displays advisories the automation currently is sending to any a/c. In addition, a so-called mileage-metering scale shows the distances between a/c on the final approach. The head-up display shows labels of a/c approaching the TMA on a timeline according to their controlled times of overfly (CTOs) at the referring IAFs. All a/c are simulated by pseudo-pilots except for three that are simulated by fully equipped simulation cockpits operated by airline pilots. Pseudo-pilots operate the non-piloted a/c according to standard procedures until they are transferred to the TMA. At the beginning of the scenario a pseudo-ACC-controller overtakes communication with the piloted a/c and handing it off to the ATCO when the a/c reaches the TMA border.

24.3.4 Dependent Variables and Procedure

The experiment was conducted in five sessions over 3 days. In total five ATCOs from Deutsche Flugsicherung GmbH (DFS) ACC Langen with valid licenses for TMA Frankfurt and 15 pilots from various airlines participated. Only ATCO data will be reported here. The ATCO sample was all male with a mean age of 32.5 years (SD: 6.42), and a job experience of 10.8 years in the mean (SD: 6.85). A session comprises four runs of a future scenario, one baseline scenario, and five training runs for the ATCOs or three training runs for pilots, respectively. A single simulation run involved one ATCO, three pilots, and two pseudo-pilots. The ATCOs were controlling the air traffic of the whole Frankfurt TMA in a one-man operation modus. Pilots were starting airborne within the lower airspace and entered TMA via three different IAFs. The ATCOs, pilots, and pseudo-pilots communicated using r/t or CPDLC/data-link transmissions.

For the ATCOs the first exercise day started with an introduction of questionnaires, calibration on the eye-tracking system, and an about 90-min in-depth walk-through the F-JAS dimensions. The second part of the first day included two initial training runs of about 1 h to get acquainted with the simulator and the 4D-CARMA AMAN interface (first training run) and to familiarize with pseudo-pilots interactions (second training run). The second day started with a recapping training session in which the merging and spacing procedures were practiced with pseudo-piloted a/c. Afterward the first joint training scenario with ATCO, pilots, and the pseudo-pilots was conducted. This was followed by a second joint training in which merging and spacing procedures were practiced with the piloted a/c. The measured traffic sample contained 24(+x) flights with destination Frankfurt, with at least six a/c entering the TMA on each of the four IAFs during the scenario run. The scenario contained a mix of a/c of heavy and medium weight class planned to land in staggered mode on the two parallel runways 25R and 25L, which have to be operated dependently. Next, the ATCOs took part in the five experimental runs, one at the end of day 2 and four on day 3. For comparison reasons, one traffic sample was used for all runs. However, to avoid familiarization with the traffic sample, all a/c signs were shuffled or substituted between each run. Furthermore, pilots switched working positions after each run to let them experience different traffic situations. Experimental run 4 and 5 were always performed consecutively (en bloc) and the baseline run was always performed first, that is, at the end of day 1. The runs 2, 3, and the block 4 + 5 were counterbalanced to avoid practice effects. Each run was performed only after a break.

Quantitative and qualitative data were gathered during and after each simulation run. As a subjective measure of workload and situation awareness, the instantaneous self-assessment (ISA; Kirwan and Flynn, 2001) was used. ATCOs were prompted four times during a scenario (appointed according to the flight phases of the piloted a/c, that is, RTA IAF, merge, follow, and turn) to give their ratings on current situation awareness and workload on two three-point scales. During the simulation runs, eye movements of the ATCOs were tracked to gain indication for participants' visual sampling behavior. Immediately after each simulator run, the participants filled out (a) a 14-Component Version of the Situation Awareness Rating Technique (SART; Taylor, 1990), (b) the NASA Task-Load Index (Hart and Staveland, 1988), a questionnaire to assess the impact on mental workload (AIM; Dehn, 2013), and the SHAPE Automation Trust Index (SATI; Dehn, 2013). At the end of the last experimental run of exercise day 2 and exercise day 3, the participants filled out a modified version of the Fleishman Job Analysis Survey (F-JAS; Fleishman and Reilly, 1995) and cognitive interviews were held directly afterwards to gain in-depth information about experienced job requirements. Additionally, a questionnaire on user satisfaction and acceptance was completed at the end of day 3.

24.4 Results

On several usability scales the ATCOs evaluated the system as easy to learn, the time needed to learn the system operation to be adequate, the simulation scenarios to be interesting and inspiring, and regarded their traffic load to be comparable to current real-life traffic demands.

24.4.1 Questionnaire on Job-Ability Requirements

The ATCOs' judgments on the abilities required for future job performance were assessed once for the presented future scenarios (overall) and compared to the ability-requirement profile the experts initially identified for the baseline scenario. These profiles are put together from a set of 37 ability-requirement scales the ATCOs filled-in, with 32 of them selected from the F-JAS, supplemented by five scales covering additional requirements developed by DLR in a format similar to F-JAS.

A paired-sample *t*-test on the single scales revealed a significant effect of future changes (versus baseline) on required problem sensitivity (cognitive scale; future increase) and number facility (cognitive scale; future decrease) abilities (see Table 24.1). With respect to practical relevance the marginal differences in required spatial orientation (cognitive scale; future decrease), perceptual speed (cognitive scale; future increase), situation awareness (interactive/social scale; future increase), resistance to premature judgment (interactive/social scale; future increase), and vigilance (future increase) abilities will also be considered (see Table 24.1). A Wilcoxon signed-rank test on the rank of means for the 38 ability requirement scales showed no fundamental change in the ability requirement profiles from current (baseline) to future scenarios. Figure 24.1 presents the means and standard errors of SMEs' F-JAS-type scale ratings related to the baseline and the future scenarios on the 37 ability requirement scales. For the baseline scenario, time sharing received the highest mean rating and is thus considered as the most important ability required for the current job.

TABLE 24.1

Mean Ratings of F-JAS Type Scale Ratings for the Baseline and Future Scenarios

F-JAS Scales	ATC			df	P (2 Way)
	Baseline	Future	T		
Problem sensitivity	4.92	6.44	−3.919	4	0.017
Number facility	4.04	2.56	2.929	4	0.043
Spatial orientation	4.20	2.46	2.400	4	0.074
Perceptual speed	4.66	5.56	−2.535	4	0.064
Situation awareness	6.56	7.00	−2.269	4	0.086
Resistance to premature judgment	3.90	5.42	−2.513	4	0.077
Vigilance	4.76	6.56	−2.126	4	0.066

Note: Only those scales are listed for which a paired-sample *t*-test yielded a significant or marginal within-group effect (N = 5).

In the resulting future ability-requirement profile, situation awareness*—which received the highest mean ratings—followed by stress resistance,[†] vigilance,[‡] problem sensitivity,[§] emotional control,[¶] selective attention,[**] auditory attention,[††] resilience,[‡‡] and perceptual speed[§§] received higher values than the ability of time sharing[¶¶] (cf., Figure 24.1). The increasing importance of situation awareness, vigilance, problem sensitivity, and perceptual speed should be considered as important developments in task requirements, as they are newly listed in the top 10 required future abilities when compared by their mean ratings (cf., Figure 24.1). Significant future decreases (see Table 24.1) were only determined for scales of minor importance (namely number facility and spatial orientation).

24.4.2 Gaze Analysis

24.4.2.1 Transitions

In Figure 24.2 for each experimental condition the state-transition matrices containing the total number of gaze transitions between a/c are displayed. The referring a/c are labeled and ordered according to their final position in the landing sequence. For each experimental condition the degree a/c were attended to in succession can be read out by first tracing the respective line and then the respective column of the a/c. For example, to determine for the baseline condition the degree the a/c in the fifth position of the landing sequence is coactivated with other a/c, the counted transitions for the preceding a/c can be read out

---

* Ability to perceive the current automation–traffic interaction, understand the reasons behind it, and project the future development.
[†] Capability of dealing with stress situations in such way that control is maintained and the objective achieved.
[‡] Ability to maintain attention and alertness over prolonged periods of time.
[§] Ability to tell when something is wrong or likely to go wrong.
[¶] Ability to stay in self-command in emotionally challenging situations, when irritating, unexpected, difficult, or stressful factors occur.
[**] Ability to concentrate on a task without getting distracted.
[††] Ability to focus on a single source of auditory (hearing) information in the presence of other irrelevant and distracting sounds.
[‡‡] Ability to rapidly recover normal energy and enthusiasm following a discouraging situation.
[§§] Involves the degree to which one can compare letters, numbers, objects, or patterns, both quickly and accurately. This ability also includes comparing a presented object with a remembered object.
[¶¶] Ability to shift back and forth between two or more sources of information.

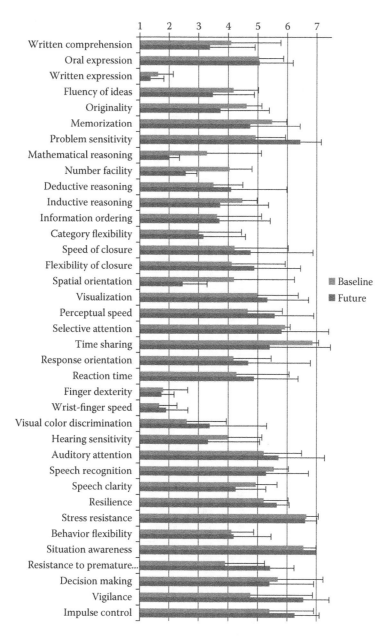

FIGURE 24.1
Mean and standard errors of SMEs (N = 5 ATCOs) F-JAS-type scale ratings on the 37 ability requirement dimensions plotted for the simulated baseline scenario and the simulated future scenarios.

in the line labeled "5th" (counts for a/c ordered in their spatial distance to the a/c under consideration are 26, 14, 34, and 11), the transition numbers for the succeeding a/c can be traced in the column correspondingly labeled "5th" (counts for a/c ordered in their spatial distance to the a/c under consideration are 37, 24, 12, 14, 5, 13, 0, 2, 0, 0, 0). To simplify the interpretation the gray shades additionally indicate the suspected strength of these relations in the ATCOs mental picture, from bright-gray with higher numbers (indicating

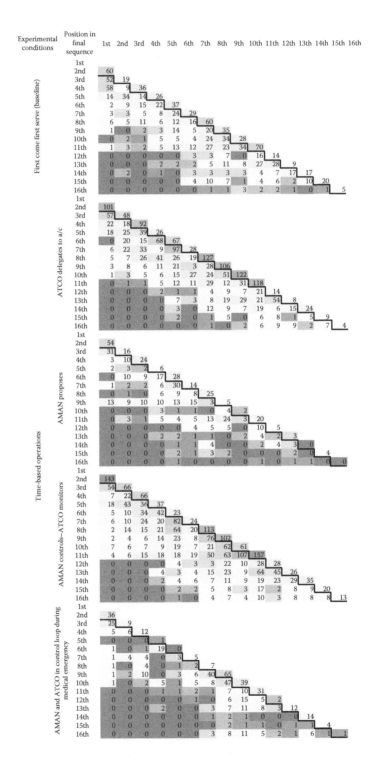

FIGURE 24.2
State-transition matrices containing the total number (summed up for N = 5) of gaze transitions between dynamic areas of interest (blip and data block) representing the single a/c on the radar monitor during an experimental run.

high strength) to white with lower numbers (medium strength) to dark-grey with low numbers to zero (minor or no strength). The top line of cells is separated by a black line to emphasize that these numbers supposedly indicate data comparisons related to distance, level, and speed between neighboring a/c at merging STARs or near to the final approach as well as strategies to sequentially scan moving objects from the inner to outer sector areas, whereas the values displayed in the cells below that line supposedly indicating the monitoring contingencies produced by other mechanisms of distributed attention monitoring (e.g., switching from a/c on final to relevant airspace structures, reacting to a/c calls via r/t, or a switch to a somehow pre-tactical work style). Obviously, for all experimental conditions the ATCOs allocated much of their attention to relations between a/c neighbors on the final approach. Maximum transition numbers between neighboring a/c are counted for experimental run 2 in which the ATCO delegated responsibility of certain tasks to suitably-equipped a/c and experimental run 4 when 4D-CARMA controlled and the ATCO monitored only. However, for both these conditions, and especially for the SC condition, even relations between distant a/c (preceding and succeeding) received more attention than in the baseline condition.

24.4.2.2 Fixations per Screen

The numbers of fixations counted for the radar display for all experimental conditions are all higher compared to the baseline (cf., Figure 24.3). This is contrary to the directed hypothesis predicting that the number of fixations counted for the radar display should decrease when trajectory management tasks are delegated to other agents (eher: closer monitoring). As the data was not normally distributed, the most appropriate way to statistically test the differences from the baseline was the Wilcoxon signed-rank test. There was a significant increase from baseline to the third condition (ATCO controls—AMAN proposes) in the number of fixations counted for the radar screen, $Z = -2.023$, < 0.05, and also to the fifth condition (AMAN and ATCO in control loop: $Z = -2.023$, $p < 0.05$). The fourth condition (AMAN controls—ATCO monitors) only marginally differed from the baseline ($Z = -1.753$; $p = 0.080$). Figure 24.3 shows that under all experimental conditions the ATCOs only rarely attended to this planning information. However, there was a significant increase from baseline to the fifth condition ($Z = -2.023$, $p = 0.042$) indicating some planning activity was involved during the emergency handling. As also can be seen in Figure 24.3, the AMAN (4D-CARMA) screen was more attended to under the fourth ($Z = -1.753$; $p = 0.080$) and fifth condition ($Z = -2.023$, $p = 0.043$).

24.4.3 Control Variables

Repeated Measures Analyses of Variance (rANOVA) were used to statistically compare the questionnaire outcomes for the various runs. Partial eta-squared (η_p^2) is given as a measure of effect size. For each analysis, $p < 0.05$ was used.

24.4.3.1 Workload

For each run workload, measures with NASA TLX were obtained (SME). An ANOVA on the NASA TLX overall scores for the five runs showed only a marginal effect [$F(4,16) = 2.694$, $p < 0.10$, $\eta_p^2 = 0.402$]. The within subject single contrasts showed *that experimental run 5 imposed significantly higher workload* compared to the baseline condition [$F(1,4) = 9.476$,

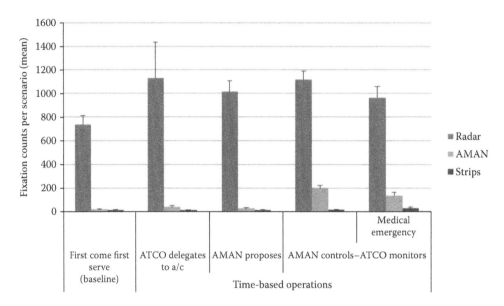

FIGURE 24.3
Number of fixations on each of the three monitors of the controller working position for each scenario run.

$p < 0.05$, $\eta_p^2 = 0.703$]. No effect for the five runs on the AIM total score was measured. Single contrasts indicate significantly higher demands on monitoring for experimental run 2 (ATCO delegates to a/c) compared to experimental run 1 (baseline), $F(1,4) = 16.000$, $p < 0.05$, $\eta_p^2 = 0.800$. A 4 (moment of measurement) × 5 (ATCO role)-factorial rANOVA for the ISA scores showed a significant effect for the factors "moment of measurement," $F(3,6) = 14.600$, $p < 0.05$, $\eta_p^2 = 0.880$. These findings contradict the hypothesis that trajectory management task delegation would release the ATCO. There is a distinctive trend for a workload increase with the scenario progressing (for all experimental conditions). The experimental factor (ATCO Role) had no significant effect on workload. Single contrasts revealed significant higher ISA workload ratings only for the experimental run 2 (ATCO delegates to a/c) compared to run 1 (baseline), $F(1,4) = 64.000$, $p < 0.05$, $\eta_p^2 = 0.970$. (Again, ATCO delegates to a/c; significant increase in workload; not for full automation.)

24.4.3.2 Situation Awareness

An ANOVA for the 5 runs showed a significant effect for the 10-D SART scale values, $F(4,16) = 4.063$, $p < 0.05$, $\eta_p^2 = 0.504$. Single comparisons with the baseline run revealed significantly lower situation awareness under the third experimental run (ATCO controls— AMAN proposes: $F(1,4) = 9.732$, $p < 0.05$, $\eta_p^2 = 0.709$), as well as under the fourth experimental run (AMAN controls—ATCO monitors: $F(1,4) = 24.500$, $p < 0.05$, $\eta_p^2 = 0.806$). This is contrary to the directed hypothesis which predicts the allocation of attentional resources released by tasks delegation to enhance situation awareness. A 4 (moment of measurement) × 5 (ATCO role)-factorial ANOVA for the ISA scores on "situation awareness" showed a significant effect for the factors "moment of measurement," $F(1.931, 3.861) = 8.826$, $p < 0.05$, $\eta_p^2 = 0.815$. There is a distinctive trend for situation awareness to decrease with the scenario progressing. There was no significant effect for the experimental factor (ATCO role).

24.5 Discussion

A prospective job analysis for the ATC task in future TMAs was performed. Potential future scenarios were simulated in a human in the loop experiment with current job holders performing the air-traffic control task in different role settings with the locus of control gradually switching to automation. The ATCOs assessed the future job-ability requirements experienced within the simulation on the F-JAS (Fleishman, 1992). Their monitoring performance was assessed by tracking their eye-gaze behavior during all experimental trials. Before coming to an interpretation of the results, some methodological reservations have to be made in advance. It may be questioned if it were future abilities we have measured, or simply the requirements to cope with the problems we have created by simulating a more or less clumsy automation setup. It may be questioned if receiving tactical advises from automation to comply with 4D-plan requirements is a means to expand the ATCOs' functional envelope (cf., Amalberti, 1999). Operators are rather compelled to adapt reactive strategies instead of anticipative strategies insuring long-term adaptation (cf., Cellier et al., 1997). It is also evident that the general problems arising with the introduction of higher levels of automation in ATC cannot be solved by focusing on personnel selection issues only. Another methodological weakness of the current study is the ATCOs' sample size, which is not large enough to reliably perform inferential statistic tests. Gaze transition data is only presented on a descriptive level. So the results presented above should not be treated as evidence, and any conclusion drawn on the results presented above should be treated as preliminary.

The ability analysis revealed the specific job requirements which occur for ATCOs when moving from the current to the monitor role. As hypothesized the ATCOs actually assessed the abilities to perceive the current automation-traffic interaction, understand the reasons behind it, and project the future development (labeled as situation awareness) as being of highest relevance when selecting future job holders. The SME also assessed that a higher sensitivity with regard to problems produced by automation will be required (problem sensitivity). It was conceived in post simulation interviews that the main demand in the system-supervising role was to continuously assess if the 4D-CARMA instructions were reasonable or not. This experience is reflected in the higher ratings for the ability to stay alert over a long period of time, that is, "vigilance." In the passive supervisor roles some ATCOs also reported they were unable to form expectations on traffic behavior on the basis of their own intentions, so they were more heavily occupied with tracing and integrating automation behavior into their mental picture to understand what is happening and project what the system is going to do. This uncertainty with regard to the actual traffic control processes they had to supervise induced stress to the monitoring ATCO, which is reflected by an increase in ratings on the stress-resistance scale, which is assessed to be an ability required at maximum for this future condition. It can be assumed that the ATCOs tried to counteract against this uncertainty by increasing their efforts in visual sampling. So, the answer to what experts regard to be the main changes in job abilities required when working with advanced automation comes as no surprise: they experience higher job requirements with regard to monitoring capabilities.

There also is some indication for changes in the monitoring behavior when the locus of control switches from ATCO to automation. ATCOs showed more transitions of fixations between distant a/c on the radar screen under high LoA, like it would be expected when they use free resources to widen or distribute their focus to attend to what is occurring around their sector (cf., Moore and Gugerty, 2010). However, this change in visual sampling behavior is accompanied by comparably lower levels of situation awareness and higher levels of experienced workload, as it is to be expected when ATCOs work with clumsy automation while they are kept out of the control loop.

The gaze transitions between dynamic AoI representing the single a/c on the radar monitor could be interpreted to reflect the relational structure of the ATCOs' current mental picture (cf., Eyferth et al., 2003) as well as the contingent flow of intentions for monitoring and control of goal-directed behavior, or in other words: the cognitive mechanisms of goal-directed behavior at work. The highest transition numbers are counted for relations between spatially close a/c. It can be assumed these are more closely monitored in relation because of comparisons of speed, level, and distance information, but also because of the strategy to recurrently move the attentional "spotlight" from the center line to the object near the sector borders, which according to an ATCO's report is performed about once in a minute. Transition counts between spatially more distant a/c may reflect scanning guided by goal directing mechanism of contingently active intentions, such as airspace structure checks or bottom-up activation (a/c calling into the sector), plus visual search for their identification.

The finding that the highest counts for transitions between all a/c (regardless of their proximity) were observed for scenarios where the locus of control for the separation task is transferred to other agents, human or automation, may indicate how monitoring behavior is changing when it is no longer guided by the ATCOs' intentions and mechanisms of goal-directed action control, but by their expectations and recurrent checks of other agents' actions.

In the baseline condition the eye-movements can be conceived to be affected by three factors: (1) the ATCOs' intentions when to give clearances to a/c, (2) their general strategies for sector monitoring, as well as (3) the salience of a/c entering the sector. The comparable higher number of fixations during the merge and space scenario could designate a closer monitoring of the decisions and actions of the three human agents within their room for maneuver. The comparable lower number of transitions counts measured under conditions of automation assistance may reflect some automation bias induced by the 4D-CARMA. Maybe 4D-CARMA's continuous provision of action proposals prevented the mechanisms of goal-directed behavior to unfold, causing that less intentions were contingently active to guide the ATCOs' monitoring. When in the SC condition the locus of control switched to 4D-CARMA, the transition numbers increased in combination with the high fixation frequency counted for the advisory window where clearances prepared by 4D-CARMA were displayed. It can be conceived that the observed changes in scanning behavior do not reflect a switch in monitoring strategies, but simply to the necessities to closely monitor the human-machine interface (HMI) element where the automation was displaying its instructions, and then for visual search to retrace the a/c to which 4D-CARMA was addressing the clearance. As an emergency occurred under the SC condition the ATCOs focused more attentional resources to the flight strips, indicating more planning activity. This came along with lower transitions numbers between a/c. As the ATCOs had to handle the emergency, gaze data suggest that they reduced their checks of the automation behavior considerably.

24.6 Conclusions

The data provide no indication for fundamental changes in the ability requirements profiles from current to the projected job setting for ATCOs. Behavioral data on the other hand once again show the problems arising when human operators are taken out of the decision loop. It simply is hard to remain involved into the events happening on a tactical level when ATCOs are not enabled to build up a state of intentionality by developing a shared understanding of the activities that are necessary to control sector events. For SC conditions the ATCOs reported they were stuck in the process of understanding the control principles the automation is following, but did not fully understand the algorithms. When ATCOs in our simulation were advised by automation when and how to act, they were hardly able to assess if the automation advisories were reasonably 4D-plan compliant. As the ATCOs actually had to decide on situational matters which they could not judge, they were only acting as a transducer of system advisories to the pilots. This might have caused their experienced decrease in situation awareness. One could argue the system could equally well directly interact with the receiver, in this case the a/c.

For follow-up studies it would be advisable to reach a participant sample large enough to draw conclusions based on statistical testing, although it is often owed to challenges in expert access that an optimal sample size is not reached, like in the current study. A further analysis of the scan paths could also shed some more light into the ATCOs' monitoring behavior. However, before starting any follow-up study an advanced automation setup that is gaining high SA ratings should be identified. This seems to be a condition sine qua non to validly carve out the requirements for working with advanced automation in ATC following a simulation-based approach. Candidate systems supposedly could be found in the area of cooperative human–machine interaction design (cf., Sarter et al., 1997; Hoc, 2000; Woods and Sarter, 2000). Systems designed according to the principles of cooperative automation supposedly would provide an ongoing understanding of the reasonability of its actions to its human team partner. Maybe it would be a fruitful approach to design a planning system that sketches out and follows plans in the way that humans do. This would give the human operator the chance to reenact, anticipate, and evaluate what the system is doing.

But whatever the design decisions may be, it is recommended here that the human performance assessment in the system validation activities should also encompass the use of job analysis tools. This is to ensure that ATCO selection will sooner rather than too much later look out for employees who are capable of filling in the roles the new ATM system assigns to the human operator.

References

Alexander, J. R., Alley, V. L., Ammerman, H. L., Fairhurst, W. S., Hostetler, C. M., Jones, G. W. et al. 1989. *FAA Air Traffic Control Operations Concepts Volume VII: ATCT Tower Controllers*. Washington, DC: Federal Aviation Administration Advanced Automation Program (AAP-110).

Amalberti, R. 1999. Automation in aviation: A human factors perspective. In D. Garland, J. Wise and D. Hopkin (Eds.), *Handbook of Aviation Human Factors*. Mahwah, NJ: Lawrence Erlbaum Associates, 173–192.

Ammerman, H. L., Becker, E. S., Bergen, L. J., Claussen, C. A., Davies, D. K., Inman, E. E. et al. 1987. *FAA Air Traffic Control Operations Concepts Volume V: ATCT/TCCC Tower Controllers*. Washington, DC: Federal Aviation Administration Advanced Automation Program (AAP-110).

Arnaldo, R., Saez, F. J., and Garcia, E. 2012. HALA!: Towards higher levels of automation in ATM. Position Paper—State of the art and research agenda. *Paper Presented at 28th International Congress of the Aeronautical Sciences*, Brisbane, Australia, pp. 1–10.

Bainbridge, L. 1983. Ironies of automation. *Automatica*, 19, 775–779.

Block, R. A. and Zakay, D. 2006. Prospective remembering involves time estimation and memory processes. In J. Glicksohn and M. S. Myslobodsky (Eds.), *Timing the Future: The Case for a Time-Based Prospective Memory*, 25–49. London: World Scientific.

Broach, D. 2013. *Selection of the Next Generation of Air Traffic Control Specialists: Aptitude Requirements for the Air Traffic Control Tower Cab in 2018* (Rep. No. DOT/FAA/AM-13/5). Oklahoma City, Oklahoma: Federal Aviation Administration, Civil Aerospace Medical Institute.

Bruder, C., Eissfeldt, H., Grasshoff, D., Gürlük, H., Friedrich, M., Hasse, C. et al. 2013. *Aviator II-Simulator-Based Research on Operational Monitoring and Decision Making for Human Operators in Future Aviation.* Cologne: German Aerospace Center.

Cellier, J. M., Eyrolle, H., and Marine, C. 1997. Expertise in dynamic environments. *Ergonomics*, 40, 28–50.

Cowan, N. 1999. An embedded-processes model of working memory. In A. Miyake and P. Shah (Eds.), *Models of Working Memory: Mechanisms of Active Maintenance and Executive Control.* Cambridge: Cambridge University Press, pp. 62–101.

Cummings, M. L. 2004. Automation bias in intelligent time critical decision support systems. In *Paper Presented at AIAA 1st Intelligent Systems Technical Conference*, Chicago, IL, pp. 557–562.

Dehn, D. M. 2013. The SHAPE Questionnaires. https://www.eurocontrol.int/sites/default/files/field_tabs/content/documents/nm/safety/safety-shape-leaflet.pdf

Dekker, S. W. A. and Woods, D. D. 1999. To intervene or not to intervene: The dilemma of management by exception. *Cognition, Technology & Work*, 1, 86–96.

Ehr, H. and Uebbing-Rumke, M. 2013. *Interface AviaSim 4D-CARMA* (Rep. No. IB 112–2010/43). Braunschweig: DLR.

Eissfeldt, H., Deuchert, I., and Bierwagen, T. 1999. *Ability Requirements for Future ATM Systems Comprising Data-Link: A Simulation Study Using an EATCHIP III Based Platform* (Rep. No. DLR Research Report 1999-15). Köln: DLR.

Eissfeldt, H., Grasshoff, D., Hasse, C., Hoermann, H.-J., Schulze Kissing, D., Stern, C. et al. 2009. *Aviator 2030: Ability Requirements in Future ATM Systems II. Simulations and Experiments* (Rep. No. DLR Research Report 2009-28). Köln: DLR.

Ellis, J. 1996. Prospective memory or the realization of delayed intentions: A conceptual framework for research. In M. Brandimonte, G. O. Einstein and M. A. McDaniel (Eds.), *Prospective Memory: Theory and Applications.* Mahwah, NJ: Erlbaum, pp. 1–22.

Ellis, S. R. and Stark, L. 1986. Statistical dependency in visual scanning. *Human Factors: The Journal of the Human Factors and Ergonomics Society*, 28, 421–438.

Endsley, M. R. 1987. The application of human factors to the development of expert systems for advanced cockpits. In *Proceedings of the Human Factors and Ergonomics Society Annual Meeting September 1987*, 31, 1388–1392. SAGE Publications.

Endsley, M. R. and Kaber, D. B. 1999. Level of automation effects on performance, situation awareness and workload in a dynamic control task. *Ergonomics*, 42, 492.

Eyferth, K., Niessen, C., and Spaeth, O. 2003. A model of air traffic controllers' conflict detection and conflict resolution. *Aerospace Science and Technology*, 7, 409–416.

Fleishman, E. A. 1992. *Fleishman Job Analysis Survey (F-JAS) Administrator's Guide.* Potomac, Maryland: Management Research Institute.

Fleishman, E. A. and Reilly, M. 1995. *Administrator's Guide to the Fleishman Job Analysis Survey (F-JAS).* Potomac, Maryland: Management Research Institute.

Freed, M. A. and Johnston, J. C. 1995. Simulating human cognition in the domain of air traffic control. In M. T. Cox and M. Freed (Eds.), *AAAI Technical Report SS-95-05: Papers from the AAAI Spring Symposium.* Palo Alto, CA: AAAI Press, pp. 39–45.

Gross, H., Friedrich, M. B., and Möhlenbrink, C. 2010. *Development of a Process for the Evaluation of Eye-Fixation Data Using Dynamic Areas of Interest* (Rep. No. HMI Laboratory Report). Braunschweig: DLR.

Harris, J. E. and Wilkins, A. J. 1982. Remembering to do things: A theoretical framework and an illustrative experiment. *Human Learning*, 1, 123–136.

Hart, S. G. and Staveland, L. E. 1988. Development of NASA-TLX (task load index). Results of empirical and theoretical research. In P.A. Hancock and N. Meshkati (Eds.), *Human Mental Workload*. North Holland: Elsevier, pp. 139–181.

Hoc, J. M. 2000. From human machine interaction to human machine cooperation. *Ergonomics*, 43, 833–843.

Kahneman, D. 1973. *Attention and Effort*. New Jersey: Prentice-Hall Inc.

Kirwan, B. 2001. The role of the controller in the accelerating industry of air traffic management. *Safety Science*, 37, 151–185.

Kirwan, B. and M. Flynn. 2001. Controller conflict resolution strategies for CORA2. In *Paper Presented at 4th USA/Europe ATM R&D Seminar*, Santa Fe, NM.

Loft, S., Sanderson, P., Neal, A., and Mooij, M. 2007. Modeling and predicting mental workload in en route air traffic control: Critical review and broader implications. *Human Factors: The Journal of the Human Factors and Ergonomics Society*, 49, 376–399.

Manning, C. A. and Broach, D. 1992. *Identifying Selection Criteria for Operators of Future Automated Systems* (Rep. No. DOT/FAA/AM-92/26). Washington, DC: Federal Aviation Administration Office of Aviation Medicine.

Miller, C. and Parasuraman, R. 2007. Designing for flexible interaction between humans and automation: Delegation interfaces for supervisory control. *Human Factors*, 49, 57–75.

Miller, G. A., Galanter, E., and Pribram, K. H. 1960. *Plans and the Structure of Behavior*. New York: Holt, Rinehart and Winston.

Mogford, R. H. 1997. Mental models and situation awareness in air traffic control. *International Journal of Aviation Psychology*, 7, 331–341.

Moore, K. and Gugerty, L. 2010. Development of a novel measure of situation awareness: The case for eye movement analysis. In *Human Factors and Ergonomics Society 54th Annual Meeting*. Santa Monica, California: Human Factors and Ergonomics Society, pp. 1650–1654.

Nickels, B. J., Bobko, P., Blair, M. D., Sands, W. A., and Tartak, E. L. 1995. *Separation and Control Hiring Assessment (SACHA) Final Job Analysis Report*. Washington, DC: Federal Aviation Aviation Administration Office of the Assistant Administrator for Human Resources Management.

Nijhuis, H. 2000. *Role of the Human in the Evolution of ATM (RHEA)*. Final Report (Rep. No. RHEA/NL/SPR/01). Amsterdam: NLR.

Norman, D. A. and Shallice, T. 1986. *Attention to Action: Willed and Automatic Control of Behavior*. New York: Plenum Press.

Parasuraman, R. and Manzey, D. 2010. Complacency and bias in human use of automation: An attentional integration. *Human Factors*, 52, 381–410.

Parasuraman, R., Sheridan, T. B., and Wickens, C. D. 2008. Situation awareness, mental workload, and trust in automation: Viable, empirically supported cognitive engineering constructs. *Journal of Cognitive Engineering and Decision Making*, 2, 140–160.

Redding, R. E. 1992. Analysis of operational errors and workload in air traffic control. In *Proceedings of the Human Factors and Ergonomics Society Annual Meeting*. London: SAGE Publications, pp. 1321–1325.

Rovira, E. and Parasuraman, R. 2010. Transitioning to future air traffic management: Effects of imperfect automation on controller attention and performance. *Human Factors*, 52, 411–425.

Sarter, N. B., Woods, D. D., and Billings, C. E. 1997. Automation surprises. In G. Salvendy (Ed.), *Handbook of Human Factors and Ergonomics*, 2nd ed. Hoboken, NJ: Wiley, pp. 1926–1943.

Schneider, B. and Konz, A. M. 1989. Strategic job analysis. *Human Resource Management*, 28, 51–63.

Schulze Kissing, D., Zierke, O., Hoermann, H. J. and Eissfeldt, H. 2010. Demonstrating new ATM concepts in a new low cost simulation environment: A human-in-the-loop study to test for future job requirements. *Paper Presented at Conference of the European Association for Aviation Psychology*, Budapest, Hungary.

Seamster, T. L., Redding, R. E., Cannon, J. R., Ryder, J. M., and Purcell, J. A. 1993. Cognitive task analysis of expertise in air traffic control. *The International Journal of Aviation Psychology*, 3, 257–283.

Senders, J. W. 1964. The human operator as a monitor and controller of multidegree of freedom systems. *Human Factors in Electronics, IEEE Transactions on*, HFE-5, 2–5.

SESAR 2012. The European ATM Master Plan—The Roadmap for Sustainable Air Traffic Management. https://www.atmmasterplan.eu/

SESAR Consortium 2007. *The ATM Target Concept—D3* (Rep. No. DLM-0612-001-02-00a). Brussels: Eurocontrol.

SESAR Consortium 2012. The European ATM Master Plan—The Roadmap for Sustainable Air Traffic Management. https://www.atmmasterplan.eu/

Sheridan, T. B. and Parasuraman, R. 2005. Human-automation interaction. *Review of Human Factors and Ergonomics*, 1, 89–129.

Sheridan, T. B. and Verplank, W. L. 1978. *Human and Computer Control of Undersea Teleoperators*. Cambridge, Massachusetts: MIT Man-Machine Systems Laboratory.

Smith, P. J., McCoy, C. E., and Orasanu, J. 2001. Distributed cooperative problem-solving in the air traffic management system. In E. Salas and G. Klein (Eds.), *Linking Expertise and Naturalistic Decision Making*. Mahwah, New Jersey: Lawrence Erlbaum Associates, pp. 367–381.

Stein, E. S. 1992. *Air Traffic Control Visual Scanning*. (Rep. No. DOT/FAA/CT-TN92/16). Atlantic City, NJ: Federal Aviation Administration.

Taylor, R. M. 1990. *Situation Awareness Rating Technique (SART): The Development of a Tool for Aircrew Systems Design* (Rep. No. AGARD-CP-478). Neuilly Sur Seine: NATO-AGARD.

Truman, T. and de Graaff, A. 2007. *Out of the Box: Ideas about the Future of Air Transport, Part 2 Advisory Council for Aeronautic Research in Europe (ACARE)*, Directorate-General for Research ACARE.

Voller, L. and Low, I. 2004. *Impact of Automation on Future Controller Skill Requirements and a Framework for Their Prediction* (Rep. No. HRS/HSP-005-REP-04). Brussels: Eurocontrol.

Wickens, C. D. 2000. *Imperfect and Unreliable Automation and Its Implications for Attention Allocation, Information Access and Situation Awareness* (Rep. No. NAG 2-1120). Moffett Field, CA: NASA Ames Research Center.

Wickens, C. D., Goh, J., Helleberg, J., Horrey, W. J., and Talleur, D. A. 2003. Attentional models of multitask pilot performance using advanced display technology. *Human Factors: The Journal of the Human Factors and Ergonomics Society*, 45, 360–380.

Wickens, C. D., Helleberg, J., Goh, J., Xu, X., and Horrey, W. J. 2001. *Pilot Task Management: Testing an Attentional Expected Value Model of Visual Scanning*. Savoy, IL, UIUC Institute of Aviation Technical Report.

Wickens, C. D., Mavor, A. S., Parasuraman, R., and McGee, J. P. 1998. *The Future of Air Traffic Control: Human Operators and Automation*. Washington, DC: National Academic Press.

Willems, B. 2002. *Decision Support Automation Research in the en route Air Traffic Control Environment*. DTIC Document.

Woods, D. D. and Sarter, N. B. 2000. Learning from automation surprises and "going sour" accidents. In N. B. Sarter and R. Amalberti (Eds.), *Cognitive Engineering in the Aviation Domain*. Hillsdale, NJ: Lawrence Erlbaum, pp. 327–353.

25

Evaluation of an Arrival Coordinator Position in a Terminal Metering Environment

Joey Mercer, Michael Kupfer, Todd J. Callantine, Vimmy Gujral, and Ashley Gomez

CONTENTS

25.1 Introduction

During peak traffic periods, the current air transportation system around terminal areas is impaired by flight inefficiencies, such as frequent vectors and level segments, often at low altitudes. This increases the noise, fuel consumption, and emissions. In consideration of forecasted increases in air traffic demand, Europe's Single European Sky ATM Research Program (SESAR) and the United States' Next-Generation Air Transportation System (NextGen) initiative focus on developing new air traffic management (ATM) technologies, systems, and procedures, such as optimized profile descents, to mitigate these inefficiencies (Federal Aviation Administration, 2013; SESAR Joint Undertaking, 2013).

The National Aeronautics and Space Administration (NASA), the Federal Aviation Administration (FAA), and industry partners are currently working jointly on a multiyear effort, called the ATM Technology Demonstration-1 (ATD-1) (Prevot et al., 2012). Researchers at NASA Ames and NASA Langley are supporting the ATD-1 efforts by conducting human-in-the-loop simulation (HITLS) with the goal of first integrating the different system components into the laboratory setting, followed by refining the system components, tools, procedures, and phraseology, and then studying the overall system performance and comparing it with current-day operations (Callantine et al., 2013; Murdoch et al., 2013; Thipphavong et al., 2013). This chapter focuses on the most-recent simulation conducted in NASA's Airspace Operations Laboratory (Prevot et al., 2014), with a particular emphasis on the strategies employed by the operator working at a position that combined the roles and responsibilities of a TRACON traffic management controller (TMC) and an arrival radar coordinator (ARC).

25.2 Background

IM-TAPSS includes three components: a traffic management advisor for terminal metering (TMA-TM) scheduler, controller-managed spacing (CMS) tools, and flight-deck interval management (FIM) cockpit automation. The TMA-TM is an extension to the currently fielded TMA, and provides a timeline graphical user interface (TGUI) displaying schedule information. The CMS tools, comprised of slot markers, timelines, early/late indicators, and speed advisories, are TRACON decision-support tools that assist controllers in managing delays and delivering aircraft in accordance with the schedule (Kupfer et al., 2011). The airborne spacing for terminal arrival routes (ASTARs) algorithm is used onboard with an FIM-equipped aircraft to support flight crews during airborne-spacing tasks.

The ATD-1 range of operation extends from the TRACON into Center airspace. While still in cruise, TMA-TM assigns runways and computes the estimated times of arrival (ETA) at various scheduling points, such as meter fixes, terminal-area merge points, and runways. Using minimum spacing standards, a scheduled time of arrival (STA) is computed for each aircraft at every schedule point. The STAs are then frozen at a particular distance (i.e., a freeze horizon), providing controllers with a stable target time. Center controllers then begin to reduce schedule delays using speed control and path assignments. FIM-equipped aircraft may receive a clearance to achieve and maintain an air traffic control (ATC)-assigned spacing interval behind a target aircraft. After transitioning into the terminal area, controllers use the CMS tools to issue speeds to an unequipped aircraft to ensure the correct interarrival spacing, while the FIM aircraft continue to manage their speed to work toward the schedule.

25.3 Experiment

The CA5 series of simulations continues with the work of prior ATD-1 simulations (Callantine et al., 2012, 2013), but comprises three separate phases to compare ATD-1 operations with current-day PHX operations. Traffic scenarios are designed to closely resemble the actual PHX traffic and wind data. Conducted in July 2013, the CA5-1 simulation served as a baseline study that only included the TMA-TM component, which was configured to operate like the deployed version of TMA. This chapter focuses on the CA5-2 simulation, conducted in September 2013, which examined an IM-TAPSS environment without FIM. Scheduled for April 2014, the third phase, CA5-3, will include FIM operations.

25.3.1 Airspace, Routes, and Scenarios

Figure 25.1 shows the test airspace, including the test sectors and routes, along with the altitude and speed restrictions associated with waypoints along the routes. The simulation's test airspace comprised three high-altitude sectors (sectors 37, 50, and 93) and five low-altitude sectors (43, 39, 38, 46, and 42) from Albuquerque Center (ZAB), four high-altitude sectors (40, 60, 35, and 36) from Los Angeles Center (ZLA), and four Phoenix TRACON (P50) sectors: two feeder sectors (206-Apache, 203-Quartz) and two final sectors (205-Freeway, 204-Verde). Several confederate ghost controllers also supported the simulation, working on the surrounding en-route airspace, PHX departures, satellite arrivals, and a PHX tower arrival position. Additionally, the simulation staffed both a ZAB TMC position and a P50 ARC position, where the latter also doubled as a P50 TMC.

The CA5 studies simulated both west- and east-flow airport configurations. The published RNAV OPD routes and non-RNAV routes (see Figure 25.1) were adapted based on the existing published RNAV OPD (MAIER5, EAGUL5, KOOLY4, and GEELA6) and non-RNAV (COYOT2, JESSE1, SUNSS8, and ARLIN4) procedures; all feeding the runway 25L and 26 approach procedures during west-flow operations, and the runway 07R and 08 approach procedures during east-flow operations. RNAV routes from the other approach directions included downwind segments that were unconnected. The non-RNAV routes were also unconnected, requiring vectors to the final approach course.

The scenarios mirrored real-world, high-demand arrival flows under realistic wind conditions, with few differences from the live traffic recordings. All scenarios included PHX arrivals and departures, as well as overflight traffic departing from and landing at other

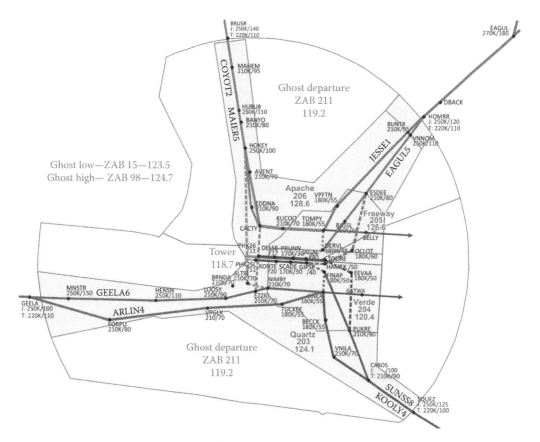

FIGURE 25.1
Simulation airspace showing the test sectors and adapted routes for west-flow configuration.

airports. The arrivals consisted of a range of aircraft-performance capabilities (i.e., jets, high- and low-performing turbo-prop aircraft, and piston aircraft), but only the west-flow scenario contained low-performing turbo-prop and piston aircraft, which did not follow any of the RNAV or non-RNAV routes. These flights approached PHX under the control of the satellite ghost controller's airspace, on routes that were not adapted in (i.e., not known to) the TMA-TM system. Copies of the two traffic scenarios were created that included an increased number of PHX arrivals (six in east flow and five in west flow) to create a period of a sustained, high demand.

25.3.2 Participants

The Center and TRACON sectors were staffed with recently retired ZAB and P50 controllers, with an average of 26.25 years of experience. CA5-2 also included a ZAB TMC, as well as a position combining the P50 TMC and the P50 ARC (referred to throughout this chapter as the ARC). The ARC, who retired in December 2011, had 29 years of experience as a tower/terminal controller; nine of those years as the TMC/ARC at P50 TRACON. Local retired controllers staffed the ZLA, KPHX tower, departure ghost, en-route ghost, and satellite ghost confederate positions. Prior to the CA5-1 simulation, none of the participant controllers had any previous experience with the IM-TAPSS concept and tools, or the multi-aircraft control system (MACS) simulation software (Prevot et al., 2014). Boeing glass-cockpit, type-rated pilots, many of whom had participated in the previous CA simulations, worked on single-pilot mid-fidelity desktop flight simulators, while local, commercial, and student pilots, who were familiar with the MACS, worked on the pseudo-pilot positions.

During the simulations, operator workstations were arranged such that, with the exception of the satellite ghost controller, all confederate ATC positions were located in one room. The TRACON controllers, satellite ghost controller, and ARC were located in a second room, while two additional rooms split the ZAB en-route controllers. The pilots were distributed over separate areas throughout the lab.

25.3.3 ARC Position and Procedures

In today's operations at P50, the person working as the ARC also performs the duties of the TRACON TMC. The ARC's primary responsibility is to provide a plan for integrating the KPHX arrivals that are not sequenced by Apache and Quartz feeder sectors (i.e., satellite arrivals), accommodating aircraft approaching from one side of the airport needing to land on the opposite side (i.e., "cross-overs"), and changing the landing runway of an aircraft whenever it is necessary (Federal Aviation Administration, 2011). The ARC coordinates with the TRACON controllers to identify and create slots in the arrival sequence for accommodating the satellite arrivals. Because current-day operations are not schedule based, the ARC does not interact with the TMA TGUI to modify STAs or sequences. The ARC simply facilitates between-sector coordinations as the P50 controllers maneuver aircraft in delivering them to the runway while maintaining a safe separation.

In contrast, the schedule-based operations of the IM-TAPSS concept require the ARC to find slots in the arrival sequence, or if necessary, create a slot by adjusting the STAs of adjacent flights while minimizing the overall impact to the schedule. In this context, the ARC's "big-picture" perspective enables him to develop a plan for unscheduled arrivals using the various TMA-TM TGUI functions for manipulating STAs: move, swap, change the runway, reset (recomputation of ETA and STA), and reschedule (recomputation of STAs). He then coordinates his plan with the Center TMU and the TRACON controllers.

Early on in the simulation, the ARC created a custom-timeline layout to help him perform his tasks (see Figure 25.2). Four timelines were configured to show a 1-h time period, with the outer-left timeline displaying meter-fix STAs (i.e., TRACON entry times) for aircraft coming over the northern meter fixes, and the outer-right timeline displaying meter-fix STAs for aircraft coming over the southern meter fixes. Furthermore, the outer-left timeline's left side displayed the northwest meter-fix STAs, while the right side displayed the northeast meter-fix STAs. Similarly, the outer-right timeline's left- and right sides displayed the southwest and southeast meter-fix STAs, respectively. The call signs displayed in these two timelines were time shared with the aircraft's assigned runway (left inset in Figure 25.2), allowing the ARC to quickly identify any crossover flights. Additionally, delay values were displayed next to the call signs (right inset in Figure 25.2).

The TGUI's two middle timelines provided schedule information for the runways. The center-left timeline displayed runway ETAs, with north-runway ETAs on the left side, and south-runway ETAs on the right side. The center-right timeline displayed runway STAs, with north-runway STAs on the left side, and south-runway STAs on the right side. Similar to the meter-fix timelines, delay values were also displayed next to the call signs in the runway-STA timeline. On the basis of the TGUI configuration in current use at P50, the ARC explained the careful arrangement of runway ETAs and STAs during one of the runs: "VFR airplanes coming from the Center [airspace], they are never scheduled, but they are always estimated." The two middle timelines, then, helped the ARC to easily assess if slots were available on one of the runways. Although VFR aircraft were not included in the CA5 series of simulations, this example demonstrates the TGUI's importance to the ARC's tasks.

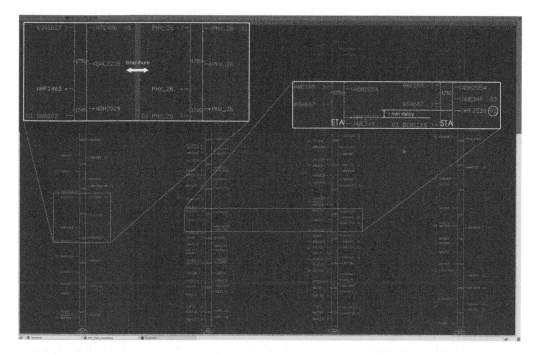

FIGURE 25.2
TGUI setup used by the ARC. Timelines for meter-fix STAs are on the outer left (northern routes) and outer right (southern routes). Runway ETAs for both runways are shown on the center-left timeline, and runway STAs (color-based frozen/unfrozen distinction not depicted here) for both runways are shown on the center-right timeline.

The simulation environment posed some limitations on the realism of the operations. In current-day operations, satellite arrivals are held on the ground until a "call for release" is approved by the ARC. However, in CA5-2, those flights began flying at prespecified times, without the option to delay their departure until a slot in the PHX arrival schedule was available. The slots identified for those arrivals by the ARC sometimes required that large amounts of delay be absorbed, often resulting in the satellite ghost controller vectoring the aircraft. Additionally, STA calculations for these aircraft were unreliable, because the tested version of TMA-TM did not know what their planned routes were. Consequently, the ARC suspended the STAs of those flights (removing them from the STA timeline), developed a new arrival plan, and coordinated with the TRACON controller team.

25.3.4 Training and Data-Collection Schedule

Controller and pilot participants received 4 days of training, including separate and joint exercises, which concluded with a debrief discussion. The data-collection period took place over 4 days during the following week, with experimental trials conducted in each of the airport flow configurations, under four realistic forecast-wind error conditions, with current-day and increased-load traffic scenarios, for a total of 16 1-h data-collection runs. A postrun questionnaire and a 20-min break followed each trial. The study concluded with a postsimulation questionnaire and an in-depth final debrief discussion.

Recorded simulation data logged various metrics such as trajectory- and flight-state information, pilot and controller entries, schedule data, etc. Voice communications between controllers and pilots were recorded using an emulation of the FAA's voice switching and communication system (VSCS). An individual voice recording was obtained from the ARC using a wearable microphone and voice recorder. Additionally, the displays of all workstations, along with the ambient audio, were captured via screen-recording software. Finally, controllers rated their current workload every 3 min, on a scale from 1 (low) to 6 (high), using the air traffic workload input technique (ATWIT) (Stein, 1985).

25.4 Results

The CA5-2 simulation's results highlight how the ARC adjusted to an IM-TAPSS environment. Objective data lend insights into which TGUI features the ARC used to accomplish his task, while subjective data provide details on tool acceptance and traffic management techniques. To emphasize the complexity of merging flights into an already-dense arrival stream, and to highlight ARC-controller coordination, two explicit examples of arrival coordination problems are presented.

25.4.1 TGUI Interactions

The analysis of TGUI log files, cross-referenced with screen recordings, identified the frequency of the ARC's interactions with the TGUI. Owing to software-stability issues identified in the first half of the simulation, this analysis only includes data from the last eight trials of the simulation, all of which were increased-load traffic scenarios with an average of 64 flights entering the TRACON.

TABLE 25.1

Frequencies of the ARC's TGUI Interaction Events (for the Eight Increased-Load Trial Runs)

Event	East-flow (inner circle)	West-flow (outer circle)	
Move STA	8	50	
Reschedule A/C	4	25	
Reschedule A/C and after	0	2	
Indirect STA changes from 'Reschedule A/C and after'	0	15	
Suspend flight	0	20	
Swap assigned runway	1	9	
Indirect STA changes from 'Swap assigned runway'	0	6	
Reset STA	0	4	
Swap two A/C STAs	0	2	
Resume STA computation	0	1	

As seen in Table 25.1, there was a clear difference between the east- and west-flow conditions, as the unscheduled satellite arrivals were only present in the west-flow scenarios. The "Move STA" function, which allows changing an aircraft's STA by a simple drag-and-drop mechanism, was the most frequently used feature, followed by the "Reschedule A/C" function, which automatically finds another slot for the selected aircraft. The "Reschedule A/C and after" function was used less frequently, since it also changed the STAs of all following aircraft. The ARC's decision to twice use the "Reschedule A/C and after" function impacted the STAs of 15 other flights. The "Suspend flight" function excludes the selected flight from being considered by the scheduler, which the ARC often used on the satellite arrivals.

25.4.2 Questionnaire Responses

After the conclusion of each run, the ARC answered questionnaires through which he provided a feedback on the procedures and tool usefulness, and also described his work strategies. In general, the ARC was very confident in using the TMA-TM [7 on Likert scale: 1 Not at all confident—7 Very Confident]. The ARC rated the initial sequence and schedule provided by the TMA as a very workable starting point [7 on Likert scale: 1 not at all workable—7 very workable], and gave ratings of "not at all frustrated" by the metering tools or the metering procedures [1 on Likert scale: 1 Not at all frustrated—7 Very frustrated]. Along with these positive ratings, he provided a suggestion on how to improve his tools: "A TMA 'what-if scenario' feature would be helpful for probing aircraft-pair runway swaps or STA schedule tweaks at the meter-fixes."

25.4.3 Sequence Coordination

The west-flow traffic scenarios contained flights departing from airports close to KPHX (typically, small turbo-prop and piston aircraft), which did not follow the routes used by large turbo-prop and jet aircraft, but instead followed the routes through airspace controlled by the satellite ghost controller. When asked which barriers the ARC saw to being able to use the IM-TAPSS metering concept in real operations, he elaborated on the problem of scheduling satellite arrivals: "In heavier traffic situations, local traffic would have to be scheduled into TMA. [For aircraft without] fixed routings (such as for VFR traffic or traffic from satellite airports), it would be difficult to determine these times. [But] it would be possible to hold slots [for those aircraft] as long as controllers can hit them." For the traffic densities simulated in the study, the ARC rated incorporating the satellite arrivals into the arrival stream as somewhat easy [5 on Likert scale: 1 very difficult—7 very easy].

The ARC indicated that the new metering environment and availability of metering tools significantly changed how frequently he coordinated with his controller team [1 on Likert scale: 1 Coordination decreased a lot—7 Coordination increased a lot]. He also reported that the operations demanded new traffic management strategies for delivering to the schedule [6 on Likert scale: 1 No changes—7 A lot of changes]. To accommodate the satellite arrivals, the ARC developed a general strategy where he first obtained a detailed picture of the arrival situation by coordinating with the TRACON controller team and by studying the TGUI. The ARC explains: "The slot markers dictated [the] scheduled aircraft sequence." Then, he suspended the STAs of the satellite arrivals so as to calm the inherent scheduling complications associated with the unadapted routes. Next, he identified any preexisting excess spacing (i.e., "gaps") between consecutive flights in the arrival stream to accommodate these suspended (i.e., now unscheduled) arrivals. If the gap was not large enough, he tried to extend the gap by modifying the STAs of neighboring aircraft. The ARC also had tools at his disposal for runway or aircraft swaps, allowing him, for example, to check for any unused space in the flow going to the other runway, and giving him the option to move another aircraft into that space to accommodate an unscheduled arrival. Finally, he then coordinated his plan with the satellite ghost controller and the TRACON controller team. This work flow was repeated throughout the simulation.

The following two examples were chosen to illustrate such interactions, and are representative of the ARC's work flow and strategy. Both stem from a west-flow run that includes satellite arrivals. Figures 25.3 and 25.4 visually support the communication excerpts that are used to frame the course of the coordination events. They provide an overview of the track plots of the flights involved in the example problems. The tracks were color coded by a relative simulation time, from between 600 s after simulation start to the end of the run at 3600 s. The information in the two examples is based on data from TGUI and controller screen recordings, as well as audio transcriptions.

Example 1 (Figure 25.3) emphasizes how the ARC coordinated an arrival sequence, focusing on flights from the south side of the TRACON, where two satellite arrivals (CFS806 and N181IS) needed to fit into the arrival stream. During this period, the TMA-TM assigned crossover runways to three aircraft. At 832 s, the ARC discussed the sequencing of two south-side arrivals with the satellite ghost controller: "I don't think [CFS806] can get ahead of the AMF2136. I have one spot behind the guy on the west side [SWA1423]; the only one. And then [AMF2837 will follow]. And after that, I am full for quite some time." At 935 s, he briefed the south-feeder controller about a crossover flight on the KOOLY4 route, coming from the south: "You should have a [runway] 26 [flight] (AWE745) coming up there." At 1146 s, he briefed the north-feeder controller about two other crossover flights

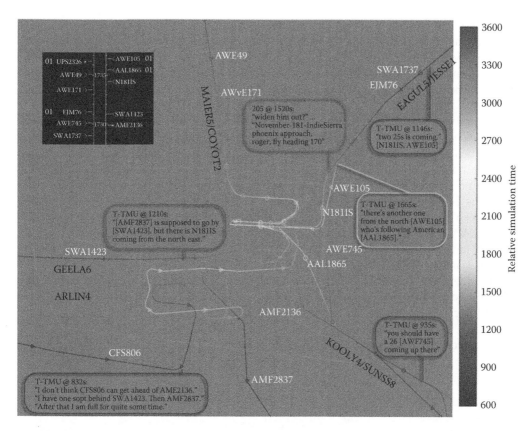

FIGURE 25.3
Overview of the track plots for the flights involved in example problem 1.

on the EAGUL5 route, coming from the north: "Two [runway] 25's are coming: one [N181IS, (pointing at the screen)] and two [AWE105, (pointing at the timeline)]." A few minutes later, at 1210 s, the ARC coordinates the arrival sequence with the south-final controller: "So, the Am-flight [AMF2837] is supposed to go behind Southwest [SWA1423], but there is a Falcon [N181IS] coming from the north-east." The south-final controller then asks for a clarification of the sequence: "Where is the Falcon [N181IS] going? Back here?" The ARC responds: "[With] the Falcon [N181IS] just go [behind the] Am-flight [AMF2837]." The south-final controller points out: "Well, if [the satellite ghost] doesn't keep him tight it's not gonna work." The ARC then verifies with the satellite ghost controller: "You'll work him right behind Southwest [SWA1423], right?" The ARC and the south-final controller agree that AMF2837 will barely fit into the slot. At 1270 s, the ARC comments: "Just eyeballing while I watched him appear, I thought this [SWA] 1423 and [AMF] 2136 is gonna be close […]."

In this first example, the ARC initially discussed the arrival sequence with the satellite ghost and later with the final controllers. He also pointed out crossover flights to the feeder controllers. Later in the same run, at 1520 s, the ARC elaborated on the crossover problem: "She's [north feeder controller] got no idea what's going [on] on the other side, unless I tell her […], and he's [south final controller] got no idea he [N181IS] is coming to his runway. So unless we widen him [N181IS] out…" The north-final controller then interrupted to verify: "Widen him out?" The ARC confirmed: "Yea." The north-final controller gave the clearance: "November 1-8-1 India Sierra, Phoenix Approach, roger, fly heading 1-7-0, vector for spacing." At 1629 s, the north-final controller pointed out the crossover flight to the south-final controller: "He's [N181IS] on the 170 heading right now." The south-final controller commented: "Seems like he fits better on your runway, but I'll take him." The ARC confirmed his plan to the south-final controller and briefs him about AWE105 crossing over from the north to the south runway: "And you don't know this yet, he [AAL1865] goes behind him [N181IS]. There's another [cross-over] from the north [AWE105] who follows American [AAL1865]." The south-final controller sought a clarification with the ARC: "Who goes behind whom?" The ARC then reiterated the sequence: "India-Sierra [N181IS] is going behind Am-flight [AMF2837 coming from the south], American [AAL1865] is going behind Sierra [N181IS]. There is another guy coming from the north [AWE105] that goes behind American [AAL1865]." The north-final controller: "Oh, he's going to... uh... oh! Cactus 1-0-5 reduce speed to 1-9-0. Alrighty then... I see that now." The embedded TGUI screen snippet in Figure 25.3 shows the final arrival sequence: for runway PHX26: SWA1737, AWE745, EJM76, AWE171, and AWE49; for runway PHX25L: AMF2136, SWA1423, AMF2837 (an unscheduled satellite arrival), N181IS, AAL1865, and AWE105.

This section highlights the difficulties that the controllers had in recognizing the intended arrival sequence on their own. They assessed the situation differently than the ARC, without his broader situational awareness. Controllers verified the order with the ARC, who then had to clearly state which flight followed which.

Problem 2 (see Figure 25.4) emphasizes the specific TGUI interactions by the ARC. The ARC was working with the TGUI trying to open slots to fit in satellite arrivals. In the

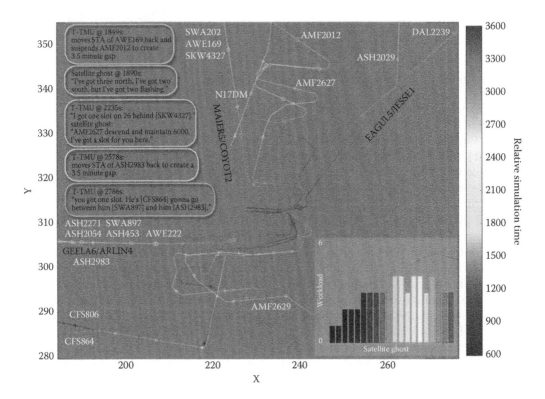

FIGURE 25.4
Overview of the track plots for the flights involved in example problem 2.

meantime, the satellite ghost controller had to issue delay vectors to several flights until any arrival slots became available.

At 1849 s, the ARC modified the arrival schedule via the TGUI interface, moving the frozen STA of AWE169 from 17:56:20 back to 17:56:50, and suspending the STA of AMF2012 to open up a gap after SKW4327. He discussed the sequencing problem with the satellite ghost controller at 1890 s: "What do you got? Three [flights in the] south and one or two [flights in the] north? Three north?" The satellite ghost responds: "I've got three [flights in the] north, I've got two [flights in the] south, but I've got two flashing [incoming, in handoff status]." The ARC replies: "Well, I'm gonna have two slots for you in a little bit." The satellite ghost answers: "Alright then, I'll bring that guy [N17DM, landing at Phoenix Deer Valley Airport] to the south, and that's probably gonna do it for me for a while. Not a lot of room here."

At 2235 s, the ARC informed the satellite ghost about an open slot: "I got one slot on [runway] 26 behind him [SKW4327]." Replying "Just one, okay… Here's my man.," the satellite ghost controller cleared AMF2627 (from the north) for the descent: "AMF2627 descend and maintain 6000. I've got a slot for you here." At 2578 s, the ARC moved the frozen STA of ASH2983 from 18:03:40 back to 18:04:30 to open up a gap after SWA897. The ARC briefed the satellite ghost at 2786 s about two more gaps available for the satellite arrivals: "He's [CFS864 from the south] gonna go between him [SWA897] and him [ASH2983] […] and the other one [AMF2012] goes behind him [DAL2239]."

This section illustrated the ARC's strategies when interacting with the TGUI. He identified and enlarged the preexisting gaps in the schedule by utilizing the allowable STA range and/or adding small amounts of delay to the flights. The satellite ghost had more flights to control than there were sequence slots available, resulting in the drastic delay vectors of the satellite arrivals shown in Figure 25.4. The workload histogram for the satellite ghost position embedded in Figure 25.4 shows a workload increase at 1800 s. At this time, he had six flights in his control, and most of them were on vectors (the workload histogram uses the same time-based color coding as Figures 25.3 and 25.4).

25.5 Conclusions

Future schedule-based arrival operations, such as the IM-TAPSS concept, rely on the integrity of an arrival schedule to provide system benefits (e.g., a sustained, high-runway throughput, as well as the fuel savings from enabling continuous descent operations). Without TMA-TM tracking the satellite arrivals, an unfortunate consequence was that the CMS tools (slot markers, timeline ETA/STA information, and speed advisories) were unavailable to assist the controllers, who, without such guidance, may potentially issue less-efficient clearances to the aircraft. However, even if most of the arrival traffic is accounted for by future scheduling systems, other issues will arise, as unforeseen events will likely still occur. Unscheduled (VFR) traffic or off-nominal flights (e.g., go-arounds), are the examples of such unexpected events. Accommodating those flights, then, will impact the arrival schedule.

The simulation described in this chapter investigated the role, responsibilities, and strategies of an operator who, during the simulation, served as both the P50 TRACON TMC and ARC. He had, relative to the individual TRACON sector controllers, a more complete picture of the overall arrival operations. Therefore, he was well suited to develop efficient plans and solutions. The simulation illustrated his function as a schedule manager and team strategist, and showed that he played an integral part in coordinating, adjusting, and instantiating the arrival schedule.

The results underline the importance of such a position in schedule-based arrival management environments. To create room in the arrival sequence, flights in current-day operations are maneuvered as necessary to maintain a safe separation from other aircraft, with no concern for meeting specific arrival times. In a schedule-based environment, however, the arrival schedule imposes additional constraints on the possible control actions. The ARC was able to identify the naturally occurring gaps to help accommodate any flight not tracked by the scheduler. When necessary, he adjusted other aircraft STAs to increase the gaps, while limiting disruptions to the overall schedule. Because the goal was to minimize the impact on the overall schedule, he looked for solutions that manipulated the schedule of only one flight, leading to his often using the "Move STA" function. When traffic densities are high though, and the available gaps are sparse, it is especially difficult to accommodate any unscheduled flights. Vectors have to be issued to delay the flights until a slot becomes available. This translates to increased fuel consumption and emissions, and increased controller workload.

Given the complexity of the arrival problems, a clear coordination between the ARC and the controller team about the intended plan is crucial to avoid any misperceptions. In the examples provided, controllers seemed to be sometimes confused about the intended arrival sequence (e.g., "Where is the Falcon going?"), or even surprised (e.g., "Oh! I see that now."). Besides the unscheduled arrivals, crossover flights to other runways also necessitated coordination among the team: controllers often had to check back with the ARC to verify the desired sequence.

Similar to the findings from Martin et al. (2012), a "what-if-feature" built into the TGUI would help the ARC to independently play out several strategies before implementing them into the active schedule. In the current software, however, each modification of aircraft STAs by the ARC immediately impacted the CMS tools, leading to unintended artifacts, such as jumps in the positions of the slot markers.

The CA5-2 study provided valuable insights on the role of an ARC position within a schedule-based terminal metering environment. Upcoming research will investigate the ARC's role when FIM operations complement the CMS tools.

References

Callantine, T., Cabrall, C., Kupfer, M., Omar, F., and Prevot, T. 2012, Initial investigations of controller tools and procedures for schedule-based arrival operations with mixed flight-deck interval management equipage, *Proceedings of the 12th AIAA Aviation Technology, Integration, and Operations (ATIO) Conference*, Indianapolis, IN, USA.

Callantine, T., Kupfer, M., Martin, L., and Prevot, T. 2013, Simulations of continuous descent operations with arrival-management automation and mixed flight-deck interval management equipage, *Proceedings of the Tenth USA/Europe Air Traffic Management Research and Development Seminar*, Chicago, IL, USA.

Federal Aviation Administration. 2011, *Order P50 TRACON 7110.1A*, Washington, DC, USA.

Federal Aviation Administration. 2013, *FAA NextGen Implementation Plan*, Washington, DC, USA.

Kupfer, M., Callantine, T., Martin, L., Mercer, J., and Palmer, E. 2011, Controller support tools for schedule-based terminal operations, *Proceedings of the Ninth USA/Europe Air Traffic Management Research and Development Seminar*, Berlin, Germany.

Martin, L., Mercer, J., Callantine, T., Kupfer, M., and Cabrall, C. 2012, Air traffic controllers' control strategies in the terminal area under off-nominal conditions, *Proceedings of the 4th International Conference on Applied Human Factors and Ergonomics*, San Francisco, CA, USA.

Murdoch, J., Wilson, S., Hubbs, C., and Smail, J. 2013, Acceptability of flight deck-based interval management crew procedures, *Proceedings of the AIAA Modeling and Simulation Technologies (MST) Conference*, Boston, MA, USA.

Prevot, T., Baxley, B., Callantine, T., Johnson, W., Quon, L., Robinson, J. et al. 2012, NASA's ATM Technology Demonstration-1: Transitioning fuel efficient, high throughput arrival operations from simulation to reality, *Proceedings of the International Conference on Human–Computer Interaction in Aerospace (HCI-Aero)*, Brussels, Belgium.

Prevot, T., Smith, N., Palmer, E., Callantine, T., Lee, P., Mercer, J. et al. 2014, An overview of current capabilities and research activities in the Airspace Operations Laboratory at NASA Ames Research Center, *Proceedings of the 14th AIAA Aviation Technology, Integration, and Operations (ATIO) Conference*, Atlanta, GA, USA.

SESAR Joint Undertaking. 2013, *SESAR Release 2013*, Brussels, Belgium.

Stein, E. 1985, *Air Traffic Controller Workload: An Examination of Workload Probe, (DOT/FAA/CT-TN84/24)*, Federal Aviation Administration Technical Center, Atlantic City International Airport, NJ, USA.

Thipphavong, J., Jung, J., Swenson, H., Witzberger, K., Martin, L., Lin et al. 2013, Evaluation of the controller-managed spacing tools, flight-deck interval management and terminal area metering capabilities for the ATM Technology Demonstration #1, *Proceedings of the Tenth USA/Europe Air Traffic Management Research and Development Seminar*, Chicago, IL, USA.

26

Identifying Markers of Performance Decline in Air Traffic Controllers

Tamsyn Edwards, Sarah Sharples, Barry Kirwan, and John Wilson

CONTENTS

26.1 Introduction

Air traffic control (ATC) is a complex and highly dynamic safety-critical environment. Within this environment, air traffic controllers (ATCOs) are responsible for flight safety and efficiency. Unlike other safety-critical industries, there are no physical barriers or defenses that protect aircraft in flight. It is therefore essential that ATCOs maintain a consistently high standard of human performance to maintain flight safety.

With a large potential for incidents, air traffic management (ATM) is remarkably reliable (Amalberti and Wioland, 1997). However, it is widely documented that human factors, such as workload and fatigue, are the "major determinants of human error" (Park and Jung, 1996, p. 330) and can negatively influence human performance in air traffic control environments (e.g., Cox-Fuenzalida, 2007; Edwards, Sharples, Wilson, and Kirwan, 2012). Traditionally within the discipline of human factors, the investigation of human performance and the association with potential performance-influencing factors has focused on a wide range of methods, from laboratory experiments to "research in the wild" (Sharples, Edwards, and Balfe, 2012). However, little research has focused on the identification and application of "signs and symptoms," or indicators, of the negative association between human factors and performance, that is, indicators that may provide a signal that a controller is reaching their personal "edge" of performance.

Previous research aligned with this topic has typically focused on the identification of control strategies utilized to support controllers in maintaining performance under demanding conditions. For example, Sperandio (1971) identified that controllers utilized adaptive strategies to manage the increases in traffic load by decreasing the amount of time processing each aircraft. This workload regulation enabled the performance to be maintained in conditions that otherwise would have led to an excessive workload (called an "overload" situation) (Wickens, Mabor, and McGee 1997). In addition, Kontogiannis and Malakis (2013) identified behavioral markers for air traffic control strategies that were applied to cope up with complex situations. This research has provided a valuable insight into the maintenance of performance under demanding conditions in an air traffic environment. However, a comprehensive set of indicators that show when controllers are reaching their edge of performance, developed based on real-world operations, has not yet been identified.

The investigation of indicators of performance decline has both theoretical and practical motivations. The identification of behavioral indicators that may be associated with performance will contribute to the relatively sparse literature on behavioral indicators of performance (Sharples et al., 2012). A practical application of the findings is a contribution to the identification of indicators of potential performance decline, enabling those in operational contexts to put mitigations in place before system performance and safety degrades. The research presented in this chapter aims to address this gap in research by identifying indicators that controllers may be reaching their edge of performance, based on data from real-world operations. The principal aim of the research presented in this chapter was to generate expert opinion regarding the indicators of potential performance decline in an ATC setting. A subaim was to identify indicators that were commonly associated with specific human factor influences (e.g., workload influences) on performance.

26.2 Method

26.2.1 Design

A total of 22, 1 hour, face-to-face semistructured interviews were conducted with en-route ATCOs. The sample was a nonprobability purposive sample. Participants were recruited by Maastricht Upper Area Control (MUAC) center managers, and selected to represent subgroups of controllers based on age, sex, experience, and role. The number and length of interviews was based on pragmatic considerations. Interviews took place at MUAC, in the Netherlands. A protocol was used to standardize the interview procedure. An interview schedule was developed to guide the semistructured interview; participants were asked predesigned lead questions that were then followed by probes.

Workload, stress, and situation awareness (SA) had been identified through previous studies (Edwards, Sharples, Wilson, Kirwan, and Shorrock, 2010) as three human factors that were important in an ATC environment, and had the potential to influence ATCO performance. Therefore, these factors were specifically selected to be included in the interviews to further explore the relationship between these factors, associated indicators, and performance decline in ATCOs. Participants were provided with definitions of each factor, developed during a previous study from a review of the literature and feedback from two ATCOs (Edwards et al., 2010). For the factor of workload, as the task of an ATCO is primarily cognitive, the interviews focused specifically on mental workload. The definition utilized for workload was "experienced demand (amount and complexity) imposed by ATC tasks, and associated subjective perception of effort to meet demands." Task demand (also referred to as taskload) and workload were differentiated during the interviews; task demand was described as the objective demands of a specific task or situation, compared to workload that refers to the individual's perception of the objective task demands and the capability of the individual to meet those demands. In relation to the factor of stress, stress was defined as "pressures imposed by situations which challenge the controller's ability to cope." Several stressors exist within the ATC environment that can influence the perception of stress, and ATCO performance. The third factor that was included in the interviews was situation awareness (SA). SA is important for a successful performance in ATC (Endsley and Rodgers, 1994). ATCOs must continually assess dynamic information sources to develop a mental representation of aircraft in three-dimensional (3D) space and projected the future aircraft locations. Controllers call this "the picture" (Endsley and Rodgers, 1994). The definition of SA used in this chapter was "maintenance of a coherent mental picture for current and future events based on continuous extraction of environmental information, which includes controller performance."

Participants were first asked about their experience of the existence and use of indicators of potential performance decline in an ATC domain. Participants were then asked to list the indicators of potential performance decline that were perceived to be associated with the specific factors of workload, stress, and SA (e.g., "What indicators of potential performance decline are associated with the influence of high workload on performance?"). Interviews were tape recorded and were then orthographically transcribed. Thematic analysis (Strauss and Corbin, 1990) was applied as the analysis strategy.

26.2.2 Participants

In total, 22 en-route controllers were interviewed. The majority of participants were male (17, or 77.27%), compared to female participants (5, or 22.73%). All participants worked as en-route controllers in the Maastricht Upper Airspace control (MUAC) center. Participants' ages ranged from 21 to 60. Participants responded to grouped age ranges and so an average age could not be calculated. A total of five participants (22.72%) were in the 21–30 age range, and a majority of nine participants (40.9%) were in the 31–40 age range. A total of seven participants (31.81%) responded to the 41–50 age range and one participant was in the 51–60 age range (4.54%). All participants were qualified ATCOs who had completed training. Years of experience as an ATCO (excluding training) ranged from 1.5 to 31 years (M = 14.55, SD = 8.68). A total of 15 participants had worked as an on-the-job training instructor (OJTI). Years of experience as an OJTI ranged from 2 to 25 years (M = 10.93, SD = 7.11). In total, six participants were also supervisors. The experience as a supervisor ranged from 1.5 to 11 years (M = 6.08 years, SD = 3.56).

26.2.3 Materials

A schedule was designed to structure the interview, which comprised of 11 open-ended lead questions relating to four areas of interest:

1. Meaning and use of indicators in ATC settings
2. Categorization of indicators of human performance limits
3. Indicators associated with specific human factors (workload, stress, and SA)
4. Awareness of own indicators and colleagues' indicators

The interview schedule was reviewed and approved by two human factors experts and two ex-controllers. The lead questions were arranged from general topics (e.g., "Could you please tell me about a time in the recent past when you identified that either yourself or a colleague was nearing the edge of your performance?") to more specific questions (e.g., "To what extent are the indicators you've specified common between controllers?"). This format was selected to lead participants through the topics in a logical progression (Millward, 2006). The interview was recorded on two Olympus DSS standard digital recorders.

26.2.4 Procedure

The participant was welcomed to the interview room and provided with a standardized brief. The participant was then asked to sign an informed consent form if he/she was happy to continue, and completed a demographic questionnaire. The interview began with an open question, followed by several probes. Once the interview was complete, the participant was thanked for his/her time and given a standardized debrief that contained the researcher's contact details.

26.2.5 Strategy of Analysis

Interviews were transcribed orthographically. The level of detail resulting from orthographic transcription was sufficient for the aims of the research and the method of analysis. Only the words that were spoken were captured in the transcription. False starts or self-corrections were included, but no paralinguistic features were captured (such as sighing, intonation) (Strauss and Cobin, 1990; Wilkinson, 2003). Thematic analysis was selected as the analysis strategy (Strauss and Cobin, 1990). In line with the thematic analysis procedure (Strauss and Cobin, 1990; Wilkinson, 2003), the transcripts were read through, and elements of participant responses that were related to the aims of the study were identified. The transcripts were then reread with the aim of categorizing the identified elements into emerging themes. No identifying information was stored in the transcription. Where quotations are used, participants remain anonymous.

26.3 Semistructured Interviews: Results

26.3.1 Controllers' Understanding and Use of Indicators of Potential Performance Decline: Does the Concept of "Performance Indicators" Have Face Validity with Controllers?

All participants (22/22) were familiar with the concept of indicators and confirmed that indicators occurred in the ATC operations room. In general, participants characterized indicators as a sign that a controller was not feeling comfortable with the control task. All controllers stated that indicators are commonly used in ATC to provide information regarding when a controller may not be controlling optimally, or a factor (such as fatigue or stress) is negatively influencing the performance. Indicators also provided information about when a controller was moving toward the edge of performance: *"The indicators are part of losing the control or going towards the limits or crossing the limits [of performance]." (Participant 21).* All controllers monitor their own personal indicators as well as colleagues' indicators: *"… We work close together, we monitor each other, you would be very aware of the person sitting beside you whether they're on the ball or whether they're slightly less, whether they're tired, whether they're distracted by whatever, it's part of the job, and you make allowances" (Participant 2).* This was perceived as a natural process that *"you don't think about … I just do it like it's a brain process that isn't conscious" (Participant 2).* Another controller stated *"I think it's a natural thing to look for signs" (Participant 11).* Controllers used indicators for the primary functions of gaining information that subsequently led to supporting own, or colleagues', performance.

26.3.2 Categorization of Indicators Used by ATCOs: How Do Controllers Conceptualize Indicators?

The participants distinguished between *internal*, subjectively experienced indicators and *external*, observable indicators. Figure 26.1 provides an initial broad categorization of indicators based on qualitative data from controllers. It should be noted that the categories are not mutually exclusive. An indicator can be both experienced internally and also overtly reflected in performance, such as a feeling of being "slow" and falling behind with traffic.

Internal indicators may alert the controller to a specific state or a negative influence on performance: *"I know that when I start thinking, 'Oh it's going fine', I've learned that I force myself to tighten the bolts and to really pay extra attention" (Participant 1).* Another said that they change their control strategy *"when I start getting a little bit nervous" (Participant 2).*

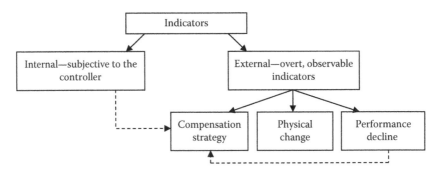

FIGURE 26.1
Categorization of indicators produced from qualitative data internal indicators.

26.3.2.1 External Indicators

In contrast, external indicators are overt and observable. Three broad subcategories of external indicators emerged from participant responses (Figure 26.1). Examples of external indicators that the controllers identified in the interviews are underlined throughout this section. Changes to a controller's performance, such as a performance decline, serve as an external marker to the controller and their colleagues that the controller may not be comfortable. One participant explained: *"When people are controlling you expect a specific performance meaning that they are not going to give too many unnecessary clearances to the aircraft...if you have people doing this, giving an alternative level for no reason then you just start to wonder, 'Why is he doing this?' Okay, he does it once. He was not paying attention on this one. He does it twice and then you start wondering, 'Well it's not his day'"* (Participant 1). Another subcategory of an external indicator is behavioral and physical changes in a controller. Examples include a face becoming red, or fidgeting, and these provide information to other controllers: *"You see it coming, you see them getting nervous, you see them talking faster"* (Participant 2).

Adaptive changes to the control strategy to mitigate negative influences on performance were also identified as a subcategory of external indicators. Employing compensation strategies provides information that a controller is feeling uncomfortable, but is aware of the present situation, and is attempting to protect and maintain the performance. As one controller summarized *"When somebody is just extra careful, I suppose that it's because they feel that they have to be extra careful"* (Participant 1).

If a controller is aware of the present internal indicators, then, a compensation strategy may be selected and applied. This is represented in Figure 26.1 by an arrow with a dashed line connecting the internal marker to the compensation strategy. This is then observable in the method of controlling. Alternatively, a decline in performance may alert the controller that a compensation strategy is required to protect performance. This is represented by an arrow with a dashed line connecting performance decline and compensation strategy. In addition, colleagues may observe the controller's performance decline and apply their own compensation strategy to support the controller's performance.

26.3.3 Subcategories of External Indicators May Provide Distinct Information: Where Are the "Hard Edges"?

Performance declines or errors may provide a more serious indication that a controller is experiencing difficulties with the task. Controllers appear to place weight on this category of an indicator and provide support in response to these indicators: *"If I see that someone is* <u>*correcting themselves*</u> *very often then I would pay a lot of attention to what he's actually doing...I really follow every single clearance. I will try to focus more on what my controller is doing and try to support as well like giving hints" (Participant 11).* A physical change (e.g., red face, yawning, and laid-back posture) may indicate a change in controllers' cognitive state (Sharples et al., 2012), although it may not be related to the feelings of discomfort. Controllers suggested that the meanings of indicators are dependent on the context. Adaptive changes in the control strategy inform colleagues that the executive controller (EC) (also known as the "tactical" controller) is experiencing discomfort with the control task (the reason for discomfort may not be observable), although the EC is aware of this and is attempting to protect and maintain performance with the application of a compensation strategy.

26.3.4 External Indicators May Mean Different Things Depending on Context: Do Controllers Moderate the Importance of the Indicators According to What Is Going On?

Indicators may mean different things to controllers also depending on the control context. Participant 1 suggested: *As a coordinator controller you follow what the executive is doing and there is a variety of complexity levels for situations and if it's an easy situation, a very crystal clear solution to a problem and then you see that the person is not applying it straightaway it triggers maybe a little alarm in your head.* This may have been interpreted differently during a high taskload. Controllers use experience and knowledge of the control situation to interpret the meaning of the indicator.

26.3.5 Indicators of Specific Factor Influences: Which Human Factors, and the Associated Indicators, Are the "Usual Suspects" for Controllers?

Participants were asked to identify internal or external indicators that they believed to be associated with a specific influencing factor (workload, stress, and SA). All participants naturally reported adaptive compensation strategies that were applied in response to the detection of a potential performance decline, listed below as one category of an external indicator.

26.3.6 Workload

When talking about workload as a potential influence on performance, controllers differentiated between high workload, low workload, and transitions between workload extremes. Each form of workload was reported to be associated with different indicators.

26.3.6.1 High Workload

Participants reported internal (Table 26.1) and external (Table 26.2) indicators of potential performance decline that were associated with high workload. The findings were grouped into categories, developed from controllers' responses. Changes to the subjective feelings and performance were reported as important indicators that a controller may be reaching the edge of performance: *"If you start to miss the things that you should be doing at certain times, it gets exponentially busier and then you can't catch up anymore" (Participant 4).* In comparison, physiological change and visible cues indicators were not interpreted to indicate that a

TABLE 26.1

Indicators Internal to the Controller

Proposed Category	Marker
Cognitive changes	Don't know the next steps
	Increased focus
	Calls are a surprise
	Reduced self-awareness
Changes to control	Reactive
	No back-up plan
	No space for an unexpected event/ working to capacity
	Future plan reduces in minutes
Physiological changes	Heartbeat is faster
	Sweat
	Red cheeks
Subjective feeling	Feeling of losing control
	More traffic than can handle
	Panic and uncertainty
	Not comfortable

controller was reaching the edge of performance or that a potential performance decline was likely.

Controllers reported using specific compensation strategies in high taskload periods if they were aware of the potential performance decline (Table 26.3). These were primarily control strategies such as reducing the efficiency to ensure safety, or going "back to basics" to ensure that all aircraft are safe. Several respondents reported that ECs became less self-aware under periods of high workload and therefore more reliant on the coordinating controller (CC) to apply compensation strategies: *"They start to swim... the planner next to them is very much paying attention and they tell them 'Okay now you do this, now you do this, now you do this'"* (Participant 14). Knowing that a high-taskload phase is coming, and the preparation for a high taskload was reported to be the most effective strategy.

TABLE 26.2

Observed Indicators

Proposed Category	Marker
Perception changes	Executive doesn't hear colleagues
Visible cues	Fidgety
	Move closer to the screen
	Colleagues not talking to one another
Changes to voice	Talking faster/more "say again"s (from pilots)
	Tone of voice
Performance changes	Miss actions
	Mixing call signs
	Can't see a simple solution
	Overlook an aircraft

TABLE 26.3

Compensation Strategies That Were Also Identified as Indicators

Category	Compensation Strategy
Control strategy	Less prioritization on efficiency and more on safety
	Back to basics
	Defensive controlling
	Keep talking so that pilots cannot interrupt
	Quicker decisions but less considered
Verbal changes	No pleasantries
	Speak slowly
Support from CC	Seek guidance from CC
Increase the field of awareness	Sitting back

26.3.6.2 Low Workload

Tables 26.4 through 26.6 list the internal, external, and compensation strategy indicators of reaching the edge of performance during low workload. In comparison to high workload, the indicators reflect the influences on performance through potential boredom or relaxation, leading to distraction: *"In low workload, there's nothing to do so you start doing other things, boredom becomes an issue and then you start talking or having a chat or doing whatever*

TABLE 26.4

Indicators Internal to the Controller

Proposed Category	Marker
Cognitive changes	Pay less attention
	Easily distracted
	Reduced awareness
	Reduced self-awareness
Changes to control	Leave situations to develop for longer
	Trying to create more complex situations
	Less safety buffer
Subjective feeling	Boredom
	Relaxed

TABLE 26.5

Observed Indicators

Proposed Category	Marker
Perception changes	Incorrect assessment of a situation
Visible cues	Sit back on a chair
	Away from the radar screen
	Talking to a colleague
Performance changes	Overlooking an aircraft
	Forgetting an aircraft
	Repeated "sloppy" mistakes
	Fall behind traffic due to distraction

TABLE 26.6

Compensation Strategies Which Are Also Indicators

Repeatedly Check Situation

Sitting forward in low-workload periods: trying to concentrate on the problem

and it's, yeah, you can miss things" (Participant 10). Interestingly, controllers reported leaving a problem to develop for longer or creating complex situations to reduce the boredom during periods of low workload, which could ultimately create a potential uncomfortable situation. This result demonstrates that it is essential to capture the context in parallel with an indicator of performance decline, to ensure an appropriate interpretation.

26.3.6.3 Workload Transitions: When Controllers Need to Shift Gear

The transitions between taskload extremes (low–high and high–low) were associated with specific indicators (Tables 26.7 through 26.9). The indicators were different depending on the direction of transition. A transition from low-to-high taskload required controllers to change their "state" to meet the speed and demands of the traffic, known as a *"gear shift"*

TABLE 26.7

Indicators Internal to the Controller

Potential Category	Marker	Transition Direction
Cognitive changes	Fall behind	Low–high
	No plan	Low–high
	Lack of awareness	Low–high
	Gear shift	Low–high
Subjective feeling	Relax	High–low
	Tiredness	High–low

TABLE 26.8

Observed Indicators

Potential Category	Marker	Transition Direction
Performance changes	Overlooking an aircraft	Low–high High–low

TABLE 26.9

Compensation Strategies Also as Indicators

Potential Category	Marker	Transition Direction
Change in control style	Lower complexity in preparation	Low–high
	"Relax" between busy periods	High–low
Subjective experience	More effort to concentrate	High–low
	Conscious internal reminder to focus	High–low

(Participant 4). Indicators that controllers may not be performing optimally during this transition included falling behind the traffic and losing awareness. Indicators associated with high-to-low taskload transitions were mostly characterized as emerging from a feeling of relaxation after the traffic peak and a resulting loss of concentration: *"… The edge of your performance, that's probably the reason why things go wrong, just after a busy period because people start relaxing and the adrenalin goes away and you lose your concentration" (Participant 16).*

Several distinct types of performance decline are associated with workload, such as an overlooking aircraft (vigilance issues) and mixing call signs (communication issues). Factors do not occur in isolation, but instead can co-occur, and interact, to produce a cumulative impact on performance (Edwards et al., 2012). For example, workload may negatively influence other factors such as vigilance, fatigue, and awareness that are then observed to be a causal factor of performance decline. This is an important finding for understanding the underlying causes of performance decline that may manifest as a result of a different factor.

26.3.7 Stress

Although respondents differentiated between stress resulting from personal situations and task-related stress, participants reported that both negatively influence performance, and suggested that the indicators and influences of stress on performance were the same regardless of the cause. Respondents also differentiated between "positive" stress and stress that results in a negative experience. Participant 11 explained: *"It's almost excited because there is more traffic coming. It's a different situation if someone is already in a complex situation, you realise he is falling behind then it's a different impression you get from the person."*

Only indicators of stress that influenced controllers negatively were discussed. Respondents emphasized the changes in a subjective feeling, such as feeling tense, uncomfortable, and anxious, as the unambiguous indicators of stress (Table 26.10). This suggests that stress may affect the subjective experience and the associated cognitive changes rather than performance directly. Several observable indicators (Table 26.11) were the manifestations of emotional responses, such as frustration and demonstrations of anger, and the associated physiological changes such as vocal changes, shaking, and fidgeting. Compensation strategies (Table 26.12) were designed to counteract the influences of stress on the controller, such as emotion regulation and practical strategies such as reducing the rate of speech. Support was sought from the CC to further protect performance.

TABLE 26.10

Indicators Internal to the Controller (Negative Stress)

Proposed Category	Marker
Cognitive changes	Start to think slower
Physiological changes	Heartbeat
	Sweat
Subjective feeling	Not coping
	Feeling of doing badly / uncomfortable (negative)
	Anxious (negative)
	Nervous
	Tense

TABLE 26.11

Observable Indicators

Proposed Category	Marker
Visible cues	Fidgeting
	Red cheeks/neck, flushed
	Sit closer to the screen
Changes to voice	Speaks faster (negative)
	Speaks higher (negative)
	Speaks louder/quieter than usual
Demeanor	Easily frustrated
	Angry/confrontational
Verbal cues	Ask to open a sector
	Communication changes
	Shouting

TABLE 26.12

Compensation Strategies Also as Indicators

Proposed Category	Compensation Strategy
Verbal changes	Speak slower
	More authoritative in instruction
Support from CC	Pay more attention to EC's actions
Emotion regulation	Reduce stress
	Sit back, reduce the anxiety
	Relax

26.3.8 Situation Awareness

Controllers referred to a decline or loss of SA as "losing the picture." The loss of SA was reported to be progressive and occurs in stages that were associated with different indicators: *"It starts off by just falling behind a bit. So you might just be a few steps behind what you're supposed to be doing and if that builds up too much then you will get to the point where you start to lose the picture" (Participant 20).* Therefore, below, indicators of a controller losing the picture, or having lost the picture, are differentiated (Tables 26.13 through 26.16).

The decline of SA was reported to be influenced by the presence of a high or low task demand. The progressive decline of SA was only reported under the conditions of a high taskload. In low traffic, the loss of awareness was more instantaneous: *"We sort of relaxed,*

TABLE 26.13

Indicators Internal to the Controller When Losing the Picture

Proposed Category	Marker
Cognitive changes	Difficulty in selecting priorities
	Thinking while giving the clearance
	Tunnel vision/hearing
Subjective feeling	Underconfident

TABLE 26.14

Indicators Internal to the Controller
Having Lost the Picture

Proposed Category	Marker
Cognitive changes	Everything as a surprise
	No plan
	Cannot see a solution
Changes to control	Reactive control
Subjective feeling	Panic

TABLE 26.15

Observable Indicators of Losing the Picture

Proposed Category	Marker
Visible cues	Slow at the task
Performance changes	Running behind
	Time working ahead degrades
	Missing calls

TABLE 26.16

Observable Indicators of Having Lost the Picture

Proposed Category	Marker
Visible cues	Zig-zagging head movement of where to look
	"Blacked out"/silent
Verbal cues	Asking for confirmation
Performance changes	Unsafe clearance
	Unexpected decisions
	Jumping from one aircraft to another
	Don't know who's calling
	Don't react correctly

'Oh, it's done now,' eating a sandwich and the fourth [aircraft] both of us had forgotten about it or not assessed it, but suddenly it's flashing and we're, 'How did we miss that?'" (Participant 4). Controllers felt that the recovery of SA was easier in periods of low traffic as compared to high traffic.

Compensation strategies (Table 26.17) protect performance when a controller is losing the picture. It was reported to be difficult to rebuild awareness after losing the picture. The compensation strategies from the EC primarily attempt to make the situation safe when awareness is degraded. Compensation strategies by the CC are tactical and appear to facilitate the EC in rebuilding the picture. For example, CCs will change the control strategy to reduce the complexity and/or traffic frequency to allow the EC to catch up and rebuild the picture. In addition, CCs will monitor the EC's instructions and step in if necessary until the EC has recovered the picture *"They [CC] tell them 'Okay now you do this, now you do this, now you do this'"* (Participant 14). The more degraded awareness is, the more reliant the EC may be on the CC to protect performance and rebuild the picture.

TABLE 26.17

Compensation Strategies Also as Indicators

Proposed Category	Compensation Strategy
Control strategy	Build the plan as go
	Conservative clearances
	Reduce the complexity
Prevention	Prevention—freeing up space
Support from CC	Get CC to decrease traffic load to allow to build up the picture again—to catch up
	CC to monitor controllers' actions
	CC provides instructions

26.3.8.1 Progression to the Edge of Performance: A Slippery Slope?

Progression to the edge of performance was spoken about in terms of "stages" and the associated experiences. Figure 26.2 provides a representation of the controllers' experience at each stage, developed from participant responses.

Performance is first represented as an uneven line; there will often be minor changes in performance although the overall performance is maintained to a consistently high standard. Within the region of safe performance, controllers experience nominal situations daily that are addressed with relative comfort. Performance is maintained. If demand increases, then, the controller may experience a subjective discomfort with the task. However, respondents suggested that these experiences are seen as "part of the job" and something that all controllers should deal with. Although experiencing a subjective

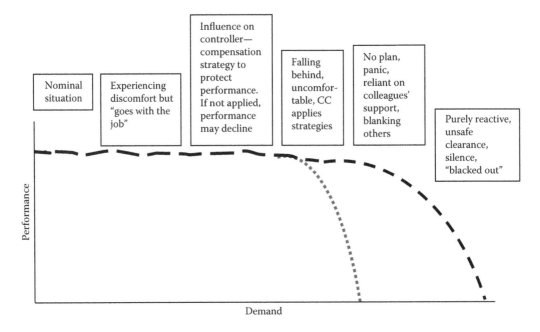

FIGURE 26.2

Diagram of performance and demand with associated indicators and controllers' subjective experience.

discomfort, controllers still complete the control task with a high level of performance. If demand increases further, then, performance may begin to be negatively affected by factor influences such as workload and fatigue. Here, performance may decline. Alternatively, if the controller identifies the threat to performance, then, a compensation strategy can be applied and performance can be maintained. The compensation strategy may not be sufficient to protect and maintain performance if task demands increase. The controller may begin to fall behind the traffic: *"It's something that will build up and you miss one…and then okay maybe you miss another one or two or you're confused as to who called you. Sometimes that happens and it'll go back down again and there's no problems and sometimes it will keep rising, starting to lose the picture"* (Participant 2). Control may become reactive and controllers may rely on colleagues' support for the maintenance of performance. If task demands are reduced, then, it is possible that performance can be recovered. However, if demand is not reduced at this point, then, the so-called "edge of performance" may be reached. Control becomes reactive, and controllers may experience panic. Unsafe clearances may be given. Severe negative reactions may occur, such as a controller shaking or becoming silent. The controller has reached performance limits and is operating outside of safe performance.

26.3.8.2 Awareness of Indicators

26.3.8.2.1 Compensation Strategies Are Dependent on Awareness

Controllers emphasized that the awareness of indicators was critical to employing a compensation strategy. One participant summarized: *"I'd say 300%, if you know that you're not being top performing today then that's fine, just adapt your working style and you'll get through the day. It doesn't really matter but if you don't know it and you're still trying to do the same then it might end in tears"* (Participant 12).

Performance may be protected by several "barriers" (created from awareness and compensation strategies) before becoming vulnerable to factor influences. If a controller becomes uncomfortable with the control task, then, internal indicators such as feelings of discomfort may alert the EC and trigger the application of a compensation strategy. Performance may then be maintained. If an internal marker did not occur or was not detected, then, another opportunity to detect the issue may occur through observable indicators. For example, *"It's getting busy … you start speaking fast and then somebody says 'Say again' and then that's it, you have a hint. 'Okay good, I have to slow down because I was not aware that I was speeding up my transitions because of the amount of traffic'. You slow down and everything's fine again"* (Participant 1). However, if the EC is not aware of indicators, then, the protection of performance is dependent on a colleague's (i.e., the CC's) awareness: *"You're not aware that you're working to the edge of your performance then you need to rely on other people to tell you or people to remind you of how you are working"* (Participant 15). If neither controller notices an issue, then, the participants suggested that performance is more likely to decline than if a compensation strategy was applied.

26.3.8.2.2 Individual Differences in Awareness and Observable Indicators: Same Indicators, Different Reactions

Participants differed in the extent of a conscious awareness of personal indicators. A minority of controllers (3/22) suggested that they personally "sense" or "just know" when they are reaching a performance limit but found it difficult to identify and describe the specific indicators. Overall, there was a consensus that, in general, indicators were generic and common between controllers. However, some indicators appeared to depend on the individual. For example, a change in voice pitch was seen as a general indication that a controller may be finding a situation difficult, but whether the pitch rose or fell would depend on the individual.

26.3.8.3 Indicators Are Learned through Experience: Learning the Hard Way?

Indicators of potential performance decline are not formally taught but are learned through experience. One respondent explained *"You start to know that you've been burning your fingers before on this kind of situation that you really have to pay attention"* (Participant 1). This has implications for trainees and newly qualified controllers. Respondents suggested that inexperienced controllers will be more vulnerable to performance decline as *"they don't know how to protect themselves"* (Participant 18).

26.4 Discussion and Conclusion

An expert opinion regarding the indicators of potential performance decline was generated by conducting 1 hour face-to-face interviews with 22 en-route controllers from MUAC. The interviews were orthographically transcribed and analyzed using thematic analysis. The results revealed that indicators were used in an ATC setting by all respondents, as an indication of when a controller was reaching the edge of performance, or a factor was negatively influencing performance. It was considered as a natural process that all controllers were used. Participants differentiated between internal indicators, representing a subjective experience, and external indicators, which were observable. Three subcategories of external indicators were identified: the changes in performance, physical signs, and application of a compensation strategy.

Participants confirmed that specific factor influences on performance were associated with specific internal and external indicators. Indicators were identified for the factors of workload, fatigue, stress, vigilance, SA, communications, and teamwork. Participants also reported adaptive changes in the control strategy that were applied to maintain performance when the influencing factor was present that resulted from the identification of indicators. This finding builds on previous research investigating controller control strategies (e.g., Kontogiannis and Malakis, 2013), by identifying different control strategies that are applied to compensate for differing performance-influencing factors or situations. Compensation strategies are an integral aspect of the use of indicators in ATC. In addition, compensation strategies were also identified to be used as external indicators themselves of controller discomfort.

The finding that controllers have, and use, indicators of when they are reaching the edge of performance, has implications for the design of dynamic automation systems. If these indicators could be measured online, it may be that tasks could be allocated between the controller and the automation depending on the needs of the controller at the time. In addition, these findings may have implications for the development and application of an observation methodology for controllers in a live operational environment. Further research will need to investigate the generalizability and validity of the identified indicators as an observation method.

Specific factors influenced performance differentially, which in turn influenced the associated indicators. The factors of workload and stress can influence other factors (e.g., communications, teamwork, SA, and vigilance) and the subsequent association with performance. The influences of workload and stress may not be visible in performance but manifest as other factor influences, such as an overlooking aircraft (a vigilance issue) due to fatigue. It is therefore important for aviation professionals to acknowledge the underlying issues of performance declines to gain a valid and comprehensive understanding about factor influences and to understand how to best protect performance. Factors such as SA influence performance directly.

A progression to the edge of performance was developed based on participant responses. The representation describes the subjective experience of controllers at each stage, and the indicators associated with the edge, and moving over the edge of performance. Although this may not be applicable in all control situations, the representation of a move to the edge of performance may provide a standardized understanding of the indicators and control situations to monitor and contributes an understanding to the wider human performance field.

Awareness emerged as an integral element in the use of indicators; controllers needed to be aware of their own or their colleagues' indicators to apply a compensation strategy. It was suggested that there were individual differences in the overall levels of awareness. In addition, controllers suggested that it was harder to be self-aware than be aware of colleagues' indicators. This was especially true for inexperienced controllers who were perceived to not have the experience to identify indicators and apply adaptive strategies. The implication of this finding is that controllers may benefit from an awareness of an initial standardized list of generic indicators. An initial knowledge of generic indicators may especially support less-experienced controllers, or controllers who work on an individual shift system—and therefore do not see their teammates regularly—while building colleague-specific knowledge. Workshops that provide standardized indicators to monitor, and support the development of the awareness of indicators for the self and colleagues, may support controllers in protecting and maintaining performance in the presence of negative influences. This may also support trainees in protecting performance while developing the required experience to identify their own indicators.

Further research may confirm the stages of progression toward the edge of performance and investigate the validity and reliability in generalization to other control centers. In addition, further research may investigate the use of standardized indicators in facilitating supervisors and controllers in identifying the potential performance decline and subsequently applying compensation strategies to prevent the performance decline. Further research should also investigate the feasibility of utilizing the identified indicators during live operations to inform the optimal dynamic allocation of control tasks, with an ultimate goal to support the controller in maintaining a high level of human performance, and preventing performance-related incidents. Finally, the utility of the human performance envelope concept for other similar tasks should be explored. Such work is now ongoing, via the European Commission-funded Future Sky Safety* program of work, which is investigating the applicability of the human performance envelope, along with indicators, edges, and compensation strategies, for airline pilots.

---

* Project P6 of Future Sky Safety aims to define the human performance envelope for cockpit operations, and identify the methods to recover crew's performance: https://www.futuresky-safety.eu/project-6/

References

Amalberti, R., and Wioland, L. 1997. Human error in aviation. In H. M. Soekkha (Ed.) *Aviation Safety* (pp. 91–108). The Netherlands: VSP BV.

Cox-Fuenzalida, L. E. 2007. Effect of workload history on task performance. *Human Factors*, 49(2), 277–291.

Edwards, T., Sharples, S., Wilson, J. R., and Kirwan, B. 2012. Multifactor combinations and associations with performance in an air traffic control simulation. *Proceedings of the 4th AHFE International Conference*, July 21–25th, San Francisco: USA.

Edwards, T., Sharples, S., Wilson, J. R., Kirwan, B., and Shorrock, S. T. 2010. Towards a multifactorial human performance envelope model in air traffic control. Presented at the *Eurocontrol/FAA Research and Development Conference*, October 19th–20th, Brétigny-sur-Orge: France.

Endsley, M. R., and Rodgers, M. D. 1994. Situation awareness information requirements analysis for en route air traffic control. In *Proceedings of the Human Factors and Ergonomics Society Annual Meeting* (Vol. 38, No. 1, pp. 71–75). Santa Monica, CA: Human Factors and Ergonomics Society.

Kontogiannis, T., and Malakis, S. 2013. Strategies in coping with complexity: Development of a behavioural marker system for air traffic controllers. *Safety Science*, 57, 27–34.

Mackieh, A., and Cilingir, C. 1998. Effects of performance shaping factors on human error. *International Journal of Industrial Ergonomics*, 22, 285–292.

Millward, L. J. 2006. Focus groups. In G. M. Breakwell, S. Hammond, C. Fife-Schaw, and J. A. Smith (Eds.) *Research Methods in Psychology* (3rd Ed.). UK: Sage Publications.

Park, K. S., and Jung, K. T. 1996. Considering performance shaping factors in situation-specific human error probabilities. *International Journal of Industrial Ergonomics*, 18, 325–331.

Sharples, S., Edwards, T., and Balfe, N. 2012. Inferring cognitive state from observed interaction. *Proceedings of the 4th AHFE International Conference*, July 21–25th, San Francisco: USA.

Sperandio, J. C. 1971. Variation of operator's strategies and regulating effects on workload. *Ergonomics*, 14(5), 571–577.

Strauss, A., and Corbin, J. 1990. *Basics of Qualitative Research: Grounded Theory Procedures and Techniques*. Newbury Park, CA: Sage Publications.

Wickens, C. D., Mabor, A. S., and McGee, J. P. 1997. *Flight to the Future: Human Factors in Air Traffic Control*. Washington, DC: National Academy Press, 14(5), 571–577.

Wilkinson, S. 2003. Focus groups. In J. A. Smith (Ed.) *Qualitative Psychology: A Practical Guide to Research Methods* (pp. 184–204). London: Sage.

27

Investigating Relevant Cognitive Abilities in the Velocity-Obstacle-Based Display for Collision Avoidance

Ursa Katharina Johanna Nagler, Peer Manske, Pierre Sachse, Marco Michael Nitzschner, Markus Martini, and Marco Furtner

CONTENTS

27.1 Theoretical Background

27.1.1 Introduction

Midair collisions of aircraft are rare, but an actual threat. For this issue, aircraft separation and systems for its support are one of the most important fields in aviation research. There are several displays to assist the pilot navigation through airspace. The purpose of this chapter is to give a short overview of the existing displays that are used in air traffic, safety standards and to present the newly developed VOD-CA (velocity-obstacle-based display for collision avoidance; Manske et al. 2013) by the German Aerospace Center in showing the advantages of comparison. VOD-CA represents a display based on planes' velocities that helps pilots to avoid conflicts by recognizing them early. Because VOD-CA represents a new kind of space with no spatial reference, potential mental problems in handling could occur. For that reason, VOD-CA and its comprehension were tested and related to the measures of spatial, figural, and numerical intelligence. It was assumed that spatial thinking might play a role. Even though velocity space is different to the familiar aerial space, it might still require similar abilities to comprehend. The purpose of this chapter was to examine the influence of mental factors on display comprehension.

27.1.2 Overview of History and Existing Displays

27.1.2.1 Safety Standards

In aviation, there are certain safety standards to avoid conflicts. Aircraft are separated from each other vertically and laterally by definite spacing (International Civil Aviation Organization [ICAO] 1996). This separation should not be undercut, otherwise there is a conflict. An aircraft converging closer to another aircraft than the mandatory minimum of separation is considered as a conflict. Thus, air traffic controllers ensure for sufficient separation between airplanes. In cases of malfunctioning of this safety layer, alternative safety systems must fill in. Such systems on board of aircraft have been developed to prevent a collision at the last moment when a conflict has already occurred.

27.1.2.2 Existing Displays

One well-established system in aviation is ACAS/TCAS (traffic alert and collision avoidance system) that advises pilots to climb or descend to avoid collisions. TCAS also shows other near-aircraft's positions within the navigation display to support an awareness of the current situation. As the display only shows the relative positions to the own aircraft without speed or heading indications, future hazards are hard to predict. Still, there is no advanced display to feature an awareness of future situations.

27.1.2.2.1 Benefits and Disadvantages of Existing Displays

TCAS has been proven to work well (Kuchar et al. 2004). It provides a picture of the current surrounding situation and gives advisories to resolve collisions when the separation loss occurs. But TCAS has one major disadvantage: it only shows the current situation. Changes of the situation can only be assessed by observing the display for a longer time. A new approach to display hazards early and make changes of the situation apparent to the pilot has been found in the representation of velocity space (Peinecke et al. 2013). In contrast to TCAS' vertical deconflicting advice, velocity space can support early lateral deconflicting. This means the actions to avoid conflicts can be taken by adapting headings or speeds. Such a concept could provide a more efficient option for deconflicting and could be more economic due to less kerosene consumption. The common velocity space displays, such as VOD-CA, show velocities of other aircraft in relation to the ownship, which are depicted by triangular shapes, called velocity obstacles (VOs). These VOs show all velocities that would lead to a conflict with other planes. Fiorini and Shiller (1993) first used a representation of VOs to actually plan motion in dynamic environments. VOs have been used for several years in different domains. Brooks (1982) already used velocity vectors for robots in motion to avoid accidents with other objects or robots. This concept has later been used for several conceptual aviation displays (Peinecke et al. 2013). All these displays were designed to show which velocities can safely be chosen and to support pilots' situation awareness. At the same time, separation zones around aircraft were included as well, so that not only the aircraft alone represented a VO but the whole separation zone. Under the assumption that other planes keep their tracks and speeds (velocities), these displays show safe and unsafe velocities not only for the moment but also for the future. If an aircraft changes its velocity, the VO on the display will change as well. Besides VOD-CA, there are several approaches to this concept of aircraft separation (Peinecke et al. 2013), for instance, the state-vector envelope (SVE) display was developed by Delft University of Technology (Van Dam et al. 2009). For more examples, see Peinecke et al. (2013).

27.1.2.2.2 Velocity-Obstacle-Based Display for Collision Avoidance

VOD-CA was consequently designed to display only velocity space from an egocentric perspective combining symbologies derived from the existing displays. First of all, the track of the aircraft in degrees is displayed on top (see Figure 27.1) similar to a navigation display. Ground speed is shown on the left side comparable to the indication in a primary-flight display. Zero speed would be at the center of the ruptured semicircle (see at the bottom of the display in Figure 27.1). The crosshairs designate their own velocity. Because there is not a spatial but a velocity space displayed, own speed can be drawn as a circle assigning the same speed in all the directions. Looking at Figure 27.1, the own speed is 217 knots, and the ownship is heading toward north. Other planes' velocities are shown as dots, and altitudes relative to the ownship are represented as white numbers next to them. With some experience, one can easily see how fast another plane is moving and

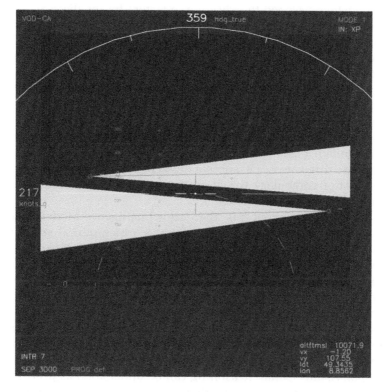

FIGURE 27.1
Velocity-obstacle-based display for collision avoidance.

where it is heading. Each triangle represents the velocities including the lateral separation zone, which have to be avoided by the crosshairs to stay clear of the conflict. Although the center line represents the actual velocity of this plane, which would lead to certain collision if the crosshairs and line were kept superposed. In conclusion, one should avoid the triangles to stay clear of the conflicts. Under the assumption that no aircraft changes its velocity, one will then stay safe at all times. Otherwise, when the own aircraft is flown within the triangle, a conflict is imminent and can be avoided by adapting speed, or heading, or both.

The advantage of VOD-CA in comparison to other displays is the absence of spatial hints and spatial relations, such as in the SVE display, for instance, as the representation of different spaces within one single display could lead to confusion. VOD-CA solely works in velocity space. Nevertheless, an analysis of the understanding of velocity space is needed to evaluate the system. Therefore, we shall examine which cognitive factors lead to a better understanding of the velocity space.

27.1.3 Psychological Concepts

27.1.3.1 Intelligence

Intelligence is a common concept in psychology. Intelligence means the human's ability to adapt to his environment, to plan, to solve problems, to learn fast and through experience, and to understand complex ideas (Gottfredson 1997). Through the history of psychology, several concepts of intelligence were developed. A common concept refers to a theory of Spearman (1927) who developed the idea of a "g-factor" for intelligence, a general factor underlying each kind of intelligence and "s-factors" for special kinds of intelligence. This theory was confirmed in empirical research more than once (e.g., Anderson and Phelps 2001). However, it is unclear as to how the g-factor influences different cognitive abilities (Floyd et al. 2009).

Another model of intelligence by Guilford (1961) distinguishes between three kinds of traits in intelligence tasks: content, product, and operation. Content can be divided as well in different kinds: figural, symbolic, semantic, and behavioral, while products can be seen as units, classes, systems, transformations, and implications. Last but not least, operations are divided in cognition, memory, divergent production, convergent production, and evaluation. Out of these three areas, 120 subsets of intelligence were combined, for example, remembering a list of vocabulary would be the section semantic contents, the product would be in units, and the operation memory.

On the basis of these two concepts, the Berlin Intelligence Model (Jäger 1982) was developed. It can be understood as a bimodal and a hierarchical structure model of intelligence, which has three underlying principles. The first principle is that each intellectual ability is needed to a different extent for every task that involves intelligence. The second principle says that each ability can be classified into two different aspects: operations and contents. The third principle means that all the abilities can be structured hierarchically and be classified on different levels (Jäger et al. 1997). A general intelligence, such as the g-factor of Spearman (1927), is made up of 12 subfactors, which can be categorized in operations and contents as well. There are four categories of operations, which are reasoning, creativity, memory, and speed, while content can be verbal, numerical, or visual–spatial. Reasoning can be understood as processing complex information, which requires getting further information, while creativity means the ability to create innovative and flexible ideas or to consider something from a different perspective. Memory describes the ability to remember items and to reproduce them for a short time. Furthermore, speed means the power of comprehension, working speed, and concentration in problem solving. For the contents, the verbal category represents the relationships between words, the numerical category describes thinking with numbers and the relationship between numbers, and the visual–spatial factor means the ability to think in spatial and pictorial terms. For understanding the velocity space and VOD-CA, visual–spatial intelligence and numerical intelligence were assumed to be the most important.

27.1.3.2 Understanding of Velocity Space

People need spatial thinking regularly in everyday life, for example, navigating through the streets to a certain destination. But understanding the velocity space as it is represented in VOD-CA may require a different skill. Participants have to understand how velocity space works and what it means to move in velocity space. There might be a problem of participants mistaking velocity space with local spatial space. Yet, there is no concept such as "understanding velocity space" in psychological research. "Understanding velocity space" is the ability to operate and perceive all aspects of the velocity space. One interesting question is whether this ability has to be learned or is inherent. Although there is evidence that participants have no problems in avoiding conflicts by using VOD-CA (Manske et al. 2013), it has not yet been investigated as to which extent participants understand velocity space and can adapt to it.

27.2 The Present Study

The representation of velocity space on VOD-CA shows information that people cannot compare to their everyday situations and thus, may have problems in understanding it. The aim of this chapter was to find the underlying relevant cognitive factors for understanding VOD-CA and velocity space. So, the common measures of intelligence were linked to the flight performance with VOD-CA and five hypotheses were formed.

First, spatial thinking should be linked to the performance with VOD-CA. This is because on the one hand, a high performance in spatial thinking might be helpful due to the missing spatial references in VOD-CA. It is assumed that building a simulated model of the current situation presented on the display might ease the adaption to the new kind of representation (Kieras and Bovair 1984). On the other hand, a high spatial intelligence might be obstructive, because the hints given to understand velocity space could be mistaken as spatial hints. Therefore, it is hypothesized that spatial thinking plays a crucial role in understanding velocity space and handling VOD-CA and may influence the flight performance.

Second, to perform well in the simulator flying with VOD-CA, it is crucial that all the numbers, heading, and speed are taken into account. Remembering all these values might be important for the right course and a good flying performance (Hardy and Parasuraman 1997). Thus, it is hypothesized that there is a relation between numerical memory and understanding velocity space.

Third, because there are several graphic representations on the display, such as the velocity obstacles as triangular shapes, and other aircrafts' velocities as dots, which have to be observed, it should be important to remember these graphic figures well and recognize the changes (Hardy and Parasuraman 1997). Hence, it is hypothesized that there is a relation between visual–spatial memory and velocity space.

Fourth, participants with a higher visual–spatial intelligence could adapt better to the new situation with the other graphical representation and perform better than participants with lower results in visual–spatial intelligence (Bühner et al. 2008). Therefore, it was hypothesized that there is a relation between visual–spatial intelligence and the understanding of velocity space.

Furthermore, the motion of the geometric figures are relevant to understand how the actual situation changes and to adapt properly to it. This is the basic requirement to identify the possible intruder aircraft and to react well to the changes. So, the fifth hypothesis was that there is a relation between visual–spatial processing and velocity space.

To sum up, the aim of this chapter was to determine which cognitive abilities are relevant for handling VOD-CA.

27.3 Methods

27.3.1 Participants

The study was conducted with 24 participants (six women and 18 men). The average age was 23.92 ($SD = 3.57$) and participants' experience with flight simulation was documented. Most of the participants were students (87.5%) from the Technical University of Brunswick, Germany. Some experience with flight simulators and sufficient German language skills were required. All participants had normal or corrected-to-normal vision. The study took about 2 h for each participant. All participants were monetarily rewarded and offered to receive a feedback of their performance in the intelligence tasks.

27.3.2 Tasks

Spatial thinking was tested with the "3-D-Würfel-Test" (Gittler 1990). The task is to rotate a cube mentally and pick the right answer out of a selection of seven choices. Every right answer counts. There was no time limit but time was recorded (range: 420–1625 s, $M = 847.38$). The test was conducted in a short version with 12 tasks.

Figural and numerical intelligence. The Berlin Intelligence Structure (BIS) test, version 4 (Jäger et al. 1997), was applied to measure intelligence referred to visual–spatial and numerical content. Visual–spatial intelligence is built up by visual–spatial reasoning and visual–spatial memory. The subtests of visual–spatial memory are "Orientierungsgedächtnis" (orientation memory; remember certain buildings on a map), "Wege–Erinnern" (route recall; remember a way through geometric figures; and draw it), and "Firmen–Zeichen" (company logo; remember a sign with its content). In addition, the subtests of visual–spatial reasoning are "Analogien" (analogy of two geometric figures), the so-called tests "Charkow" (the task is to complete a pattern) and "Bongard" (search a picture which fits to six others), "Figuren–Auswahl" (figure selection; select two pieces which build a presented figure), and "Abwicklung" (unwinding; select a figure which is built by a flat pattern). Numerical memory is made up of "Zahlen–Paare" (a pair of numbers; remember 12 pairs of numbers), "Zweistellige Zahlen" (double-digit numbers; remember 16 binary numbers),

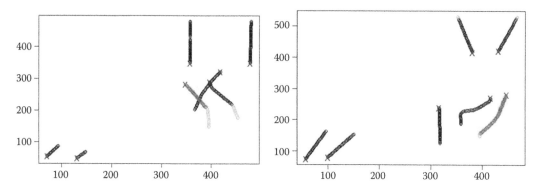

FIGURE 27.2
Scenarios 1 and 2.

and "Zahlen–Wiedererkennen" (recognize nine numbers in 66 numbers) (see Nagler et al. 2014).

An understanding of VOD-CA and velocity space was evaluated by the flying performance. The participants flew six scenarios in a flight simulator with VOD-CA. Figures 27.2 through 27.4 show examples of flight tracks for all the scenarios. The dark gray dots mark times of the own plane without conflicts, light gray shows the separation losses, and the cross marks the last position of each plane. Each scenario was tested beforehand (see Manske et al. 2013) and each ended after resolving all the conflicts.

27.3.3 Hardware

The used hardware for steering was a *Saitek* "Pro Flight Yoke System"®. There were a yoke and a thrust lever. To display the simulation content, two monitors represented the scene and the displays. They had a frequency of 60 Hz/75 kHz, a screen size of 20.1″, and a resolution of 1600 × 1200 pixels. Both monitors were from *Fujitso Computers Siemens*®. The computer itself had an Intel Core i7-2006 3® processor with 4 GHz and the graphic card was from NVIDIA (GForce GTX 570®).

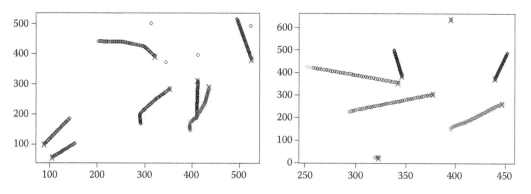

FIGURE 27.3
Scenarios 3 and 4.

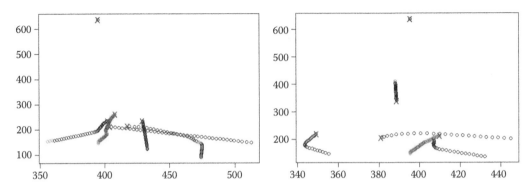

FIGURE 27.4
Scenarios 5 and 6.

27.3.4 Software

The simulation was done with *X-Plane10*® and a self-written program to edit the scenarios was programmed by the German Aerospace Center. The flight data were computed with R. The following variables were generated for each scenario: the duration of conflict, duration of separation loss, number of conflicts, and the closest point of approach. The duration of conflict was the time that the own velocity was in the marked triangle that would have led to a conflict if no action occurred, while the duration of the separation loss was the time that a person spent in a highly endangered area. This means that the minimum separation of three miles was undercut. The number of conflicts was the counted number of times, where a person flew in the marked triangle. The closest point of approach was the closest distance between the own and another plane. The overall means of the variables were processed across all the scenarios. The statistical analyses and correlations were conducted using IBM SPSS Statistics 21®.

27.3.5 Procedure

First, the participants had to complete the BIS 4 (Jäger et al. 1997). Second, the "3-D-Würfeltest" (Gittler 1990) was executed. Afterward, there was a short instructive presentation about VOD-CA and its functions (see Figure 27.2 for an example). As can be seen in Figure 27.5, it was explained as to how the representation of VOD-CA is made up of and how to interpret the represented VOs. At the end of the presentation, two examples were provided that showed the potential and actual conflicts. Keywords such as the separation zone or conflict were explained as well.

The participants had to confirm their understanding in the first three exercise flights of the simulation with additional hints given by the instructor. Thereafter, the participants flew six testing scenarios, which were tested before (see Manske et al. 2013), without any comment from the instructor. For all runs, participants were not only instructed to fly north, but also to avoid conflicts with other aircraft. The participants should not act if it was not necessary to do so. Staying conflict free had the highest priority. The autopilot held the airplane at the height of 10,000 ft to ensure that all planes were on the same level to eliminate the option of vertical deconflicting. TCAS as an additional tool was deactivated. Each scenario was finished when all the potential conflicts were resolved. Finally, the participants filled out the questionnaire and the demographic data sheet.

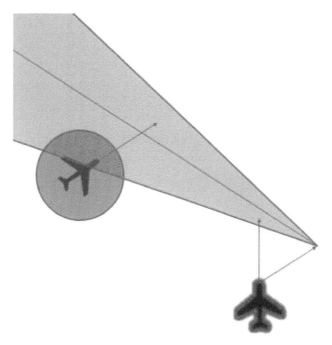

FIGURE 27.5
Picture of the presentation for the understanding of VOD-CA.

27.4 Results

All variables were distributed normally that was tested with the Shapiro–Wilk test. Flight experience, simulation experience, and experience with handling a yoke did not correlate significantly with the flight data. The intercorrelations among all the variables can be found in Table 27.1.

Spatial thinking: There was no significant relation between the "3-D-Würfeltest" and the duration of conflict ($r = -0.36, p = 0.09$), duration of the separation loss ($r = -0.12$, $p = 0.58$), number of conflicts ($r = -0.40, p = 0.05$), and the closest points of approach ($r = 0.05, p = 0.81$).

Numerical memory: Furthermore, numerical memory was not related to the duration of conflict ($r = 0.07, p = 0.75$), duration of the separation loss ($r = -0.19, p = 0.38$), number of conflicts ($r = -0.09, p = 0.68$), or the closest points of approach ($r = -0.33, p = 0.11$).

Visual–spatial intelligence: There was no significant relation between visual–spatial intelligence and the duration of conflict ($r = -0.25, p = 0.24$), duration of the separation loss ($r = 0.04, p = 0.85$), or the closest points of approach ($r = 0.01, p = 0.65$). However, there was a negative relation between visual–spatial intelligence and the number of conflicts ($r = -0.54, p = 0.01$).

Visual–spatial memory: Moreover, there was no significant relation between visual–spatial memory and the duration of conflict ($r = -0.28, p = 0.18$), duration of the separation loss ($r = -0.01, p = 0.97$), or the closest points of approach ($r = 0.01, p = 0.97$).

TABLE 27.1

Intercorrelations with p-Values among All Variables

	3DW	VM	VR	VI	NM	Con_G	Seploss_G	N_G	CPA_G
VM	0.31 (0.139)								
VR	0.35 (0.097)	0.33 (0.112)							
VI	0.40 (0.053)	0.85*** (0.000)	0.78*** (0.000)						
NM	0.21 (0.318)	0.28 (0.182)	0.38 (0.071)	0.40 (0.055)					
Con_G	−0.36 (0.086)	−0.28 (0.178)	−0.11 (0.619)	−0.25 (0.241)	0.07 (0.751)				
Seploss_G	−0.12 (0.580)	−0.01 (0.971)	0.08 (0.697)	0.04 (0.847)	−0.19 (0.379)	−0.38 (0.064)			
N_G	−0.40 (0.052)	−0.46* (0.025)	−0.43* (0.036)	−0.54** (0.006)	−0.09 (0.681)	0.54** (0.007)	−0.16 (0.453)		
CPA_G	0.05 (0.811)	0.01 (0.968)	0.17 (0.436)	0.10 (0.647)	−0.33 (0.113)	−0.36 (0.089)	0.52** (0.010)	−0.42* (0.043)	

Note: 3DW = 3-D-Würfeltest; VM = visual–spatial memory; VR = visual–spatial reasoning; VI = visual–spatial intelligence (global); NM = numerical memory; Con_G = average of the duration of conflicts over all the scenarios; Seploss_G = average of all separation losses over all the scenarios; N_G = average of all numbers of conflicts over all the scenarios; CPA_G = average of each closest points of approach over all the scenarios.

*$p = 0.05$, *$p = 0.01$, **$p = 0.001$.

However, visual–spatial memory was related to the number of conflicts ($r = -0.46$, $p = 0.03$) as well.

Visual–spatial reasoning: The duration of conflict ($r = 0.11$, $p = 0.62$) was not linked to visual–spatial reasoning, such as the duration of the separation loss ($r = 0.08$, $p = 0.70$) and the closest points of approach ($r = 0.17$, $p = 0.44$). There was a significant finding of visual–spatial reasoning with the number of conflicts ($r = -0.43$, $p = 0.04$).

27.5 Discussion

The aim of this chapter was to examine the relationship between different facets of intelligence and the understanding of velocity space. While some expected effects could be observed, the overall majority of the participants could handle VOD-CA and resolved the potential dangers and conflicts before they occurred. There was a relation between visual–spatial intelligence, visual–spatial memory, and visual–spatial reasoning with the overall number of conflicts, while spatial thinking and numerical memory showed no such effect. Furthermore, no other flight data such as the duration of conflict, duration of the separation loss, and the closest point of approach showed a relation to the measurements of intelligence. Flight experience, experience with handling a yoke, or simulation experience did not interfere and showed no correlation with the flight data. This missing relation could be a hint that VOD-CA can be interpreted correctly even without any experience with flying in a simulator. This could mean that people can adapt to the representation of VOD-CA easily, although velocity space is not common in everyday life. The single relation between the number of conflicts and the missing relations with the other flight variables with the measurements of intelligence could be explained by the assumption that some participants accepted the temporary future conflicts to stay safe and conflict free in the long term. For this plan, they could have accepted passing the velocity obstacles to fly with a safe velocity again. This could be the reason for the missing links between the measurements of intelligence and the closest point of approach, duration of the separation loss, or duration of the conflict. There was no relation between spatial thinking and flight performance. Participants seemed not to be confused because of the representation, for example, taking the dots as the actual position of other planes. The tendency to a positive relation (participants with a better spatial thinking skill could show fewer conflicts with other planes) might be a hint that persons with high scores in spatial thinking find it easier to handle velocity space. In addition, there was no effect between the flight data and numerical memory. This could mean that the overriding graphical representation played a more important role than the numbers displayed. Hence, there were just ground speed and a track, which seemed not to be challenging. This could be the reason as to why there was no connection between the flight data and numerical memory. In contrast, there was a negative correlation of the number of conflicts with visual–spatial intelligence. This means that the better participants performed in visual–spatial intelligence, the fewer potential conflicts were generated. The relationship might be explained by participants remembering the presentation better, identifying the proper conflicting triangles in time, and therefore, reacting to them quickly. That result implies better performance, the better one's visual–spatial memory, and visual–spatial reasoning, because visual–spatial intelligence consists of those two constructs. The study showed the same result with a link between visual–spatial memory and the understanding of velocity space. This connection could be explained by the recall of the given presentation. Furthermore, there was a link between visual–spatial reasoning and the number of conflicts. That finding could be explained through the assumption that participants who are better in processing geometric figures might also be better in anticipating the consequences of their flight behavior. Because the flight experience showed no effect, it can be assumed that this applies to both experienced and unexperienced pilots.

Another factor that might have impacted flight performance is the simulation time. The whole study took about 2 h and the measurements of intelligence were applied first. This could have had an effect on fatigue. But, just as Manske et al. (2013) could show, all the participants could handle VOD-CA well. Even if it seemed that not everybody understood velocity space, the participants could use VOD-CA properly to avoid conflicts and collisions. In sum, VOD-CA can successfully be used in aviation for separation by everyone, because no complex skills seem to be required and people can adapt to the representation of velocity space.

27.6 Limitations and Conclusion

In the future, deeper investigations should show more options and the limitations of VOD-CA. A comparison of VOD-CA and TCAS could lead to interesting findings. In addition, it could be tested whether the disadvantage of potential confusion evoked by the unfamiliar display may outweigh its advantages. More studies should focus on all the underlying aspects of understanding and using VOD-CA. But so far, it seems to be clear that velocity-based displays, especially VOD-CA, have a great potential to help pilots' situation awareness and increase the flight performance and safety. In the future, displays such as VOD-CA could be a great supplement in the cockpit and support pilots' situation awareness.

References

Anderson, A. D., and Phelps, E. A. 2001. Lesions of the human amygdala impair enhanced perception of emotionally salient events. *Nature* 411:305–309.

Brooks, R. A. 1983. Solving the find-path problem by good representation of free space. In: *Proceedings of the 2nd Annual Conference on Artificial Intelligence* (pp. 381–386). Pittsburgh, PA: IEEE.

Bühner, M., Kröner, S., and Ziegler, M. 2008. Working memory, visual–spatial-intelligence and their relationship to problem-solving. *Intelligence* 36:672–680.

Fiorini, P., and Shiller, Z. 1993. Motion planning in dynamic environments using the relative velocity paradigm. In: *Proceedings of the Conference on Robotics and Automation* (pp. 560–565). Atlanta: IEEE.

Floyd, R., McGrew, K., Barry, A., Rafael, F., and Rogers, J. 2009. General and specific effects on Cattell-Horn-Carroll broad ability composites: Analysis of the Woodcock–Johnson III Normative Update CHC factor clusters across development. *School Psychology Review* 38:249–265.

Gittler, G. 1990. *Drei-dimensionaler Würfeltest (3DW). Ein Rasch-skalierter Test zur Messung des räumlichen Vorstellungsvermögens.* Weinheim: Beltz.

Gottfredson, L. S. 1997. Mainstream science on intelligence: An editorial with 52 signatories, history, and bibliography. *Intelligence* 24:13–23.

Guilford, J. P. 1961. Factorial angles to psychology. *Psychological Review* 68:1–20.

Hardy, D. J., and Parasuraman, R. 1997. Cognition and flight performance in older pilots. *Journal of Experimental Psychology: Applied* 3:313.

International Civil Aviation Organization. 1996. Procedures for air navigation services: Rules of the air and air traffic services. pp. 37–57.

Jäger, A. O. 1982. Mehrmodale Klassifikation von Intelligenzleistungen: Experimentell kontrollierte Weiterentwicklung eines deskriptiven Intelligenzstrukturmodells. *Diagnostica* 28:195–225.

Jäger, A. O., Süß, H.-M., and Beauducel, A. 1997. *Berliner Intelligenzstruktur-Test: Form 4*. Göttingen: Hogrefe.

Kieras, D. E., and Bovair, S. 1984. The role of a mental model in learning to operate a device. *Cognitive Science* 8:255–273.

Kuchar, J., Andrews, J., Drumm, A., Hall, T., Heinz, V., Thompson, S., and Welch, J. 2004. A safety analysis process for the traffic alert and collision avoidance system (TCAS) and see-and-avoid systems on remotely piloted vehicles. In: *AIAA 3rd Unmanned Unlimited Technical Conference*. Chicago, IL.

Manske, P., Meysel, F., and Boos, M. 2013. VOD-CA—Assisting human flight performance and Situation awareness in lateral deconflicting. In: D. de Waard, K. Brookhuis, R. Wiczorek, F. di Nocera, R. Brouwer, P. Barham, C. Weikert, A. Kluge, W. Gerbino, and A. Toffetti (Eds.), *Proceedings of the Human Factors and Ergonomics Society Europe Chapter Annual Meeting* (pp. 303–313). Turin, IT.

Nagler, U. K. J., Manske, P., Sachse, P., Nitzschner, M. M., Martini, M., and Furtner, M. 2014. VOD-CA—Testing a velocity-obstacle based display for collision avoidance in aviation. In: T. Ahram, W. Karwowski, and T. Marek (Eds.), *Proceedings of the 5th International Conference on Applied Human Factors and Ergonomics AHFE 2014* (pp. 1480–1487). Krakow.

Peinecke, N., Uijt de Haag, M., Meysel, F., Duan, P., Küppers, R., and Beernink, B. 2013. Testing a collision avoidance display with high-precision navigation. In D. D. Desjardins, and K. R. Sarma (Eds.), *Proceedings of SPIE 8736, Display Technologies and Applications for Defense, Security, and Avionics VII*. Baltimore, MD.

Spearman, C. 1927. *The Abilities of Man*. New York: Macmillan.

Van Dam, S. B. J., Mulder, M., and van Paassen, M. M. 2009. The use of intent information in an airborne self-separation assistance display design. In: *AIAA Guidance, Navigation, and Control Conference*. Chicago, IL.

28

Aircraft Seat Comfort Experience

Naseem Ahmadpour, Jean-Marc Robert, and Gitte Lindgaard

CONTENTS

28.1 Introduction

The design of the aircraft seat for commercial aircrafts is becoming progressively important for the airline and aerospace industry. On one hand, there are manufacturing considerations with regard to meeting standards and safety regulations, the choice of lightweight material that contributes to reducing aircraft fuel consumption and consequently environmental impacts, as well as the size and form of seats which dictate the number of passengers on board. On the other hand, seat design has a significant influence on passenger comfort and well-being (Vink et al., 2012) and impacts their purchasing decisions and choice of airline (Brauer, 2004). Therefore it is becoming increasingly challenging to design seats that are safe, comfortable, and offer a pleasurable experience to the occupants.

Seat comfort is often defined as personal and subjective, pertaining to a harmony between physical and psychological aspects of the experience (De Looze et al., 2003). Moreover it has been linked to the experience (Ahmadpour et al., 2014a) and perception (Vink and Hallbeck, 2012). The term "comfort experience" was coined as the result (Vink et al., 2005), characterized as an experience beyond the physical interaction (Helander, 2003) and with added hedonic qualities (Hancock et al., 2005). Design for comfort experience is subsequently proposed as an approach that goes beyond ergonomics and toward delivering enhanced pleasing experiences. Several studies provided conceptual information about the experiential aspects of seat comfort. These aspects are commonly subjective and described in terms to the users' feelings or affective responses. For example, Helander and Zhang (1997) discussed the office chair comfort experience in association to feeling relief and relaxed.

In the field of transportation, Coelho and Dahlman (2002) disclosed a relationship between car seat comfort and pleasure. Kamp (2012) showed that a perfect car seat elicits pleasant emotions with a minimal level of activation (e.g., excitement) and highlighted the occupant's experience of "relaxedness." Hiemstra-van Mastrigt et al. (2015) suggested that car seats with integrated active seating systems in the back rest results in occupants feeling more "refreshed" and therefore more comfortable.

The above studies highlight the importance of the pleasurable aspects of the comfort experience for the seat design. However, some researchers have expressed concern about the applicability of those results to the design of the aircraft seat due to the highly different nature of the flight context and diversity of passenger activities (Hiemstra-van Mastrigt, 2015). Therefore, further investigation is critical in order to characterize the experiential aspects of the seat in the context of the flight and their impact on passengers' overall comfort. Acquiring such knowledge could inform designers about various aspects of passenger reactions to the design elements, enabling them to develop innovative concepts that not only respond to the ergonomics requirement of the seat (e.g., physical fit) but also promote positive experiences such as relief, pleasure, etc. One objective of this chapter is to identify those aspects of passengers' experience that are linked to seat comfort in the economy class of commercial flights. This however cannot be achieved before addressing a theoretical issue that follows.

Several researchers (Helander and Zhang, 1997; De Looze et al., 2003; Helander, 2003) raised concern about the differences between the seat comfort and discomfort experiences. Helander (2003) speculated that the interaction of a user with a seat in a use context results in a number of affective responses (i.e., feelings), some of which are associated with the comfort experience while the others are linked to the discomfort experience. De Looze et al. (2003) added that discomfort results from the physical impact of the seat on the occupant's body whereas comfort is linked to emotions. The relationship among those variables is illustrated in Figure 28.1.

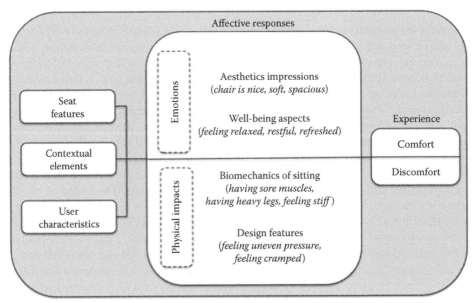

FIGURE 28.1
A model of seat comfort and discomfort experience based on Helander (2003) and De Looze (2003).

As shown in Figure 28.1, comfort experience is characterized in terms of positive affects pertaining to aesthetic impressions (e.g., luxurious, plush) and well-being aspects (e.g., feeling refreshed, feeling at ease). Discomfort experience is described in relation to the biomechanics of sitting (e.g., having sore muscles or heavy legs, feeling stiff) and design features (e.g., feeling uneven pressure, feeling cramped). Helander (2003) followed that seat comfort is experienced independently of discomfort and thus should be evaluated on the basis of a different set of criteria, as noted above.

Given that aircraft passengers spend the majority of their flight time seated, an investigation into the above hypothesis is necessary as it may have implications for the design of both the cabin environment and the seat. Therefore, in this chapter, we will first assess passengers' experiences of comfort and discomfort with the objective of establishing any potential differences among the underlying components of those.

The study presented in this chapter follows the results of a previous inquiry (Ahmadpour et al., 2014a) into the experiential aspects of passenger comfort in commercial aircrafts. A summary of that study follows.

28.1.1 The Thematic Components of Passenger Comfort

In an empirical study, Ahmadpour et al. (2014a) collected data from 155 passengers of commercial flights in the economy class who gave written accounts of their comfort experiences. A content analysis of those reports was performed and eight subjective themes were identified in order to describe passengers' experiences of the aircraft cabin interior features (e.g., seat). Subsequently participants' concerns in relation to each of those themes were identified. The themes and concerns are shown in Figure 28.2. The themes "peace of

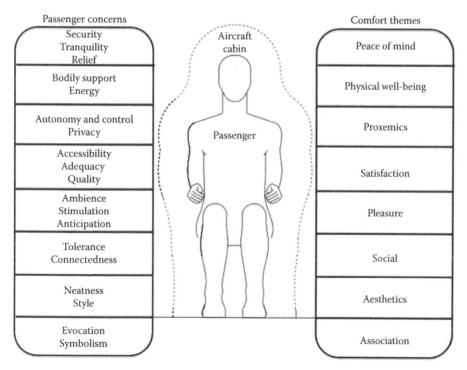

FIGURE 28.2
Aircraft passenger concerns and comfort themes.

mind" and "physical well-being" were identified as the most important themes, based on the frequency with which participants had mentioned them. The theme "association" was mentioned least frequently.

"Peace of mind" signifies the psychological aspect of passenger comfort and concerns for security (e.g., having everything needed), tranquility (e.g., feeling calm when not exposed to any excessive noise), and relief (e.g., not hitting the head on the luggage bin when getting in and out of the seat). "Physical well-being" exemplifies passengers' experience of the physical impacts (e.g., pain, soreness) on their body and the level of energy they feel (e.g., the air feels fresh).

"Proxemics" in the context of flight subscribes to one's experience of personal space and concerns for control (e.g., to make adjustments, to perform activities) and privacy (e.g., not disturbed when resting). "Satisfaction" is described in terms of the cabin's usability and whether it fulfills passengers' expectations with regard to accessibility (to cabin element, e.g., entertainment unit, seat control), adequacy (of cabin features e.g., seat recline function), and quality (e.g., of announcement sound). The theme "pleasure" concerns the extent to which passenger anticipations are exceeded (e.g., by surprising service elements), the cabin ambience (e.g., warm, welcoming) and stimulation (e.g., entertainment). The theme "social" concerns the inter-personal interactions and a balance between those that one has to tolerate (particularly those that are disturbing) and those connections that are favorable (e.g., empathetic human contacts). "Aesthetics" aspects concern neatness (e.g., cleanliness of the cabin, its maintenance) and style (e.g., colors). The theme "association" concerns the representational elements of the environment in terms of familiarity (e.g., a seat that resembles a comfortable lounge chair) and desirability (e.g., service as good as in first class).

Furthermore, the study uncovered 22 cabin interior elements. It was shown that 80% of participants identified the seat as the central determinant of their flight comfort experience. This was followed by six other cabin elements, as shown in Figure 28.3, all of which had been mentioned by at least 20% of participants. Those were legroom (64%), in-flight entertainment—IFE (37%), temperature (33%), activity (28%), noise (28%), and service (22%).

The above results necessitate further research into the comfort experience associated with the aircraft seat. The study presented in this chapter aims to provide an insight into that experience by investigating two main questions as follows:

Question 1: What are the differences between the aircraft passengers' comfort and discomfort experiences in the cabin environment?

Question 2: What types of experience are associated with aircraft seat comfort? The previous study (Ahmadpour et al., 2014a) gave indications about the eight themes of overall passenger comfort experience. The aim of the study presented in this chapter is to identify the themes that are relevant to the comfort experience of the seat.

28.2 Method

28.2.1 Participants

A convenience sample of 27 participants (15 males) was obtained. Of those, 20 were aged 18–34 and the rest 35–55. The mean height was 174 (150–193) cm and none of the participants had disabilities. All participants had more than five flight experiences in the past.

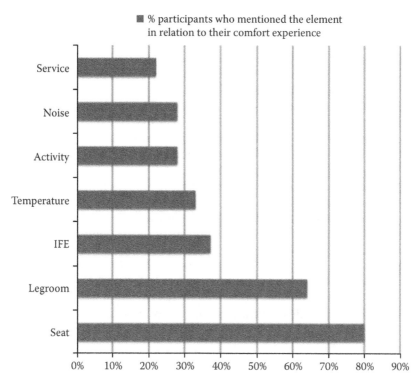

FIGURE 28.3
Seven cabin elements, which have been mentioned by above 20% of participants in relation to their comfort experience in the aircraft cabin.

They were informed that they would be asked to share details about an economy class flight experience that had taken place within the past 2 years. A total of 54 reports were then obtained, of which 44 concerned long flights (>4 hour long) and the rest were short flights (<4 hour long).

28.2.2 Questionnaire

Respondents received a link to an online questionnaire via email. The questionnaire consisted of questions inquiring about age, gender, height, disability, previous flights experiences (never/1–5 times/more than 5 times), an open-end question about details of a comfortable flight experience followed by a similar question about an uncomfortable flight experience in the economy class. Then the respondents were requested to give ratings on the influence of each of the eight themes (as summarized in Introduction) on both of those experiences using a 5-point scale ranging from slightly influential to highly influential. A short description of each theme and the types of concerns associated with each were also provided. Respondents were instructed to leave the rating section blank if a theme had no impact on their experience. A comment section was provided at the end of the questionnaire for the respondents to specify any aspect of their experience that was not covered by the eight proposed themes. Finally, respondents were informed that the principle investigator would contact them for a follow-up interview in order to collect more in-depth account of their responses.

28.2.3 Interview

Interviews were conducted within 14 days from the date an online report was submitted. At the beginning of the interviews, respondents were provided with an operational definition of comfort as "a pleasant state of well-being and ease whereby there is a physical, physiological and psychological harmony between a person and the environment." This was followed by a definition of discomfort as "a state whereby one experiences hardship of some sort which could be physical, physiological or psychological." Those descriptions were formulated based on the discussion of Ahmadpour et al. (2014a). Next, respondents were asked to specify whether they reported long or short flights in their submitted reports, as this question was missing from the initial questionnaire. During the interview, each respondent was first asked to read their responses to the open-end questions and then give more details about them including their feelings, attitudes, concerns, and reactions to the environmental elements such as the seat. A laddering technique (Jordan, 2000) was adopted to achieve a better understanding of the respondents' experiences. This included the investigator repeatedly asking "why" following each statement. The probing continued until the investigator believed the respondent had revealed their concerns and reactions associated with different aspects of their comfort and discomfort experiences. The interview ended with a review of the ratings given on eight themes. The respondents were then requested to provide justifications for each of those ratings.

28.3 Analysis and Results

The data analysis included identifying the descriptions and environmental elements that influenced passenger comfort and discomfort to some extent. However, we only report those related to the seat in this chapter. The full report is published by Ahmadpour et al. (2016a).

28.3.1 Differences between Comfort and Discomfort Experiences

To examine the differences between the ratings on the eight themes in relation to comfort and discomfort experiences, a Wilcoxon signed-rank test was performed. Significant differences ($p < 0.001$) were found between ratings given on themes "physical well-being" and "pleasure" in the comfort and discomfort reports. The ratings on all other themes exhibited no significant differences. Next the average ratings and standard deviation on each theme were computed for comfort and discomfort reports separately as summarized in Table 28.1.

The theme "pleasure" was rated the highest in reports of comfort experience whereas the themes "physical well-being" received higher ratings in reports of discomfort experience. The ratings placed the theme "peace of mind" second to the highest in both comfort and discomfort reports, confirming the importance of psychological well-being for both experiences. The theme "association" was similarly rated the least influential in both groups.

Given that six out of the eight themes did not demonstrate any significant differences in the ratings, it was decided to consider the eight themes as representative of both the comfort and discomfort experiences throughout the rest of this study. This means the overall passenger comfort is viewed as a phenomenon that is stretched across a range of

TABLE 28.1

The Mean Ratings (SD) of Eight Themes (1-Slightly Influential to 5-Highly Influential) Associated with Comfort and Discomfort Experiences (N = 27) Separately and Combined

	Pleasure	Peace of Mind	Proxemics	Physical Well-Being	Satisfaction	Social	Aesthetics	Association
Mean ratings (SD) in comfort reports	3.0(1.9)	2.9(2.2)	2.3(2.2)	2.1(2.1)	1.6(1.8)	1.5(1.9)	1.2(1.8)	0.8(1.4)
Mean ratings (SD) in discomfort reports	0.6(1.3)	3.4(1.5)	1.6(2.0)	4.0(1.5)	2.1(2.1)	1.9(2.2)	0.5(1.3)	0.3(0.8)
Mean (SD) of comfort and discomfort reports together	1.8(1.2)	3.2(0.3)	2.0(0.4)	3.1(1.0)	1.9(0.3)	1.7(0.2)	0.9(0.4)	0.6(0.3)

experiences, from a negative state of discomfort to a positive state of comfort with a neutral state in between. This phenomenon could be described in terms of eight experiential themes, for example, physical well-being, pleasure, peace of mind, etc. For this reason, comfort and discomfort experiences are not differentiated through the rest of this chapter.

Subsequently, the overall influence of the eight themes on the overall comfort experience was calculated based on combining all ratings obtained for each theme and then calculating the mean (SD) value. This is reported in the last row of Table 28.1. The results confirmed the rank order of comfort themes reported by Ahmadpour et al. (2014a) (see Figure 28.2) introducing "peace of mind" and "physical well-being" as highly influential on passenger comfort followed by "proxemics," "satisfaction," "pleasure," "social," "aesthetics," and "association," respectively.

28.3.2 Impact of the Seat on Comfort Experience

Next, a content analysis was conducted on respondents' reports. The procedure included first eliciting travel descriptions in relation to the seat. Second, those descriptions were inspected carefully and the passenger comfort themes relevant to each seat description were identified.

The first step revealed that 18 respondents had mentioned the seat at least once in their reports in connection to an experience of comfort (or discomfort). Among those, N = 11 were male, N = 13 were aged 18–34 years old, and N = 5 were aged 35–55. The average height of those participants was 174 (152–193, *SD* = 10) cm. The number and demographic information of those respondents who mentioned the seat in their reports is summarized in the first row of Table 28.2. The information about those who did not mention the seat at all is given in the second row of the table.

In the second step, the themes relevant to those seat-related descriptions were identified from the verbatim interview transcripts and then counted. A description was counted once if the seat was mentioned in relation to only one comfort theme throughout a report (comfort and discomfort experiences combined). A description was counted several times if it was mentioned in relation to different themes in a report. As an example, when a respondent mentioned the seat once in relation to its social aspect and another time in relation to its physical [well-being] aspect, it was counted twice. The results of this step yielded 52 seat-related descriptions.

The seat was described most frequently in relation to "physical well-being" (21 description, 40% of all seat descriptions) and least frequently in relation to "pleasure" and "social" (1 description, 2% of all seat descriptions) themes. The themes "physical well-being,"

TABLE 28.2

Number and Descriptions of Respondents Who Mentioned the Seat in Relation to Their Comfort Experience as Opposed to Those Who Did Not Mention It At All

	N	Male	Female	Age (Years Old) 18–34	35–55	Height (cm) Mean (SD)
Seat mentioned in relation to at a comfort theme	18	11	7	13	5	174(10)
Seat not mentioned in relation to the comfort experience	9	4	5	7	2	172(12)
Total	27	15	12	20	7	174(10)

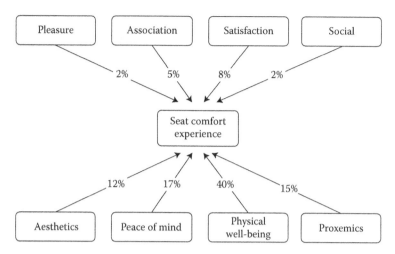

FIGURE 28.4
The overall influence of the eight comfort themes on seat comfort experience.

"peace of mind," "proxemics," and "aesthetics" accounted for 84% of all seat descriptions combined. These results are summarized in Figure 28.4. The result also revealed that 46 (88.5%) of the seat descriptions were related to long-haul (>4 hour) flights while only 6 (11.5%) were linked to short-haul (<4 hour) flights. The percentages of long- versus short-haul flights descriptions were calculated for each theme and shown in Figure 28.5 in different colors. The themes "proxemics," "social," and "pleasure" were only mentioned in relation to long-haul flights.

A Pearson product–moment correlation was performed to highlight the relationship between the heights of respondents who had mentioned the seat in relation to their comfort

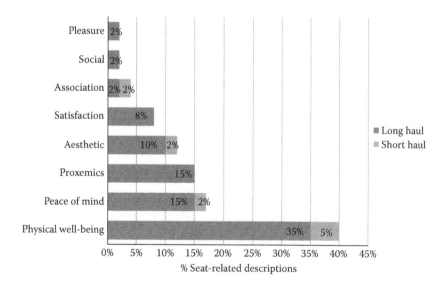

FIGURE 28.5
The seat descriptions (in total 52) that are linked to the eight passenger comfort themes on long- and short-haul flights based on flight comfort and discomfort reports of 27 participants.

experience and the ratings they had provided on the eight comfort themes. Significant correlation ($p < 0.05$, $r = 0.4$) was found between the height and ratings on the theme "physical well-being." This implies that a person's height is mainly linked to their perceived physical comfort of the seat however it may not influence other aspects of their seat experience. Moreover, it suggests that the bodily fit of the seat has a higher value for taller passengers.

Finally the respondents' concerns in relation to the elicited seat descriptions and the themes associated with those were identified. These are summarized in Table 28.3. Several differences were observed between the list of concerns generated for the seat and the overall comfort initially introduced by Ahmadpour et al. (2014a) as follows:

1. It was shown that the "physical well-being" experience of the seat was mainly concerned with the bodily support, leaving out concerns for energy. Recurrent descriptions relevant to that theme included the fit of the backrest to the spine curve, absence of cramps and pains, no sharp corners or edges, etc.

2. The concern for symbolism and experiences of higher values was not observed in the "association" theme. Participants' concerns in relation to this theme were limited to "evocation," that is, being reminded of familiar situations such as sitting in the car or a hotel lounge.

TABLE 28.3

Recurrent Passenger Concerns and Comfort Themes in Relation to the Seat Experience

Theme	Common Seat Descriptions	Concern
Physical well-being	No pain/cramp, back curve fits to the backrest, not to have to stack pillows to adjust to backrest curvature, easy to move, seat pan is not slippery, no sharp edges.	Bodily support
Peace of mind	Feeling at ease, no worry, not feeling confined, feels airy, able to store personal stuff, not feeling irritated and fidgety.	Security
	Supporting the head and neck (headrest ears) so that head does not fall off in sleep.	Tranquility
	Smooth recline, easy to position for sleep, able to lean against something to sleep.	Relief
Proxemics	Have an arm rest for myself, freely recline and control position with no worry, able to adjust and personalize the headrest.	Control
	No physical contact with neighbors, proper separation under armrest, not feeling squeezed by neighbors, feeling of having a personal space, like a cocoon.	Privacy
Aesthetics	Looks new/refurbished (vs. old/worn out), clean (no food crumbs, nor tears on covers).	Neatness
	Nice seat covers, bright colors, seat cover feels nice to touch.	Style
	Working well, solid, functioning design (no broken parts, no malfunction).	Quality
Satisfaction	Good recline system, well-designed and makes sense, confirms to a non-paying child needs, sufficiently enables comfortable eating and working.	Adequacy
	Buttons on the seat are well-placed, radio button could not be pushed accidently.	Accessibility
Association	Feels like sitting in a hotel lounge, feels like sitting in the car.	Evocation
Pleasure	Pleasant surprise, seat feels wider than usual. Cozy and inviting, modern (wood parts).	Anticipation Ambience
Social	Ability to hold hands over the low armrest.	Connectedness

3. The social experience of the seat was solely concerned with connectedness to others, for example, holding hands with one's spouse. However, the concern for intolerable interactions was not mentioned in relation to the seat.

28.4 Discussion

An in-depth inquiry was performed in order to obtain insight into the passenger comfort experience in the economy class of commercial aircrafts. Ratings on the eight thematic components of passenger comfort showed good fit to both comfort and discomfort experiences, implying that passengers may perceive those two experiences on the basis of the same components. That contradicts the proposition that comfort and discomfort, when they involve the seat, are independent entities underlined by different sets of variables (De Looze et al., 2003; Helander, 2003). However, the contradictory result obtained from our study may be due to the influence of the contextual elements of the flight, for example, legroom, IFE, temperature, activity, noise, service, etc. In addition, the social setting of the flight and the proximity of passengers to one another are different from the office environment used in Helander's study.

The differences between the ratings on themes "pleasure" and "physical well-being" suggest that enhancing the pleasant aspects of the flight highly improves passenger comfort experience whereas lack of physical well-being diminishes that experience more than any other theme. Therefore, an emphasis must be put on these themes in the design of the cabin environment. Pleasant aspects of the flight subscribe to exceeding passengers' anticipations, providing a nice ambience, entertainment, and stimulation in the flight context. Physical well-being is determined by the physical fit of the space to the passenger body and improving their level of energy.

Once the issue of differentiating comfort and discomfort was resolved, the average ratings obtained from the combined effects of the experiential themes confirmed their rank order from the previous study (Ahmadpour et al., 2014a). The outcome highlighted "peace of mind" and "physical well-being" as the most influential aspects of the experience for passenger comfort.

Previously, Ahmadpour et al. (2014a) suggested that the seat is central to most experiential aspects of passengers' comfort experience. This was confirmed by the study in this chapter. Four themes namely "physical well-being," "peace of mind," "proxemics," and "aesthetics" were mentioned more frequently when seat comfort was reported (accounting for 84% of seat comments). The theme "pleasure" was mentioned least frequently in those reports confirming the arguments of Ahmadpour et al. (2014a) that the entertainment units and service are the main determinants of the pleasure aspects of passenger experiences. The "social" aspects of the seat experience were similarly minimal in the reports. This was also suggested by Ahmadpour et al. (2015b), who examined those differences in a study that compared the social aspects of passenger comfort experience to the proxemics aspects (i.e., concerned with control and privacy). They revealed that sitting habits and sitting preferences are mainly important for the experience of privacy especially for those traveling alone (i.e., without companions).

The result indicated that physical comfort only accounts for 40% of seat comfort (see Figure 28.5). This result has implication for the seat design in that it suggests a focus on the non-physical aspects of the seat experience, particularly "peace of mind," "proxemics," "aesthetics," and passengers concerns associated with those aspects as shown in Figure 28.6.

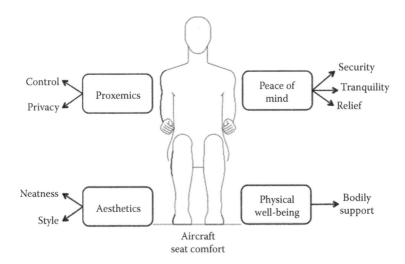

FIGURE 28.6
The four main themes of aircraft seat comfort experience and passenger concerns associated to each.

As an example, the seat headrest could be linked to several types of concern as follows: (1) concern for a better bodily support and fit to the working posture particularly when using electronic devices such as laptops and tablets (physical well-being), (2) concern for relief enabling one to sleep or relax whilst leaning against "wings" of the headrest (peace of mind), (3) concern for controlling one's position by adjusting the headrest (proxemics), (4) concern for the neatness and feel of the headrest cover, for example, softer material (aesthetics).

Among the four themes of aircraft seat comfort experience shown in Figure 28.5, the literature on "proxemics" is scarce and therefore demands further research. The theme was operationally defined as passengers' impression of the personal space and was connected to concerns for control over their space (e.g., freedom to adjust) and privacy (Ahmadpour et al., 2014a). Earlier, Hall (1966) argued that a distance of 45 cm from others in all direction yields an optimum personal space. While the average medial distance of 71–81 cm in commercial aircrafts complies with Hall's specification, the lateral distance is usually less than 45 cm and falls short on offering passengers a sense of privacy (Ahmadpour et al., 2014a). Our interviews similarly revealed a general complaint about privacy issues such as physical contact with neighbors in the areas of arms and legs. Increasing the space between adjacent seats in the economy class may significantly reduce the number of seats on the fleet and consequently it may not be feasible. Therefore, other solutions such as proper separation of passengers in those areas should be pursued in order to improve their sense of privacy. Moreover, respondents in our study encouraged the introduction of non-verbal means of communication with others, for example, "do not disturb" sign.

The above propositions do not diminish the importance of physical ergonomics for seat design. Vink and Brauer (2011) previously highlighted a need for better fit of current aircraft seats to the occupant's body and made recommendations accordingly. However, characterizing seats on the basis of comfort themes suggested by our study provides an opportunity to address a wider range of experiences by seat design. The value of that approach is demonstrated by Kamp (2012) in relation to car seats. She revealed that even when physical support is lacking on the sides of a car seat, we could still generate an experience of relaxedness through other favorable emotional characters.

Aircraft passengers' emotional responses to the cabin interior were examined by Ahmadpour et al. (2014b). It was shown that passengers' seat experience is linked to three groups of emotions. The first group consists of emotions such as disappointment, satisfaction, and relief in response to how passengers' expectations (for the seat) were met. Those expectations were similar to concerns for peace of mind (e.g., security) and satisfaction (e.g., adequacy of the design for working, sleeping, and other activities) that are presented in this study. The second group included emotions such as joy, resulting from positive experiences with the seat when passengers did not expect them. Concerns for peace of mind and physical well-being were linked to emotion joy. The third group was attraction emotions such as liking or disliking, elicited due to positive visceral impressions of the seat similar to aesthetics concerns. Ahmadpour et al. (2014b) suggested that information about passengers' concerns is essential for improving the emotional experience of the seat. The study in this chapter contributes to a better understanding of passengers' concerns with regard to the seat and the outcomes provide context for the emotional model of passengers introduced by Ahmadpour et al. (2014b).

The study presented in this chapter was carried out with the intention of exploring the content and nature of the seat comfort experience in economy class. However, some limitations of the study such as the disproportional number of participants on long-haul flights prevent us from drawing any firm conclusions in that regard. Future research should overcome this limitation by employing a larger sample of participants.

28.5 Conclusion

Aircraft passenger comfort and discomfort experiences are not two independent entities but rather they are understood as opposite sides of a holistic experience. Passenger comfort could be described as an experience ranging from extreme discomfort to extreme comfort, underlined by eight themes. The relationship between the seat and the themes of the passenger comfort experience was examined. It is concluded that four themes namely "physical well-being," "peace of mind," "proxemics," and "aesthetics" are the most prominent aspects of aircraft seat comfort. Applying this insight into the seat design is expected to improve the seat comfort experience.

References

Ahmadpour, N., Kühne, M., Robert, J-M., Vink, P. 2016b. Attitudes towards personal and shared space during the flight. *WORK: A Journal of Prevention, Assessment & Rehabilitation*, Special Issue on Environmental Design, in press.

Ahmadpour, N., Lindgaard, G., Robert, J.-M., Pownall, B. 2014a. The thematic structure of passenger comfort experience and its relationship to the context features in the aircraft cabin. *Ergonomics*, 57(6), 801–815.

Ahmadpour, N., Robert, J.-M., Lindgaard, G. 2014b. Exploring the cognitive structure of aircraft passengers' emotions in relation to their comfort experience. *Proceedings of KEER'14* (Kansei Engineering and Emotion Research), June 11–12, 2014, Linköping, Sweden.

Ahmadpour, N., Robert, J.-M., Lindgaard, G. 2016a. Aircraft passenger comfort experience: Underlying factors and differentiation from discomfort. *Applied Ergonomics*, 52, 301–308.

Brauer, K. 2004. Convenience, comfort and cost: The Boeing perspective on passenger satisfaction. [PowerPoint slides] In: *The Aircraft Interior EXPO'* 04, March 30–April 1, 2004, Hamburg.

Coelho, D.A., Dahlman, S. 2002. Comfort and pleasure. In: P.W. Jordan, W.S. Green, (eds.), *Pleasure with Products: Beyond Usability* (pp. 321–331). London: Taylor & Francis.

De Looze, M.P., Kuijt-Evers, L.F.M., Van Dieen, J. 2003. Sitting comfort and discomfort and the relationships with objective measures. *Ergonomics*, 46(10), 985–997.

Hall, E.T. 1966. *The Hidden Dimension*. New York: Anchor Books.

Hancock, P.A., Pepe, A., Murphy, L. 2005. Hedonomics: The power of positive and pleasurable ergonomics. *Ergonomics in Design: The Quarterly of Human Factors Applications*, 13(1), 8–14.

Helander, M.G. 2003. Forget about ergonomics in chair design? Focus on aesthetics and comfort. *Ergonomics*, 46(13–14), 1306–1319.

Helander, M.G., Zhang, L. 1997. Field studies of comfort and discomfort in sitting. *Ergonomics* 40(9), 895–915.

Hiemstra-van Mastrigt, S. 2015. Comfortable passenger seats: Recommendations for design and research. PhD Delft University of Technology.

Hiemstra-van Mastrigt, S., Kamp, I., van Veen, S.A.T., Vink, P., Bosch, T. 2015. The influence of active seating on car passengers' perceived comfort and activity levels. *Applied Ergonomics*, 47, 211–219.

Jordan, P.W. 2000. *Designing Pleasurable Products*. London: Taylor & Francis.

Kamp, I. 2012. The influence of car-seat design on its character experience. *Applied Ergonomics*, 43(2012), 329–335.

Vink, P., Bazley, C., Kamp, I., Blok, M. 2012. Possibilities to improve the aircraft interior comfort experience. *Applied Ergonomics* 43(2012), 354–359.

Vink, P., Brauer, K. 2011. *Aircraft Interior Comfort and Design*. Boca Raton, FL: CRC Press.

Vink, P., Hellbeck, S. 2012. Editorial: Comfort and discomfort studies demonstrate the need for a new model. *Applied Ergonomics*, 43(2), 271–276.

Vink, P., Overbeeke, C.J., Desmet, P.M.A. 2005. Comfort experience. In: P. Vink (ed.), Comfort and design: Principles and good practice (pp. 1–12). Boca Raton, FL: CRC Press.

29

Rotorcraft-Pilot Couplings Caused by Biodynamic Interaction

Vincenzo Muscarello, Pierangelo Masarati, and Giuseppe Quaranta

CONTENTS

29.1 Introduction

Aircraft or rotorcraft-pilot couplings (A/RPCs) are adverse, unwanted dynamic phenomena caused by an anomalous and undesirable interaction between the pilot and the vehicle. Adverse A/RPCs have been always a critical issue for flight since the early days—the earliest recorded examples of such phenomena date back to the Wright Brothers' glider of 1902 (Pavel et al., 2013).

Aircraft pilots act on the vehicle through control inceptors in order to perform the desired mission tasks. The main interaction with control inceptors is driven by voluntary actions of the pilot to change the status of the aircraft. In some cases, the pilot can be deceived by erroneous or misleading cues, thus voluntarily producing control efforts that are erroneously phased with respect to the vehicle motion, resulting in undesired response. In such cases, adverse phenomena may occur. These phenomena have been initially called pilot-induced oscillations (PIO).

In other cases, the pilot may involuntarily inject controls through the inceptors as a consequence of cockpit vibrations, producing what have been termed pilot-assisted oscillations (PAO). Specific attention must be dedicated to such involuntary interaction between the vehicle and the pilot. This transmission of signals as control commands through the inceptors at frequencies above the band of voluntary action is called biodynamic feedthrough (BDFT). The BDFT is the involuntary consequence of external disturbances.

Such involuntary commands may further excite the dynamics of the vehicle, causing a degradation of the flight dynamics qualities, difficulties in achieving the desired performance, and may ultimately produce an unstable closure of the control feedback loop. This problem may affect all kinds of vehicles whose pilot is accommodated within the vehicle and is thus subjected to its motion.

Research on A/RPC has mainly focused on fixed wing aircraft; it is reviewed in this classical report (McRuer et al., 1997). However, it is arguable that rotorcraft should be more susceptible to RPCs than fixed wing dynamics, in particular to PAO, due to the prominent role of high-frequency servo-aeroelastic dynamics and to the significant number of extreme task demands, which require very high pilot gain. This speculation has been, in fact, verified thanks to the project ARISTOTEL (Aircraft and Rotorcraft Pilot Couplings—Tools and Techniques for Alleviation and Detection); the main findings of such project can be found in Pavel et al. (2013, 2015).

This study discusses two important helicopter aeroelastic RPC events caused by the interaction of the vehicle with the biodynamics of the pilot: the collective bounce and the aeroelastic roll/lateral instability. Collective bounce is the consequence of an adverse interaction of the pilot with the vertical motion of the helicopter. In it, the highly damped main rotor coning mode plays an essential role by introducing enough phase delay in the vertical acceleration of the vehicle in response to collective control input to reduce the phase margin of the vehicle when the collective control loop is closed by the pilot's involuntary response to the vertical acceleration of the seat. Roll/lateral instability is caused by the interaction of the regressive lead-lag motion of the main rotor blades, which causes a dynamic unbalance of the rotor that interacts with the motion of the vehicle about the roll axis. Such motion, in turn, causes an involuntary action of the pilot on the lateral cyclic control, which unintentionally closes the control loop. It is shown, both numerically and experimentally, that for specific vehicle configurations some pilots, based on their BDFT, may drive the closed-loop system unstable. The description of the phenomena is addressed by identifying the appropriate modeling requirements for all the components involved in the process: the aeromechanics of the vehicle and its components, the dynamics of the control system and, specifically, the biomechanics of the pilot to describe the BDFT. This is a formidable task, since it involves the requirement of modeling biodynamics, aerodynamics, structural elasticity, and servo-systems altogether, leading to the development of what have been called bio-aero-servo-elastic (BASE) models. Modeling and analysis tools are developed, including detailed multibody dynamics and linearized analysis formulations for both the vehicle and the pilot. Typical adverse couplings are investigated and discussed, with reference to numerical models of representative vehicles interacting with pilot models resulting from identification.

Those RPC mechanisms have also been investigated experimentally in the flight simulator, with professional test pilots controlling the models of the helicopters through specially crafted mission task elements (MTE). In particular, it has been possible to investigate how the BDFT changes while the pilot is performing different MTE, showing that there is a correlation between pilot workload and the BDFT.

The modeling approaches are briefly recalled, along with the experimental setup used to investigate the problem. The dependence of the pilots' BDFT transfer functions on the task is discussed. Specifically, it is shown how the closure of the involuntary control loop affects the pilot's transfer functions, motivating the need to conduct specific experiments in closed-loop configuration.

29.2 Analysis of RPCs Problems

29.2.1 Helicopter Inceptors Layout

Helicopters are controlled by pilots using control inceptors whose layout evolved into a standard configuration, with a few notable exceptions. Figure 29.1 illustrates a typical layout.

A central stick, analogous to that of many fixed-wing aircraft, is usually held using the right hand. The center stick can be moved fore and aft, and sideways. Such commands are transduced into appropriate motion of a mechanical component, called the "swashplate," which in turn induces the periodic pitching of the rotor blades, thus causing periodic changes in blade loads that make the main rotor disk tilt. The tilting of the disk ultimately changes the direction of the thrust, thus producing control moments about an arbitrary axis orthogonal to the axis of rotation of the rotor. Control moments about the pitch and the roll axis are obtained by moving the stick fore-aft and sideways, respectively. The role of this control is essentially analogous to that of the alternative arrangements of center stick, side stick, or wheel that are in use in conventional fixed-wing aircraft.

Another control, specific of helicopters, is the so-called collective control inceptor, which is operated by the pilot using the left hand. Such control produces a collective change of the pitch of the main rotor blades, thus controlling the amount of thrust the rotor generates. A change in thrust is usually accompanied by a change in torque required to keep the rotor angular velocity at the desired value, so that the collective control inceptor is used to incorporate a throttle control, much like the throttle of a motorbike. Modern helicopters use an automatic control system, called FADEC that acts on the fuel flow to keep the angular velocity of the rotor as close as possible to the nominal value regardless of the amount of collective pitch, and ultimately of thrust, requested by the pilot.

Finally, pedals are used to control the collective pitch of the tail rotor, producing a moment about the yaw axis. This control is analogous to the deflection of the rudder in fixed-wing aircraft. Pedals are not considered in this discussion.

29.2.2 Bio-Aero-Servo-Elastic Feedback Loop

When the pilot is subjected to accelerations, involuntary control actions can be injected into the flight control system. Indeed, accelerations are transmitted from the seat to the pilot's spine and then to the rest of the body. The suspended mass of the pilot's arms is also subjected to inertia forces. This transmission of vibrations through the body is often

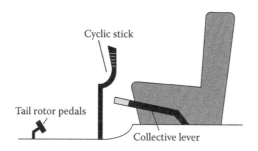

FIGURE 29.1
Typical helicopter control inceptors arrangement.

called BDFT. BDFT in general may produce annoying disturbances, increased vibratory level and discomfort; it may increase the workload of the pilot, for example because it disturbs the vision of the displays, or because it requires more attention of the pilot to perform the required task. It may also produce direct effects on the pilot's performance, by inducing fatigue into the pilot's muscles; ultimately, for long enough exposures, it might impair the pilot, something that may need to be taken into account along the lines discussed in ISO 2631 (Griffin, 1996). The problem becomes more critical when there is a direct relationship between the direction of the accelerations that affect the pilot and the control forces and moments produced by a motion of the controls caused by such accelerations. In those cases, the stability of the closed-loop pilot–vehicle system (PVS) may be jeopardized.

As shown in Figure 29.2, from a topological point of view voluntary and involuntary controls may be interpreted as two independent feedback control loops (although the problem is more complex). Voluntary control action stems from the perception of the vehicle behavior, which is acquired by the pilot using several sensors: visual, inertial (low-frequency accelerations), vestibular and proprioceptive, just to mention the main ones. The cognitive level of processing of such information and, to some extent, the reflexive level produce voluntary (and sometimes reflexive) actions on the control inceptors. As anticipated, BDFT (that is the direct, mechanical effect of vibrations on the pilot's limbs) produces additional, involuntary control inputs that are added to the voluntary ones. Their combination produces the actual motion that is commanded by the control inceptor to the control system.

Considering z the trajectory followed by the aircraft (in terms of positions and attitude as perceived at the cockpit's location) and z_d the desired trajectory, θ_i, the commands transmitted by the pilot through the inceptors, are equal to the combination of the contributions produced by the voluntary and involuntary pilot models, namely

$$\theta_i = \theta_{ap} + \theta_{pp} \tag{29.1}$$

where the voluntary pilot model is represented by the transfer function between the trajectory error $e = (z - z_d)$ and the pilot commands θ_i, and the involuntary pilot model is represented by the BDFT transfer function between the cockpit accelerations and the commands θ_i. Consequently,

$$\theta_i = H_{ap}(s)(z - z_d) + H_{pp}(s)s^2z \tag{29.2}$$

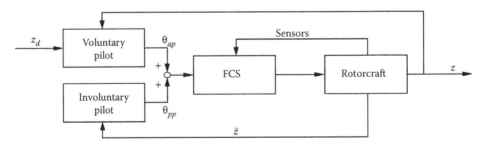

FIGURE 29.2
Feedback loops between vehicle and pilot.

The vehicle can be represented as well by a transfer function between the control input θ and the trajectory z

$$z = H_R(s)\theta \tag{29.3}$$

that could be supplemented by a flight control system (FCS) which may modify the control transmitted to the rotorcraft θ with respect to the pilot's commands θ, that is,

$$\theta = H_{FCS}(s)\theta_i \tag{29.4}$$

The simplest voluntary pilot, denominated "crossover model," was proposed by McRuer (McRuer and Jex, 1967) and represents the behavior of a trained pilot that is able to control the aircraft by adapting to its low-frequency dynamics. McRuer's model prescribes that the product of the pilot and the vehicle models is equal to an integrator and a time delay

$$H_{ap}H_R H_{FCS} = e^{-s\tau_e} \frac{\omega_c}{s} \tag{29.5}$$

This means that, to some extent, the pilot adapts himself/herself to the characteristics of the piloted vehicle. Adding all the elements together it is possible to derive the loop transfer function of the vehicle

$$H_L = H_{ap}H_R H_{FCS} - H_{pp}H_R H_{FCS}s^2 = e^{-s\tau_e} \frac{\omega_c}{s} - H_{pp}H_R H_{FCS}s^2 \tag{29.6}$$

This expresses the important fact that the element related to the voluntary behavior of the trained pilot is not explicitly affected by changes in the aircraft or flight control system dynamics, while the involuntary pilot can be strongly affected by them. So changes in the aircraft or flight control system behavior may be used to influence the coupled pilot–vehicle system during experiments to assess the bio-aero-servo-elastic stability.

The phenomena involved in helicopter RPCs present several peculiarities and require some careful study to point out the root causes. Understanding the phenomenon is essential to support the design of vehicles and human–machine interfaces (HMI) that mitigate its insurgence. It has been shown that to investigate these kinds of RPCs the analysis involves the detailed modeling of the pilot, of the vehicle, and of their interaction. Such analysis is important to understand what factors play an important role, and to define the modeling requirements for an appropriate analysis, in support of the design of new vehicles and HMIs. Pilot models have been developed according to two distinct approaches:

1. Linear, frequency domain models have been obtained by identifying the results of BDFT experiments (Figure 29.3). The experiments involved human subjects sitting in a flight simulator pod and holding the control inceptors and subject to vibration of the flight simulator base at several frequency between 0.5 and 8.0 Hz at several amplitudes and along all three directions, vertical, longitudinal and lateral, see, for example, Masarati et al. (2013) and Muscarello et al. (2016). Most of the experiments involved in this work were performed at the flight simulation laboratory of the University of Liverpool (UoL), within the joint research efforts mentioned earlier.

   Such experiments allowed to identify an envelope of possible transfer functions that are shown in Figure 29.4. In all cases the identified transfer functions showed

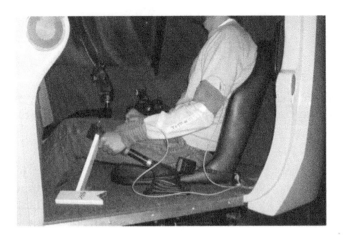

FIGURE 29.3
BDFT experimental setup for collective control inceptor.

an inter-subject variability but also an intra-subject variability, due to changes in the mental and physical status of the subject related to the level of workload and attention to the task to be performed. Overall, the transfer functions showed a reasonable agreement with those already identified in the past literature, see Mayo (1989) and Parham et al. (1991).

2. Detailed numerical models of the pilot have been developed within a general-purpose multibody dynamics environment (Figure 29.5), based on biomechanical data available from the open literature, see, for example, Masarati et al. (2013). The models were extremely detailed, including four rigid bodies representing the humerus, radius, ulna, and hand, all articulation and 25 muscles based on a Hill-like force model. Those models have been either directly used in "monolithic" multibody simulations of the coupled PVS, or used to produce numerical experiments and synthesize transfer functions to be used in linearized analysis, much like the models obtained directly from experiments.

It is worth stressing that the capability to produce linearized models of the pilot's involuntary behavior from numerical analysis paves the way for predicting BDFT within yet untested HMI configurations and cockpit layouts.

Experiments were also used to verify the predicted adverse interaction between the pilot and the vehicle. The two problems of vertical bounce and roll were considered using rather different approaches. In fact, in the case of vertical bounce, a purely biomechanical adverse interaction is postulated, when appropriate system parameters are considered, whereas in the case of roll axis instability it can only occur when the pilot is required to perform a high-gain task.

29.3 Vertical Bounce

One may speculate that the minimal vehicle model capable of describing the problem is a rigid-body model. In such cases, the motion of the helicopter along the vertical axis is all

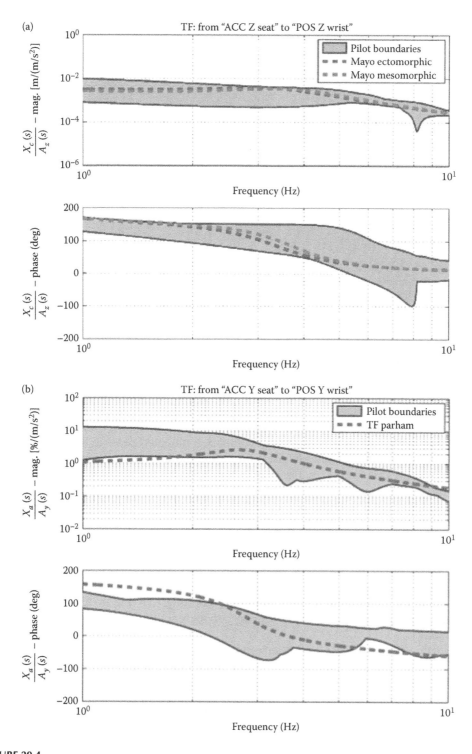

FIGURE 29.4
Envelopes of experimentally identified BDFT transfer functions for the collective lever due to vertical acceleration of the cockpit (a), and of the lateral cyclic stick due to lateral acceleration of the cockpit (b).

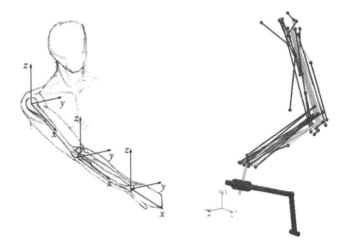

FIGURE 29.5
Pilot arm detailed multibody model. A sketch of the arm bones and articulations on the left and the representation of the 25 muscles used in the multibody model with the attachment points on the right.

one needs to describe the vehicle. In this case, one can easily show that no instability is possible within realistic vehicle parameters when using realistic models of the pilot BDFT.

Figure 29.6a shows that helicopter types that significantly differ in size and characteristics of the main rotor system present a nearly identical behavior in terms of vertical acceleration resulting from collective control input when the main rotor coning motion is neglected. Figure 29.7a shows that the corresponding loop transfer function (LTF), obtained by closing the feedback loop with a realistic involuntary pilot model, that is, the one identified in Mayo (1989), characterized by two complex conjugated poles at about 3.5 Hz with about 30% damping, is not going to circle around point (−1), thus always complying with Nyquist's stability criterion no matter of how much the feedback gain is increased. In this context, the feedback gain can be interpreted as the ratio between the rotation of the collective control inceptor and the actual pitch rotation of the main rotor blades. A realistic nominal value is considered in the curves, leading to stable and reasonably behaving curves for all the helicopters considered in the plots.

On the contrary, when the main rotor coning mode (Figure 29.7) is considered, a rather different behavior can be observed. Figure 29.6b shows some amplification for some helicopters, and none for others; all vehicles show significant phase lag in the vicinity of the frequencies of the pilot model's poles. The corresponding LTF (Figure 29.7b) now crosses the negative imaginary axis; as a consequence, an increase of feedback gain could lead to loss of stability.

It is worth stressing that pilots often prefer high feedback gain for several reasons: high gain means that the same amount of thrust change can be obtained with a smaller motion of the controls. As a consequence, the whole range of blade pitch can be obtained with less overall motion of the hand, which can be held about the most comfortable position for most of the time. Moreover, high gain is felt as a more prompt response to control inputs. Pilots can easily adjust themselves to gain changes—and in general to response changes—thanks to training—since the response can be somehow predicted by the pilot as long as it conforms to the mental model he/she has of the vehicle—and to adaptation that is intrinsic in humans when acting as regulators of predictable processes. As a consequence, from the point of view of voluntary control, a request for change of feedback gain

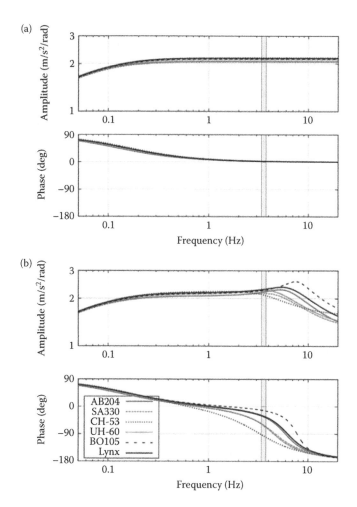

FIGURE 29.6
Vertical acceleration as a function of collective control: (a) without and (b) with rotor coning.

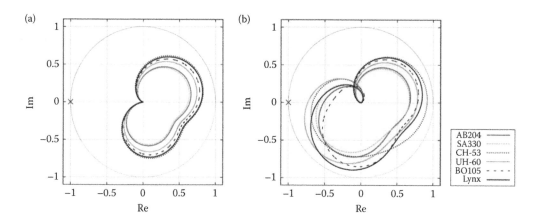

FIGURE 29.7
Loop transfer function of involuntary collective control: (a) without and (b) with rotor coning.

is perfectly legitimate and, within reasonable bounds, compatible with the behavior of human operators. However, as shown by this simple analysis, a change of feedback gain changes the involuntary contribution to the control input and can lead to loss of stability.

Experiments performed in the flight simulator considering a closed-loop setup, in which the force applied to the analytical model of the vehicle was determined by the motion of the control inceptor, showed a loss of stability of the coupled PVS as the feedback gain was increased (Masarati et al., 2014). In that case, engineer (i.e., non-professional) pilots as well as professional test pilots were considered in the experiment. Experiments were performed considering a simple, two-degree-of-freedom model that describes the motion of the vehicle along the vertical axis, in analogy with the simplified analytical model shown in Figure 29.8, with an additional degree of freedom that was tuned to produce some desired dynamics, not necessarily those of rotor coning. The dynamics associated with this second mode were used to produce the desired characteristic frequency and damping, to investigate the interaction of the involuntary pilot with specific vehicle dynamics.

The results highlight several interesting aspects:

1. The insurgence of the instability is very subjective, both inter-subject and intra-subject variabilities were experienced; different tests yielded different stability and post-stability threshold of the feedback gain, although common trends appeared.

2. The instability is dominated by the biodynamic characteristics of the pilot; this was assessed by modifying (specifically by detuning the structural dynamics from their nominal value) and noticing that the frequency of the instability is characterized by a frequency close to that of the pilot.

3. No pilot was RPC-free; pilots that reached instability with some vehicle configuration did not reach instability with other configurations, and vice versa.

4. Closed-loop experiments clearly show an adaptation of the voluntary behavior of the pilot to changes in the feedback gain, as expected, but also a dependence of the involuntary behavior. This dependence is somehow subtle, and the type and amount of tests performed did not allow to fully understand it, but it is clear that as the instability is approached, the voluntary behavior of the pilot changes; this has an impact on the involuntary dynamics, modifying the proneness to RPC. It is conjectured that by acting more cautiously, the pilot actually modifies the biomechanical properties of the limbs, reducing their equivalent stiffness, thus decoupling the biomechanical poles from those of the vehicle. This aspect needs further analysis to be adequately quantified.

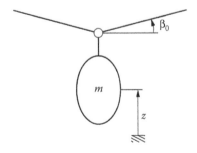

FIGURE 29.8
Rigid-body heave and main rotor coning model.

29.4 Roll-Axis Instability

Roll-axis dynamics has been prone to adverse pilot–vehicle interaction for both rotorcraft and fixed-wing aircraft. In fixed-wing aircraft, the source of adverse interaction has been often identified either in the flexibility of the vehicle, for example, low-frequency skew-symmetric wing bending that interacts with rigid-body roll dynamics to yield sufficient phase delay in the roll response, or other sources of delay, for example, input processing by a digital flight control system, or insufficient bandwidth or saturation of control system actuators.

In helicopters, a possible source of adverse interaction is associated with the lead-lag motion of the main rotor blades. Specific combinations of blade lead-lag motion displace the center of mass of the rotor from the axis of rotation. Those combinations are known as "progressive" and "regressive" lead-lag modes. When such motions occur, the center of mass of the rotor moves along a trajectory that processes about the rotation axis either in the same direction (the progressive mode) or in opposite direction (the regressive mode). Such dynamics are damped by lead-lag dampers that are usually present to prevent such dynamics from coupling with pitch and roll of the vehicle when standing on the landing gear. The corresponding dynamic phenomenon is well-known as "ground resonance" (Johnson, 2013).

A similar phenomenon may occur, when the helicopter is airborne, as long as the roll motion is controlled by some form of feedback that reacts to the roll motion with a cyclic change of blade pitch that produces a counteracting control moment of the main rotor. Such phenomenon is known as "air-resonance" when the feedback loop is closed by a stability (and control) augmentation system (S(C)AS). However, the pilot himself may close such a control loop through involuntary interaction with the vehicle dynamics. This phenomenon requires some specific circumstances to appear:

1. Acceleration about the roll axis must produce an involuntary control input.
2. The regressive lead-lag motion must have low damping; this is usually the case in helicopters that do not suffer from ground resonance, and thus have little if any blade lead-lag damping.
3. Lead-lag motion must be sufficiently coupled with roll; this is usually the case, since the main rotor is well offset vertically from the roll axis.
4. The lead-lag regressive mode, in the non-rotating frame, must be in the vicinity of the pilot's biomechanical modes; typical figures for the latter, according to the open literature and to experiments performed within this work, are between 1 and 2.5 Hz.

Figure 29.9 presents the pilot BDFT related to lateral cyclic control input as a function of lateral acceleration for three professional helicopter test pilots as measured during a dedicated test campaign performed during the project ARISTOTEL (Muscarello et al. 2016).

The figure clearly shows significant differences between the three pilots, which are explained partly by their attitude toward aircraft control, and partly by their build (pilot 1 was significantly taller than the other two). Notice how pilot 1 shows a significant occurrence of phase delay at about 2 Hz, whereas the other two pilots show a comparable amount of delay at higher frequency, together with less pronounced amplification. The combination of these two factors, the phase delay and the amplification at about 2 Hz, produces the Nyquist plots of the LTFs shown in Figure 29.10. Figure 29.10a shows the LTF

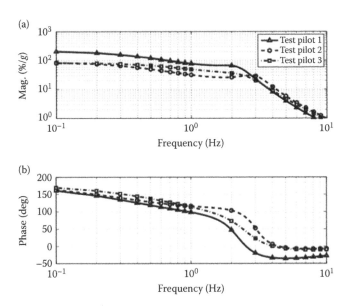

FIGURE 29.9
Biodynamics feedthrough of the lateral cyclic control as function of lateral acceleration measured on three test pilots during a test campaign in Liverpool: (a) amplitude, (b) phase. (Adapted from Muscarello, V. et al. 2016. *AIAA Journal of Guidance Control and Dynamics*, 39(1):42–60.)

of the three pilots for a helicopter with hingeless main rotor (5% damping of the regressive lead-lag mode). The helicopter is modeled using a linear state-space model that includes essential airframe and main rotor dynamics (all rigid-body and several aeroelastic modes well above the bandwidth of the pilot; control system actuator dynamics, SCAS).

The three pilots show a rather different behavior. The feedback gain is 2.5 the nominal value, a value that does not prevent the pilots from performing several high-gain tasks,

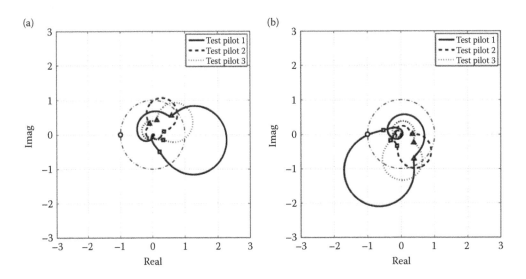

FIGURE 29.10
LTF of roll for a hingeless helicopter in forward flight, 80 kts: (a) no; (b) 140 min time delay.

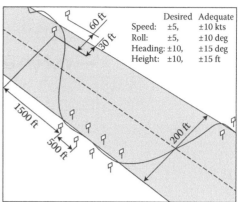

	Desired	Adequate
Speed:	±5,	±10 kts
Roll:	±5,	±10 deg
Heading:	±10,	±15 deg
Height:	±10,	±15 ft

FIGURE 29.11
Roll-step maneuver course layout.

but makes the vehicle oversensitive. Figure 29.10b shows the same plots with 140 min of digital time delay between the lateral cyclic control input and the actual blade pitch change. As one can clearly appreciate, now pilot 1 is at the verge of instability, because the time delay basically "rotates" clockwise the curve of Figure 29.10a, without impacting the magnitude.

An instability of this type only appears when the pilot triggers it by performing some specific task. In fact, usually cyclic lead-lag motion, being at least slightly damped, is not present unless triggered by specific abrupt maneuvers, like those about the pitch and the roll axes. The results of this simple analysis have been assessed by performing a dedicated test campaign in the flight simulator. The test pilots were requested to fly a "roll step" maneuver (Figure 29.11), that is, to fly along the left edge of a runway, perform a sharp right turn to line up with the right edge, fly straight a little bit more, and then return on the left edge with a sharp left turn. The sharp turns are intended to excite the rotor dynamics, significantly the lowly damped regressive lead-lag mode.

Indeed, it was shown by the experimental results (Figure 29.12) that pilot 1 was able to incur into a limit cycle oscillation (LCO) for a feedback gain of three times the nominal and a time delay of 100 ms (i.e., similar to those of Figure 29.10b). The oscillation was repeatable, at about 2.2 Hz, which is the frequency of the regressive lead-lag mode. Inspection of the state of the helicopter aeromechanics model implemented in the flight simulator confirmed that the LCO was associated with the regressive lead-lag mode.

Figure 29.12 shows that pilot 1 had to abort the task shortly after lining up after the first roll step. Such a repeatable pattern seems to indicate that the pilot could manage to fly the helicopter as long as the regressive lead-lag mode was not yet excited; he was also able to fly the helicopter throughout the roll step itself (i.e., the roll maneuver initiation and completion, which is essentially based on anticipating the response of the vehicle to compensate for the high time delay, and thus is essentially a feedforward maneuver), but failed to perform the subsequent "capture" phase, that is required to level the helicopter with respect to the roll axis. Such maneuver is a high-gain tracking task (essentially a feedback task), which requires firm pilot action; the pilot involuntarily "stiffens" while performing high-gain tracking, and thus increases the feedback gain of the involuntary response LTR, driving the coupled system to become unstable. The instability immediately reveals itself because the just performed roll step maneuver excited the regressive lead-lag mode. Thus, the combination of the maneuver, which excites the mode, and the stiffening of the pilot,

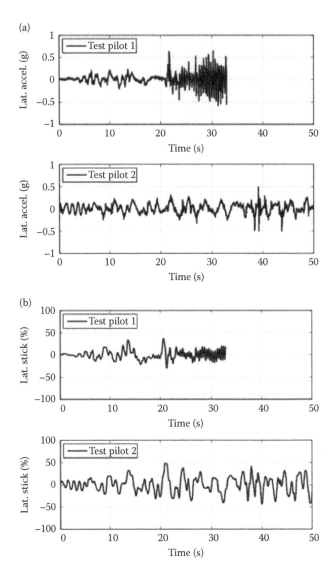

FIGURE 29.12
Lateral acceleration (a) and lateral cyclic stick deflection in percentage (b) for test pilots #1 and #2 measured during Liverpool test campaign. The measures were taken with a feedback gain three times nominal and 100 ms time delay. (Adapted from Muscarello, V. et al. 2016. *AIAA Journal of Guidance Control and Dynamics*, 39(1):42–60.)

which moves from a "pursuit" (i.e., feedforward) to a tracking (i.e., regulatory feedback) task, act as triggers of the instability.

As a further check, a similar aeromechanics model, with the regressive lead-lag mode switched off, could be flown in the same configuration, safely bringing the task to completion. Curiously enough, pilot 2, who did not incur into LCOs with the model that included the regressive lead-lag as predicted in Figure 29.10b, did incur into "conventional" rigid-body PIO associated with flight mechanics (i.e., rigid-body roll).

29.5 Conclusions

Adverse aeroelastic rotorcraft-pilot couplings may jeopardize the stability of helicopters in specific operating conditions. This work describes recent research activities aimed at understanding specific problems, determining the modeling requirements for their analysis, and investigating them both numerically and experimentally. The vertical bounce and roll axis problems are analyzed. The vertical bounce involves the vertical motion of the helicopter and the collective control inceptor, which is held by the pilot's left hand. It is discussed how the main rotor coning motion plays a fundamental role by introducing phase delay between the control action and the vertical acceleration of the vehicle at the frequencies that characterize the involuntary action of the pilot. The roll axis instability involves helicopter roll and lateral displacement and the lateral cyclic control, which is held by the pilot's right hand. It is discussed how the main rotor regressive lead-lag motion plays an essential role by introducing a lowly damped mode in the band of frequency that is characteristic of the involuntary action of the pilot. Although outside the scope of this work, alleviation of the phenomena can hardly be obtained by modifications to the aeromechanics of the vehicle, since they are the result of performance optimization. Typical means involve modifications of the dynamics of the control system, including modifications to the control inceptors and digital filtering for augmented and fly-by-wire vehicles.

References

Griffin, M. 1996. *Handbook of Human Vibrations*, London: Academic Press.

Johnson, W. 2013. *Rotorcraft Aeromechanics*, Cambridge: Cambridge University Press.

Masarati, P., Quaranta, G., and Jump, M. 2013. Experimental and numerical helicopter pilot characterization for aeroelastic rotorcraft-pilot couplings analysis, *Proc. ImechE, Part G: J. Aerospace Engineering*, 227:124–140, doi: 10.1177/0954410011427662.

Masarati, P., Quaranta, G., Lu, L., and Jump, M. 2014. A closed loop experiment of collective bounce aeroelastic rotorcraft-pilot coupling, *Journal of Sound and Vibration*, 333:307–325, doi:10.1016/j.jsv.2013.09.020.

Masarati, P., Quaranta, G., and Zanoni, A. 2013. Dependence of helicopter pilots' biodynamic feedthrough on upper limbs' muscular activation patterns, *Proc. IMechE Part K: J. Multi-body Dynamics*, 227:344–362, doi: 10.1177/1464419313490680.

Mayo, J. R. 1989. The involuntary participation of a human pilot in a helicopter collective control loop, in: *15th European Rotorcraft Forum*, pp. 81.1–12, Amsterdam, The Netherlands, September 12–15, 1989.

McRuer, D. T., Droste, C. S., Hansman Jr., R. J., Hess, R. A., Le Master, D. P. et al. 1997. *Aviation Safety and Pilot Control Understanding and Preventing Unfavorable Pilot–Vehicle Interactions*, Washington, DC: ASEB National Research Council, National Academy Press.

McRuer, D. T. and Jex, H. R. 1967. A review of quasi-linear pilot models, *IEEE Transactions on Human Factors in Electronics*, 8:231–249, doi: 10.1109/THFE.1967.234304.

Muscarello, V., Quaranta, G., Masarati, P., Lu, L., Jones, M., and Jump, M. 2016. Prediction and simulator verification of roll/lateral adverse aeroservoelastic rotorcraft–pilot couplings, *AIAA Journal of Guidance Control and Dynamics*, 39(1):42–60, doi: 10.2514/1.G001121.

Parham, T., Popelka, D., Miller, D. G., and Froebel, A. T. 1991. V-22 pilot-in-the-loop aeroelastic stability analysis, in: *Proceedings of the American Helicopter Society 47th Annual Forum*, Phoenix, Arizona, May 6–8, p. 25.

Pavel, M., Jump, M., Dang Vu, B., Masarati, P., Gennaretti, M. et al. 2013. Adverse rotorcraft pilot couplings—Past, present and future challenges, *Progress in Aerospace Sciences*, 62:1–51, doi: 10.1016/j.paerosci.2013.04.003.

Pavel, M. D., Masarati, P., Gennaretti, M., Jump, M., Zaichik, L. et al. 2015. Practices to identify and preclude adverse aircraft-and-rotorcraft-pilot couplings—A design perspective, *Progress in Aerospace Sciences*, 76:55–89, doi: 10.1016/j.paerosci.2015.05.002.

Subject Index

Author Index

A

Abbink, D. A., 192, 193, 194
Åberg, L., 191, 193
Accolla, A., 89
Adelt, F., 214, 222
Adler, N., 273, 293
Afacan, Y., 74, 79
Agichtein. E, 51, 59
Ahmadpour, N., 399, 401, 402, 404, 406, 408, 410, 411
Alderson, J. C., 310
Alexander, H., 294
Alexander, J. R., 271, 349
Alley, V. L., 349
Allen, J., 233, 241
Amalberti, R., 347, 349, 368, 384
Amdits, A., 214, 221
Ammerman, H. L., 336, 349
Anderson, A. D., 183, 396
Anderson, J., 51, 193
Andrews, J., 386, 397
Annett, J., 209
Anund, A., 196, 210
Araújo, D., 138, 149
Armstrong, A., 259, 269
Arnaldo, R., 335, 349
Asvin, G., 225, 241
Atkinson, J., 269
Augustine, G., 58
Avineri, E., 241
Ayyalasomayajula, P., 108, 118

B

Baber, C., 211, 215, 222, 261, 269, 270, 307, 310
Bacchus, A., 222
Bainbridge, L., 335, 350
Baker, C. C., 13
Baker, D. P., 215, 222
Bakker, G. J., 313, 314, 328, 329
Baldauf, M., 44
Balfe, N., 368, 384
Bandini Buti, L., 27
Banister, D., 229, 241
Banks, A. P., 222
Banks, V. A., 213, 214, 219, 220
Barkenbus, 186

Barnard, Y., 214, 221
Barry, A., 388, 396
Barshi, I., 297, 299, 309, 310, 311
Barsoux, J., 294
Bashir, M., 125, 133, 134
Batteau, A., 274, 294
Batt, R., 294
Baxley, B., 353, 365
Bazley, C., 399, 412
Beanland, V., 159, 160
Beardsall, J., 50
Beauducel, A., 397
Bechervaise, N. E., 226, 241
Becker, E. S., 336, 349
Beernink, B., 386, 387, 397
Beilinsson, L., 191, 193
Bekier, M., 214, 220
Bengler, K., 193
Bergen, L. J., 336, 349
Bernstein, D. A., 244, 253
Bessell, K., 138, 149
Bianchi Piccinini, G. F., 243, 244, 251, 252, 253
Bieg, H., 51
Bierwagen, T., 336, 350
Billings, C. E., 298, 299, 304, 310, 332, 349, 351
Bingham, C., 181, 193
Birenheide, M., 320, 328
Birrell, S. A., 192, 193, 198, 211, 223
Bisantz, A. M., 219, 221
Blair, M. D., 336, 351
Blanker, P. J. G., 313, 314, 328
Block, R. A., 333, 350
Blok, M., 399, 412
Blom, H. A. P., 313, 314, 328, 329
Bobko, P., 336, 351
Boer, P. C., 271, 274, 293, 295
Bonapace, L., 27
Bonsall, P., 226, 241
Bonsall, P. W, 226, 241
Boole, P. W., 298, 310
Boos, M., 385, 389, 391, 392, 396, 397
Bosch, T., 400, 412
Bovair, S., 389, 397
Bowen, J. T., 241, 226
Boyatzis, R. E., 260, 261, 269
Bradshaw, J., 215, 222
Brauer, K., 399, 410, 412

Printed and bound by CPI Group (UK) Ltd, Croydon, CR0 4YY

01/11/2024

01782600-0013